S0-BXC-154

THE HISTORY OF THE EARTH'S CRUST

CONTRIBUTIONS TO A CONFERENCE

HELD AT THE GODDARD INSTITUTE FOR SPACE STUDIES

NOVEMBER 10-11, 1966

THE HISTORY OF THE EARTH'S CRUST

A SYMPOSIUM

EDITED BY

ROBERT A. PHINNEY

PRINCETON, NEW JERSEY
PRINCETON UNIVERSITY PRESS
1968

Library of Congress Catalog Card Number: 68-20875
International Standard Book Number: 0-691-08063-1

Printed in the United States of America by
Princeton University Press

This book has been composed in linotype Times Roman.

Second Printing, 1970

PREFACE

The contents of this volume are the result of a two-day conference held at the Goddard Institute for Space Studies, New York, N.Y., sponsored by the Department of Geology at Columbia University and the Goddard Institute for Space Studies. The conference sessions were scheduled as follows:

Thursday morning, November 11, 1966	Bulk chemical and physical properties of the planet Earth
Thursday afternoon	Paleomagnetic and paleoclimatic studies
Friday morning, November 12, 1966	Submarine geology and crustal history
Friday afternoon	Intercontinental correlations in time and space

The results presented at the conference, as well as many subsequent results, have demonstrated rather clearly that the crust under most of the earth's oceans is geologically young and that the positions of the continents have changed very significantly in the last 150 million years. More than one-third of this information had not yet been reported when the conference was first planned in the spring of 1966. The timing of the conference was fortuitous in this respect; it happened to come at a critical period in the history of earth science when scattered evidence bearing on continental drift was available but had not been published. The occasion of the conference brought together in one place a variety of arguments whose cumulative impact on those present was very great.

The implications of these developments are just beginning to be understood. The erasure of the record of the earth's early history renders their interpretation more difficult, but it may be hoped that the exploration of the moon, Mars and Venus may supply some of the information regarding planetary history which has been irretrievably lost on our own planet.

The organizers of the conference are deeply indebted to Professor Robert Phinney, Department of Geology, Princeton University, who has invested a substantial amount of his time in the preparation of the published proceedings. Mr. Leon Thomsen of Columbia University provided valuable assistance to Professor Phinney. Finally, the organizers and participants are grateful to Miss Ruth McCarthy of the Institute for Space Studies for contributing her administrative talents and charm to the arrangements of the meeting.

PAUL GAST
ROBERT JASTROW

LIST OF CONTRIBUTORS

W. B. N. Berry, University of California, Berkeley, California
A. J. Boucot, California Institute of Technology, Pasadena, California
James C. Briden, The University of Birmingham, Birmingham, England
E. C. Bullard, University of Cambridge, Cambridge, England
Allan Cox, Stanford University, Stanford, California
G. B. Dalrymple, U.S. Geological Survey, Menlo Park, California
John Dewey, Sedgwick Museum, Cambridge, England
R. R. Doell, U.S. Geological Survey, Menlo Park, California
Paul W. Gast, Lamont Geological Observatory, Palisades, N.Y.
James R. Heirtzler, Lamont Geological Observatory, Palisades, N.Y.
P. M. Hurley, Massachusetts Institute of Technology, Cambridge, Massachusetts
E. Irving, University of Leeds, Leeds, England
J. G. Johnson, California Institute of Technology, Pasadena, California
Marshall Kay, Columbia University, New York, N.Y.
Robert A. Phinney, Princeton University, Princeton, New Jersey
X. LePichon, Lamont Geological Observatory, Palisades, N. Y.
Robert K. McConnell, Jr., Arthur D. Little, Inc., Cambridge, Massachusetts
Dan P. McKenzie, University of Cambridge, Cambridge, England
H. W. Menard, University of California, San Diego, California
Neil D. Opdyke, Lamont Geological Observatory, Palisades, N.Y.
J. R. Rand, Massachusetts Institute of Technology, Cambridge, Massachusetts
W. A. Robertson, Dominion Observatory, Ottawa, Canada
Francis G. Stehli, Case Western Reserve University, Cleveland, Ohio
Lynn R. Sykes, ESSA, Lamont Geological Observatory, Palisades, N.Y.
F. J. Vine, Princeton University, Princeton, New Jersey

CONTENTS

Preface v

List of Contributors vi

Introduction, Robert A. Phinney 3

1. THE UPPER MANTLE 13

Upper Mantle Chemistry and Evolution of the Earth's Crust, PAUL W. GAST 15

The Geophysical Importance of High-Temperature Creep, DAN P. MC KENZIE 28

Viscosity of the Earth's Mantle, ROBERT K. MC CONNELL, JR. 45

2. EVIDENCE FROM THE OCEAN BASINS 59

The Paleomagnetism of Oceanic Cores, NEIL D. OPDYKE 61

Magnetic Anomalies Associated with Mid-Ocean Ridges, F. J. VINE 73

Evidence for Ocean Floor Spreading across the Ocean Basins, J. R. HEIRTZLER 90

Time Scale for Geomagnetic Reversals, A. COX, R. R. DOELL, G. B. DALRYMPLE 101

Some Remaining Problems in Sea Floor Spreading, H. W. MENARD 109

Heat Flow through the Ocean Floor and Convection Currents (Abstract), X. LE PICHON 119

Seismological Evidence for Transform Faults, Sea Floor Spreading, and Continental Drift, LYNN R. SYKES 120

3. EVIDENCE FROM THE CONTINENTS 151

Review of Age Data in West Africa and South America Relative to a Test of Continental Drift, P. M. HURLEY, J. R. RAND 153

Appalachian and Caledonian Evidence for Drift in the North Atlantic, JOHN DEWEY, MARSHALL KAY 161

The Distribution of Continental Crust and Its Relation to Ice Ages, E. IRVING, W. A. ROBERTSON 168

Paleoclimatic Evidence of a Geocentric Axial Dipole Field, JAMES C. BRIDEN 178

A Paleoclimatic Test of the Hypothesis of an Axial Dipolar Magnetic Field,
FRANCIS G. STEHLI 195

The Crust of the Earth from a Lower Paleozoic Point of View,
A. J. BOUCOT, W. B. N. BERRY, J. G. JOHNSON 208

4. CLOSING REVIEW 229

Conference on the History of the Earth's Crust,
E. C. BULLARD 231

Author Index 237

Subject Index 241

THE HISTORY OF THE EARTH'S CRUST

INTRODUCTION

ROBERT A. PHINNEY

Unraveling the history of the earth's crust has been a principal objective of geologists and geophysicists from the very start. Various views have held sway over the past century, each in turn reflecting the strong points and limitations of the available means of observing and interpreting the earth. The papers collected in this volume discuss certain critical lines of new evidence which have, taken together, revolutionized and revitalized this inquiry, setting the science of the earth into new directions. This turning point has come, appropriately enough, as the culmination of a period of vigorous controversy, fueled by an unprecedented wave of exploration of the oceans and continents. A brief, selective historical sketch is appropriate to set the scene for the papers which follow.

The classical basis of geology is the study of the major sedimentary systems of Europe and North America. Out of this work grew all modern studies of the earth, with their shared dependence on the uniformity of processes and the role of long time scales in the incremental accumulation of major phenomena. A synthesis of ideas could be built upon the data of Phanerozoic sedimentation by Schuchert in 1923. What emerged was a world of uplifts and geosynclines, of borderlands and shallow seas. The vertical instability of continental surfaces and the importance of mountain belts as sources of sediment are among the pieces of the puzzle contributed by this tradition. On the other hand, many inferences, such as the existence of now-vanished borderlands, have continued to puzzle by their lack of compatibility with results in other fields.

Field study and mapping of structure of the continental crust has for a century and a half remained the test of truth in geology. If the analysis of sedimentation gave a connected, but indirect history of the movements of the crust, then the observation of structure in the mountain belts of the world has provided direct, but less coherent evidence of quite spectacular kinds of crustal deformations and displacements. The discovery and detailed study of the beautifully exposed overthrust structures in the Alps in the last century has made European workers on the whole more congenial to ideas involving major crustal mobility than their colleagues across the Atlantic. Structural geology, however, suffered a major difficulty of scale, which is even now only being partially overcome. Mountains, sediment wedges, fold belts, overthrusts, and other units which a geologist may spend decades studying, are one or two orders of magnitude smaller than the scales of the major driving forces connected with the evolution of continents and oceans. At the same time, they are normally much larger than one's own back yard, and it requires a major intellectual synthesis to understand the field data. It is therefore not particularly surprising that the synthesis of a coherent crustal history from the data of field geology alone has been at best a very provisional one. Wegener's proposal (1915) that the continents facing the Atlantic had drifted apart during Cretaceous time met bitter and nearly unanimous opposition from his colleagues. Although based on a datum with an appropriately large length scale—the fit of the continents—the proposal could not be discussed in the existing framework of geological data. Wegener's ideas of a driving mechanism did not survive scrutiny by Jeffreys; he did not propose convection until the 1928 edition of his book.

With the publication of *The Earth* in 1924,[1] Harold Jeffreys brought together the results of over half a century of geophysics. The spirit of this classical age is exemplified by the prodigious work of Jeffreys and his predecessors, G. H. Darwin, Lord Rayleigh, Sir Horace Lamb, A. E. H. Love, Sir George Airy, and many others who brought the power of mathematical physics to bear on the natural world. We can identify two principal outcomes of this era which hold a central place in the problem of crustal history. Studies of the deflections of a plumb line established that mountainous areas are supported by rocks of an anomalously low density. This led to the concept of isostasy and the recognition of the existence of a basic structural distinction between continental and oceanic crust. Study of near-earthquakes in Europe led Mohorovicic to identify the elastic discontinuity at the base of the crust and set the style for all subsequent geophysical models of the crust. Second was Jeffreys's demonstration that the stress differences required for the support of mountains and continents could be computed and that, even in the acknowledged presence of isostasy, stresses of the order of one-third the strength of granite were inferred for the base of a continent. He felt that this evidence, combined with the undoubted great age of many features being thus supported, pointed to a static arrangement of continents held in place by the finite strength of the substrate. He made a point-by-point rebuttal of all suggestions by Wegener and his followers that large-scale motions might be possible. He also has come out against convection in a viscous mantle, but offers no direct rebuttal on physical grounds.[2]

Let us arbitrarily take 1930 as the start of the "modern era" in earth science and summarize the "big picture" seen at that time. The meager geophysical data, combined with the theoretical analysis by Jeffreys, gave a snapshot of the two different kinds of crust, oceanic and continental, supported by isostatic forces and the strength of the rocks. Isostatic adjustment was a geophysically approved response to the movement of material by erosion, intrusion, or volcanism, and seemed to account for enough manifestations of crustal unrest. One suspects that geological opinion went along with this in loyalty to Occam's razor. Modern ideas of gravity tectonics achieved impetus from this particular world-view, as geology attempted to explain the embarrassing field evidence for horizontal displacement in mountain belts. Certainly the rich, contradictory, but incomplete data of geology presented a major problem; bolstered by the authority of physics, one could take the least complex model and explain away inconsistencies in the customary manner. The only other option was to build models according to fancy, but without serious test, due to absence of relevant data.[3] Although the static view prevailed at that time, descendants of the two schools reopened the battle in the 1950's with the discovery of paleomagnetic data favoring continental drift.

[1] The date of the first English edition of Wegener's book!
[2] For a spirited account of this issue, see p. 364 of the latest edition of *The Earth* (1962).
[3] These are the same options faced by students of the UFO problem.

INTRODUCTION

The past two decades have witnessed the proliferation of productive new geophysical methods of studying the crust, following the pioneering work of Ewing, Woollard, Bullard, and Hess and their colleagues in the 1930's and 1940's. Refraction and reflection seismology, precision echo sounding, heat flow determination, and deep ocean sampling have unlocked the ocean basins to science. The secrets of basement rocks and mountain belts on the continents are now being systematically unraveled through a combination of field mapping, radiometric age determinations, and a soundly based understanding of metamorphic petrology. The opening up of the oceans, the fruition of geophysics as a field method, the advent of the airplane as a field vehicle, and the wide availability of government support have combined to bring about this new age of exploration.

There is a logarithmic quality about the way relevant new information has been piling up; one finds the quantity and sophistication of data collected ten years ago to be hopelessly meager in comparison with what is now being acquired. This will continue to be true. At such a time, our desire to find a synthesis of crustal history is frustrated by the embarrassment of riches. By its very nature, most geophysical information is static in character; the outcome of a geophysical study is an idealized snapshot of the spatial behavior of some measurable property. Since the picture drawn from geophysical studies of the crust and upper mantle seems fairly consistent with a static crustal history, it is not surprising that most geophysicists have felt ill at ease with the rich wine of continental drift. In contrast, the more numerous geologists have never been in general agreement as to even a provisional set of hypotheses, despite intimidation by Jeffreys. Stimulated by the multifold data of the field, geology has provided haven and comfort for an incredible variety of theories to explain the major facts of continental history, while maintaining a quite static earth as the orthodox view.

The Modern Debate on Continental Drift

It would seem that the more radical ideas, lying under the geophysical interdict against continental drift, did not achieve the status of serious proposals until, in the 1950's, the new field of paleomagnetism produced data which split the geophysical camp and brought the ideas of Wegener back into serious contention. Strong objections made to these ideas were based on modern data on the distribution of heat flow and the non-equilibrium component of the earth's gravity field (MacDonald, 1963; MacDonald, 1966). Many of the assumptions and techniques of paleomagnetism appeared to be on dubious grounds; attempts to demonstrate the fallacy of the paleomagnetic method, however, apparently stimulated the kind of definitive laboratory and field work required to clarify many of these questions (Cox and Doell, 1960).

In 1960 Hess (1962) and Dietz (1961, 1962) proposed that convection may occur in the form of ocean floor spreading. This stimulated a host of further proposals which set the stage for the modern debate on crustal history. Most of this discussion has given emphasis to the role of the newly mapped mid-ocean ridge system. Runcorn (1962) believed that convection is a cellular phenomenon involving the entire mantle and used stability theory to connect the radius of the (presumed growing) core with the order of harmonic of the convection. Elsasser (1963) and Orowan (1964) advocated convection mechanisms with a relatively thin zone of flow in the upper mantle. The respectability accorded these speculations comes of the improved knowledge we now have of the phenomenology of creep in crystalline solids. It is also due in part to the realization that large displacements measured on circum-Pacific strike-slip faults may reflect directly the large-scale motions of the mantle. The use of surface wave and free oscillation methods in seismology has focused attention on the upper mantle as a region of low seismic velocity and high attenuation (Anderson, 1967); this revival of the Gutenberg low-velocity zone has stimulated a revival of the Daly asthenosphere, or weak zone, as the most plausible locus of flow in the mantle.

The maritime view of the Anglo-Saxon world, which nurtured all this radicalism, has not been shared by Soviet workers. Beloussov (1966) has developed a synthesis based on tectonic analysis of the history of geosynclines and mountain belts in Eurasia. He finds sufficient explanation of the structural data in purely vertical tectonics, characterized by uplift and foundering of crustal blocks.

A major sociological phenomenon of this decade has been the frequency with which continental drift enthusiasts gather in conference to elaborate upon their latest schemes. If one of these meetings must be picked out of the many for citation, the nod should go to the 1964 Royal Society Symposium (Blackett, Bullard, and Runcorn, 1965) at which Bullard unveiled a computer-aided fit of the continents across the Atlantic (Bullard, Everett, and Smith, 1965). This was an objective test of the first modern reconstruction of the jigsaw-puzzle match by Carey (1958), in which the continental shelf edge instead of the coastline was recognized as the structural boundary of the continental plate. The very remarkable agreement thus achieved must stand as the most direct evidence for continental drift in the geological record.

Objection to these ideas took two forms: criticism of specific data purporting to show drift, and general arguments showing the inconsistency of the drift process with known geophysical data. Space cannot permit mention of the data "proving" drift and all the counter-arguments demolishing the same proofs. Experience here clearly shows that individual lines of geological evidence are too ambiguous to settle anything without some convincing independent test. The contributions in paleontology by Briden, Stehli, and Boucot in this volume are a certain demonstration of this. The general objections may be mentioned, since it is of some interest to evaluate them in light of the new evidence of sea floor spreading.

1. Radioactive heat-producing elements are strongly concentrated upward in the continental crust, but only slightly so in the oceanic crust. By comparison with chondritic or solar system abundances, it seems probable that the upper mantle is enriched at the expense of the core and lower mantle. For time scales of the order of 100 million years, appropriate to continental drift, heat flow is shown by calculation to be in approximate equilibrium with these sources. What is singular here is the equality of the average

continental and oceanic heat flow—about 1.2 μcal/cm²/sec. If the crustal source concentrations are so different, this result indicates that the upper mantle under oceans must be richer in sources than that under continents in order for the heat flows to be so similar. The continents thus appear to be "tied" to the mantle beneath them, since drift would presumably bring continents above previously oceanic mantle and produce high heat flows of a sort not commonly found (MacDonald, 1963). This model implies a horizontally static evolution of the crust, with upward migration of mantle and lower crustal differentiates and downward migration of material in sedimentary basins as the major evolutionary mechanisms.

2. Of all the convection models proposed, some variant of sea floor spreading seemed most likely. Ewing and his collaborators at the Lamont Geological Observatory, pioneering the use of the seismic reflection profiler in the ocean basins, obtained detailed cross-sections of the upper layers of oceanic crust which do not show certain expected phenomena of the spreading hypothesis. This predicts that the ocean basins are no older than Cretaceous, or possibly Jurassic, at any point, with the age varying monotonically with the distance from a mid-ocean ridge. Let it be noted first that the distribution of sediment thickness in the ocean basins seems to bear out this model in a general way. The objections, however, are not the less cogent. Sedimentary layers can be traced continuously from the normal ocean floor into the bottom of a typical oceanic trench and seem to end discontinuously at the trench axis against a steep facing wall (Ludwig et al., 1966). While this is compatible with the trench being a major vertical dislocation, it conflicts with the expected structure if the trench is the locus of convergence and sinking. One expects the sediments to be piled up and folded at the trench axis, since they are too fluid to be stuffed back into the ground like toothpaste into a tube. A second problem posed by the sediments comes from the observation of Cretaceous sediments at the surface in the northwest Pacific (M. Ewing et al., 1966) and the Atlantic (J. Ewing et al., 1966). These sediments appear to represent "reflector A," a ubiquitous and easily identified seismic interface; the observed thickness of sediments below "A" is sufficient to infer an age of Triassic or older for the crust at certain places. In addition to there being further objections in detail, based on interpretations of sediment structure, fossils of apparent Miocene age (ca. 20 million years) found in a core from the axis of the Mid-Atlantic Ridge conflict with the predicted age of under 5 million years (Saito et al., 1966).

3. Analysis of the gravity field of the earth from satellite orbits (Cook, 1967) shows harmonics which cannot be accounted for unless the mantle has the strength to support density variations. In particular, the values of harmonics determined in the range 10-20 require that the related density variations reside in the upper mantle and be supported by stresses in the range of 10-100 bars. MacDonald's (1966) discussion of the lower-order harmonics concludes that similar long-term stresses must reside in the lower mantle, due either to density fluctuations or to variations in the radius of the core-mantle boundary. He interprets the non-hydrostatic component of the P_2 harmonic as due to the inability of the bulge of the earth to respond to the slowing rate of rotation over

times of 10^8 years. This seems to preclude convection in the lower mantle, a conclusion concurred in by McKenzie (1966) in a discussion of the viscosity.

In the case of both the heat flow and the gravity harmonics, Runcorn (1962) starts by *assuming* that convection is occurring. In the first place, this eliminates the objection based on the continental-oceanic equality of heat flow, since one can assume convection rates and geometries which invalidate the equilibrium between sources and heat flow. Then, he infers from his model, the maps for heat flow and geoid height, and finds a broad-scale correspondence with the actual data. This particular exercise depends on having a good convection model and having data which are complete up to orders well beyond 20. It is not clear, therefore, where this analysis will lead, but one can see the role of circular reasoning, which is as central to this argument as to the static interpretation of heat flow and gravity.

It should be pointed out that much of the polarization of opinion over these issues was aroused by some of the liberties taken by enthusiasts of drift at the expense of established laboratory and field data. The literal acceptance of Rayleigh-Bénard convection in the earth with respect to cell dimensions, time scale, stationarity, etc. flew in the face of the known phenomenology of deformation in solids and the evidence against widespread existence of the liquid phase in the mantle. While the contradiction may only be apparent, it is apparent to many, and no discussion appeared in the literature until very recently evaluating the probable values of relevant parameters in the mantle (Gordon, 1965; McKenzie, 1966; McKenzie, this volume). It may very well turn out that the Rayleigh stability criterion is quite relevant, but this fairly persuasive, yet conservative, point is frequently lost in a welter of undiscussed assumptions. Rightly or wrongly, theoretical and laboratory geophysics is ultimately looked to to provide the writ of legitimacy by its approval of geological models. The overabundant enthusiasm shown for continental drift by its supporters drove many creative workers in geophysics into opposition, and the chance for collaboration was lost. This unfortunate polarization appears to be a thing of the past, in light of the convincing evidence for sea floor spreading that has come out in the last two years.

The Magnetic-Reversal Time Scale

Probably the most important fundamental development of this decade is the continuing development of a time scale for the Quaternary and Cenozoic, based on reversals of the earth's dipole magnetic field preserved in volcanic rocks (Cox et al., 1967). In work which must serve as a model for integrating several disciplines, Cox and his colleagues demonstrated in a series of papers that lavas of a given K-Ar age consistently give identical magnetic pole positions regardless of geographic position. Reversals of the geomagnetic polarity are found to form a time sequence which shows a typical polarity epoch somewhat under a million years (Cox et al., this volume). The ability to read this scale in some sediments and sedimentary rocks as well as in volcanics is having a revolutionary impact on studies in all areas—paleontology, climatology, regional structure, stratigraphy, sedimentology, and so on. The existence of a scale for the reversals over the past two million years led directly to the possibility of a convincing proof of sea floor spreading.

Sea Floor Spreading—the Revolution

The Hess-Dietz sea floor spreading hypothesis had an earlier, probably premature, birth in an idea of Holmes (1928). By the present decade the time was more propitious. Vine and Matthews (1963) noticed that the very puzzling striped magnetic anomalies in the northeast Pacific (Raff and Mason, 1961; Vine, this volume, Fig. 4) might be due to the freezing in of the recent history of the earth's magnetic field reversals along an axis of spreading. Wilson (1965) introduced the transform fault to the model as an explanation of the fracture zones found offsetting the mid-ocean ridge system. Both analyses contained predictions which could be quantitatively tested: the Vine-Matthews hypothesis by comparing the magnetic anomaly patterns with the magnetic-reversal time scale, and the Wilson transform fault model by studying the epicenter distributions and focal mechanisms of earthquakes on the mid-ocean ridge system. The convincing confirmation of the predictions in 1966 (Vine, 1966; Pitman and Heirtzler, 1966; Sykes, 1967) must be regarded as a major turning point in studies of the earth. At this time two closely related endeavors also bore fruit. Opdyke et al. (1966) were able to find the magnetic-reversal history in oceanic sediments, thus tying the sediment history to the same time scale as the sea floor spreading. Hurley et al. (1967) found radiometric basement age dates in facing portions of Africa and South America which offer perhaps the best field geological evidence to date of the former contiguity of these continents.

These workers and many others who have made important contributions are represented in the chapters of this volume, based on the November 1966 conference at the Goddard Institute for Space Studies. The material brings the reader up to approximately that date, and brings together the evidence which has so revolutionized our understanding of the history of the earth's surface. Two very interesting papers, one by Don L. Anderson on "Recent Evidence Regarding the Elastic and Anelastic Properties of the Mantle" and the other by Maurice Ewing entitled "Sediment Cover of the Deep Sea Floor," which were presented at the conference are not included in this collection at the wish of the authors. The papers by Francis G. Stehli and Allan Cox and his colleagues were solicited by the editor, and were not given at the original meeting. Since the editor has somewhat of a time advantage with respect to his authors, the opportunity is taken to review here the more important developments in 1967.[4] Three further results stand out:

1. The extrapolation of the magnetic-reversal time scale back into the Cretaceous, using oceanic magnetic anomalies (Heirtzler et al., 1968). Tentative assignment of the age of various areas of the oceans has been made, and the time sequence of continental motions can be developed in fairly interesting detail.

[4] In July 1966, the results of Vine and Pitman and Heirtzler were known to only a few colleagues. Word got around during the fall, by preprint, and at the Goddard conference, slightly preceding formal publication in *Science*. By January 1967 the impact was such that nearly 70 papers on sea floor spreading had been submitted for the April meeting of the American Geophysical Union.

2. The demonstration by Morgan (1968) that continental-sized plates of the lithosphere act more or less as rigid units, moving on paths which can be referred to an instantaneous pole of motion. LePichon (Heirtzler et al., 1968) and Morgan have developed global patterns of displacement which integrate a great deal of tectonic and earthquake data quite successfully. Earthquake mechanisms (Sykes, this volume) are proving to be useful dynamic indicators of the motions of these plates (McKenzie and Parker, 1967; Sykes, Oliver, and Isacks, 1968).

3. The discovery, by seismic methods, of structural heterogeneity under trenches which is readily reconciled with the notion that trenches are loci of convergence and sinking of the lithosphere. Oliver and Isacks (1967) found a zone of anomalously low S-wave attenuation under the Tonga-Kermadec arc, coinciding with the postulated plate of depressed lithosphere. Cleary (1967) found that teleseismic P-waves traveling (north) through a comparable region beneath the Aleutian arc from a shallow source arrived anomalously early by up to 2 seconds.

In other areas, the implications of the "new tectonics" are just being felt. Displacement rates of 1 to 5 cm/yr, when compared with observed heat flows, suggest that equilibrium with respect to sources is achieved in about 10 million years (McKenzie, 1967). This both corroborates the estimate of lithosphere thickness (50-100 km) obtained from theoretical considerations of high-temperature creep and eliminates any practical possibility that present-day heat flow retains any "memory" of the early thermal history of the earth. The state of affairs with respect to the vertical and horizontal arrangement of heat sources is not at all settled, however. Equilibrium models can be constructed (Gast, this volume) which are more or less in accord with the observations. The principal features of such models are: (1) lower K + U in the continental crust, arising from the recognition that average crustal rock is more dioritic or granodioritic in composition than granitic, and (2) use of the 50 to 100 km thick lithosphere as the mechanically stable unit of the crust in place of the chemical crust as defined by the M discontinuity at 10 to 40 km. Both modifications eliminate the contradiction between heat flow and crustal spreading data.

The role of the major strike-slip fault systems is now understandable, if not yet proved in detail. It becomes inevitable, from the rigid-plate analysis by Morgan, that relative lateral motion does occur between adjacent blocks. One is led to the realization that lines of convergence will normally display transcurrent faulting at rates comparable to those of sea floor spreading. Only three or four years ago, transcurrent faulting was considered by many to be the dominant mode of tectonics in the circum-Pacific belt; viewed in this light the data seemed to indicate that the entire Pacific basin is rotating counterclockwise. Although unanimity is hardly possible on this matter, it now appears that the San Andreas Fault is a pure transform fault, and quite inappropriate as a prototype circum-Pacific fault, if such a thing can be defined.

No interpretation of the gravity field has appeared since the demonstration of sea floor spreading. A number of interesting approaches are indicated. One can take the bottom of the lithosphere to be effectively stress-free and can assume stress values on the margins of the lithospheric plates appropriate to the sense of relative displacement and the earthquake mechanisms. A gravity continentality function may be constructed for these lithospheric plates and comparison made with the actual distribution of gravity harmonics. The aim would be to estimate the depth to which the stress-free zone extends. There is reason to feel, however, that construction of the lithospheric density model in any unique sense may not be possible at this time. The problem of constructing a model for the density perturbation field connected with the convection process itself stands as an example of a limited, yet still ambitious objective of future work in this area.

It is worth pointing out the importance of stress measurements in the crust to future experimental work on crustal evolution. The lithospheric density field, which is the object of interest, gives rise not only to a measurable gravity field, but also to a stress field, produced by the action of gravity (principally the zero-order term) on the mass distribution. Contributions to the stress are produced in addition by the boundary stresses, for which data are not totally lacking. It might be hoped that stress measurement in boreholes can become as important a part of the experimental effort as is heat flow measurement today.

Finally, the body of work on paleomagnetic pole positions done over the past decade finds vindication and becomes highly pertinent to the reconstruction of crustal history on time scales of 10^7 to 10^9 years. It is fortunate that at this time many longstanding problems of technique and interpretation have yielded to the experience of a decade, providing a tool which can be used with some confidence to reconstruct the structural evolution of continents.

ACKNOWLEDGMENTS

A. G. Fischer and F. J. Vine offered helpful criticism of my more sweeping generalizations. In a selective review such as this, I take full responsibility for the choice of literature citations in an area where so many have contributed.

Princeton University January 1968

REFERENCES

Anderson, D. L. (1967), Latest information from seismic observations, pp. 355-420 in *The Earth's Mantle*, T. F. Gaskell, ed., Academic Press.

Beloussov, V. V. (1966), *The Earth's Crust and Upper Mantle of the Continents*, Nauka, Moscow, p. 58.

Blackett, P. M. S., E. C. Bullard, and S. K. Runcorn (1965), A symposium on continental drift, *Phil. Trans. Roy. Soc. A* 258:1-323.

Bullard, E. C., J. E. Everett, and A. G. Smith (1965), The fit of the continents around the Atlantic, *Phil. Trans. Roy. Soc. London, A* 258:41-51.

Carey, S. W. (1958), A tectonic approach to continental drift, pp. 117-355 in *Continental Drift: A Symposium*, University of Tasmania.

Cleary, J. (1967), Azimuthal variation of the Longshot source term, *Earth Planet. Sci. Letters* 3:29-37.

Cook, A. H. (1967), Gravitational considerations, pp. 63-87 in *The Earth's Mantle*, T. F. Gaskell, ed., Academic Press.

Cox, A., G. B. Dalrymple, and R. R. Doell (1967), Reversals of the earth's magnetic field, *Scientific American* 216 (1):44.

Cox, A., and R. R. Doell (1960), Review of paleomagnetism, *Bull. Geol. Soc. Amer.* 71:645-768.

Dietz, R. S. (1961), Continent and ocean basin evolution by spreading of the sea floor, *Nature* 190:854-857.
Dietz, R. (1962), Ocean-basin evolution by sea-floor spreading, in *The Crust of the Pacific Basin*, Amer. Geophys. Union, Monograph No. 6.
Elsasser, W. M. (1963), Early history of the earth, pp. 1-29 in *Earth Science and Meteoritics*, E. D. Goldberg and J. Geiss, eds., North Holland Publishing Co.
Ewing, J., J. L. Worzel, M. Ewing, and C. Windisch (1966), Ages of horizon A and the oldest Atlantic sediments, *Science* 154:1125-32.
Ewing, M., T. Saito, J. I. Ewing, and L. H. Burckle (1966), Lower Cretaceous sediments from the northwest Pacific, *Science* 152:751-55.
Gordon, R. B. (1965), Diffusion creep in the earth's mantle, *J. Geophys. Res.* 70:2413-18.
Heirtzler, J. R., G. O. Dickson, E. M. Herron, W. C. Pitman III, and X. LePichon (1968), Marine magnetic anomalies, geomagnetic field reversals, and motions of the ocean floor and continents, *J. Geophys. Res.* (in press).
Hess, H. H. (1962), History of ocean basins, pp. 599-620 in *Petrologic Studies: A Volume to Honor A. F. Buddington*, Geol. Soc. Amer.
Holmes, A. (1928), Radioactivity and earth movements, *Trans. Geol. Soc. Glasgow* 18:559-606.
Hurley, P. M., F. F. M. de Almeida, G. C. Melcher, U. G. Cordani, J. R. Rand, K. Kawashita, P. Vandoros, W. H. Pinson, Jr., and H. W. Fairbairn (1967), Test of continental drift by comparison of radiometric ages, *Science* 157:495-500.
Jeffreys, H. (1924), *The Earth*, Cambridge University Press (4th edn., 1962).
Ludwig, W. J., J. I. Ewing, M. Ewing, S. Murauchi, N. Den, S. Asano, H. Hotta, M. Hayakawa, K. Ichikawa, and I. Noguchi (1966), Sediments and structure of the Japan trench, *J. Geophys. Res.* 71:2121-38.
MacDonald, G. J. F. (1963), The deep structure of continents, *Rev. Geophys.* 1:587-665.
MacDonald, G. J. F. (1966), The figure and long-term mechanical properties of the earth, pp. 199-245 in *Advances in Earth Science*, P. M. Hurley, ed., MIT Press.
McKenzie, D. P. (1966), The viscosity of the lower mantle, *J. Geophys. Res.* 71:3995-4010.
McKenzie, D. P. (1967), Some remarks on heat flow and gravity anomalies, *J. Geophys. Res.* 72:6261-73.
McKenzie, D. P., and R. L. Parker (1967), The North Pacific: an example of tectonics on a sphere, *Nature* 216:1276-1280.
Morgan, J. (1968), Rises, trenches, great faults and crustal blocks, *J. Geophys. Res.* 73:1959-82.
Oliver, J., and B. Isacks (1967), Deep earthquake zones, anomalous structures in the upper mantle, and the lithosphere, *J. Geophys. Res.* 72:4259-75.
Opdyke, N. D., B. Glass, J. D. Hays, and J. Foster (1966), Paleomagnetic study of Antarctic deep-sea cores, *Science* 154:349-57.
Orowan, E. (1964), Continental drift and the origin of mountains, *Science* 146:1003-10.
Orowan, E. (1966), Age of the ocean floor, *Science* 154:413-16.
Pitman, W. C., III, and J. R. Heirtzler (1966), Magnetic anomalies over the Pacific-Antarctic ridge, *Science* 154:1164-71.
Raff, A. D., and R. G. Mason (1961), Magnetic survey off the west coast of North America, 40° N-52° N latitude, *Bull. Geol. Soc. Amer.* 72:1267-70.
Runcorn, S. K. (1962), Convection currents in the earth's mantle, *Nature* 195:1248-49.
Saito, T., M. Ewing, and L. H. Burckle (1966), Tertiary sediment from the Mid-Atlantic Ridge, *Science* 151:1075-79.
Schuchert, C. (1923), Sites and nature of the North American geosynclines, *Bull. Geol. Soc. Amer.* 34:151-229.
Sykes, L. (1967), Mechanism of earthquakes and nature of faulting on the mid-oceanic ridges, *J. Geophys. Res.* 72:2131-53.
Sykes, L. R., J. Oliver, and B. Isacks (1968), Seismological evidence for motions of the lithosphere, 2, worldwide synthesis of surface motions, *A.G.U. Trans.* 49:292.
Vine, F. J. (1966), Spreading of the ocean floor; new evidence, *Science* 154:1405-15.
Vine, F. J., and D. H. Matthews (1963), Magnetic anomalies over oceanic ridges, *Nature* 199:947-49.
Wegener, A. (1915), *Die Entstehung der Kontinente und Ozeane*, Friedr. Vieweg & Sohn, Braunschweig.
Wegener, A. (1966), *The Origin of Continents and Oceans*, Dover, trans. of 4th revised German edition (1929).
Wilson, J. T. (1965), A new class of faults and their bearing on continental drift, *Nature* 207:343-47.

1. THE UPPER MANTLE

UPPER MANTLE CHEMISTRY AND EVOLUTION OF THE EARTH'S CRUST*

PAUL W. GAST

The objective of this review is to identify and discuss certain chemical parameters that place significant restrictions on the way in which the mantle and crust of the earth evolved during geologic time. The arguments and geochemical data will be considered from three different aspects: (1) the bulk composition of the present and primeval upper mantle, (2) the relevance of lead and strontium isotope compositions, and (3) the constraints resulting from the observed heat fluxes in oceans and continents.

The Bulk Composition of the Mantle

ABUNDANCE OF Sr^{87} IN THE EARTH AND METEORITES

The mineral composition and major element chemistry has been reviewed in a number of recent papers: Ringwood (1966), Clark and Ringwood (1964), McQueen et al. (1964). The present discussion will be largely confined to an attempt to infer limits and the abundance of certain trace elements in the earth.

The proposition that the abundance of many minor elements in the earth is significantly different from that observed in chondritic meteorites will be developed. Chondritic abundances are, nevertheless, a very useful framework in which to consider the inferred differences.

The non-chondritic composition of the earth's mantle suggested by Gast (1960) was based mostly on the terrestrial and meteoritic Sr^{87} abundance. Implications of the evidence presented in 1960 were subsequently developed by Wasserburg et al. (1964) and MacDonald (1964) with regard to abundance of radioactive elements and with regard to abundance of volatile elements.

Many additional and more precise determinations of Sr isotope abundance have been published in the interval since 1960. A partial summary of these measurements is given by Gast (1967). Results published subsequent to completion of this study, e.g., Bence (1966), MacDougall and Compston (1965), and Moorbath and Walker (1965) do not materially alter the results shown in Figure 1. Several generalizations made on rather limited observations in 1960 can now be made on a much firmer basis.

It is seen that oceanic volcanic rocks have a very narrow distribution of Sr^{87}/Sr^{86} ratios, ranging from 0.702 to 0.705 with a median value of 0.703. Secondly, the distribution of Sr^{87}/Sr^{86} ratios in continental volcanic rocks is significantly different from that observed in oceanic volcanic rocks. This difference is probably the result of contamination of the continental volcanics by continental granitic and metamorphic rocks that are

* Lamont Geological Observatory Contribution No. 1206

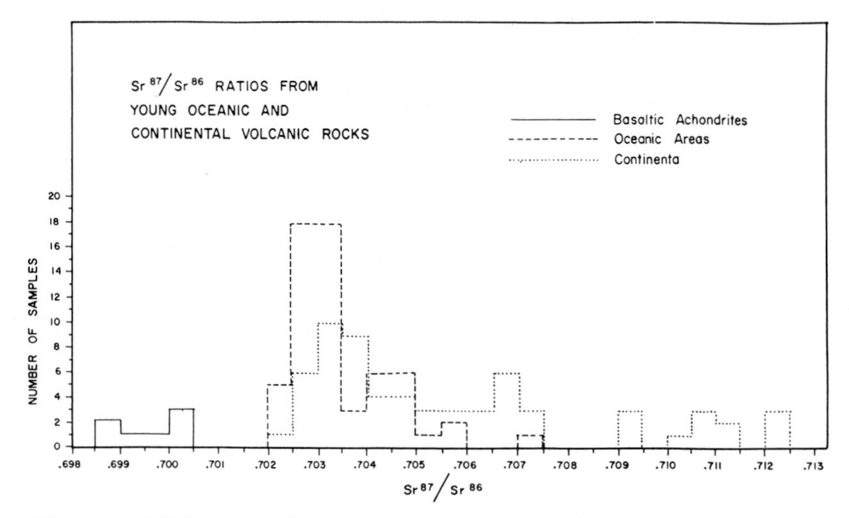

Figure 1. Sr^{87}/Sr^{86} ratios in volcanic rocks from oceanic and continental areas.

significantly enriched in Sr^{87}. Under conditions of partial fusion of mantle material, it is assumed that the melted fraction has the same isotopic ratio as the whole rock, due to the likelihood that equilibrium conditions prevail at the high temperatures. Thus the results summarized here indicate that the mean Sr^{87}/Sr^{86} ratio of that part of the mantle sampled by oceanic volcanic rocks is 0.703 ± 0.001. A second generalization resulting from the now extensive data on the abundance of Sr^{87} in granitic and metasedimentary crustal rocks is the observation that the initial Sr^{87}/Sr^{86} ratio for such rocks is almost always less than 0.710; cf. Hurley et al. (1962), Hedge and Walthall (1963). No attempt at a detailed summary of these initial ratios will be made here. The estimated value of such initial values will be taken as 0.705 ± 0.005 (error is at the 2-sigma level). (It is recognized that the distribution will probably be skewed toward the lower values.) This value, with the average age of the crust (the age here is the time measured by a Rb-Sr isochron) and the average crustal Rb/Sr ratio, readily permits estimation of the mean Sr^{87}/Sr^{86} ratio of the earth's crust. The abundance of Rb and Sr in the crust will be discussed in the following section. The results of that discussion suggest a Rb/Sr ratio for the crust of 0.2 or less (the most probable composition, Model C, gives a ratio of 0.13). The average age can be estimated from extensive geochronologic data (cf. Engel, 1963; Goldich et al., 1966; Muehlberger et al., 1966). I will use an average of 1500 m.yr. Combining these estimates results in a mean Sr^{87}/Sr^{86} ratio of 0.719 or less for the earth's crust. If the crustal Sr makes up 10% of the earth's Sr (cf. discussion below) and the ratio derived from oceanic volcanic rocks is taken as representative of all mantle Sr, the mean terrestrial Sr^{87}/Sr^{86} ratio is 0.7045 ± 0.001 or less. The major uncertainty in this number arises from our inability to sample the deep mantle adequately.

The average Rb/Sr ratio of the earth can now be estimated if the initial Sr^{87}/Sr^{86} ratio for the entire earth is known. The Sr^{87}/Sr^{86} ratios for a number of achondritic meteorites are summarized in Figure 1. When calculated back to 4.5 b.yr. B.P., these results suggest an initial Sr^{87}/Sr^{86} ratio for the earth of 0.699 ± 0.001. In addition, we may conclude from several Sr^{87}/Sr^{86} ratios reported for terrestrial samples that the terrestrial Sr^{87}/Sr^{86} ratio was 0.699 or lower. The mean increase for the terrestrial Sr^{87}/Sr^{86} ratio over 4.5 b.yr. is thus 0.0055 ± 0.0014 or less; this corresponds to a Rb/Sr ratio of 0.031 or less. The average Rb/Sr ratio of chondrites, based on the work of Gast (1962), Pinson et al. (1965), Murthy and Compston (1965), and Smales et al. (1964), is 0.25. Thus, the Rb/Sr ratio of the earth is depleted by more than a factor of eight compared to chondritic meteorites. This difference can obviously result from (1) depletion in Rb, (2) enrichment in Sr, or a combination of (1) and (2). In order to delimit these possibilities further, it will be fruitful to consider the geochemical properties of the alkaline metals, the alkaline earths, and uranium.

CRUSTAL ABUNDANCE AND TERRESTRIAL ABUNDANCE OF LITHOPHILE ELEMENTS

It has long been recognized that certain elements have been highly concentrated into the earth's crust. Birch (1958) showed that if the abundance of uranium in the earth is chondritic, the degree of concentration of this element in the crust approaches 100%. Gast (1962) suggests that a similar situation exists for barium. The probability of such great enrichment into the crust seems to me questionable. I should like to explore this question further by comparing the crustal abundance with abundances derived from a hypothetical chondritic abundance for a number of additional elements. (Abundance in this discussion will designate the total amount of a given element in the crust or in the earth.) The mass of the earth's crust used in the calculations that follow is summarized in Table 1. Even though the oceans make up a large area, the oceanic crust accounts

Table 1. The mass of the earth's crust

Shield areas		1.08×10^{25} gm
Continental shelves		0.48×10^{25} gm
Young folded belts		0.53×10^{25} gm
Oceanic crust		0.48×10^{25} gm
	Total	2.61×10^{25} gm

Total = 0.44 percent of the mass of the entire earth; 0.65 percent of the mass of the present mantle.

for less than one-third of the mass of the crust. Thus our ignorance of this part of the crust is not a major limitation in determining a crustal abundance. A much more significant limitation arises from the inability to specify the composition of the lower continental crust. Poldervaart (1955), Gast (1960), and Taylor (1964) have used the concentrations determined for common surface or upper crustal rocks (basalts, diorites, etc.) in estimating the abundance of the elements in the entire crust. Heier (1965) and Lambert and Heier (1966) have suggested that there may be very significant and systematic differences in the trace element complement of the lower and upper continental crusts. In particular, these authors find that pyroxene-granulite subfacies rocks are depleted in Th, U, and Rb relative to amphibolite facies rocks. Heier (1965) suggests that this may be the result of anatexis or partial melting of the lower crust. R. Lambert (oral communication, 1967) has made similar observations on the pyroxene-granulites of the Scottish Highlands. Thus, the previous estimates of crustal abundances may have overestimated the abundance of those elements enriched into the liquid during partial melting. The concentrations of a number of elements in common rocks are summarized in Table 2. The crustal abundances, based on several different assumptions, are given in Table 3. Ignoring the Sr^{87} evidence for the moment, these crustal abundances are com-

Table 2. Trace element content of some common rocks

	Parts per million								
Sample	U	Ba	Sr	Pb	Tl	Zn	Cd	K	Rb
Composite "granite"[1] > 70% SiO_2 (221)	4.2	—	132	28.7	—	—	—	3.70%	186
Composite "granodiorite" 60-70% SiO_2 (191)[1]	2.60	—	371	19.2	—	—	—	3.11	140
Composite "diorite"[1] < 60% SiO_2 (85)	1.63	—	635	14.6	—	—	—	2.41	93
Composite granite I[2] (27)	3.14	570	348	22.2	—	—	—	3.29	133
Composite granite II[8] (50)	2.3	685	283	—	—	—	—	3.04	125
Composite basalt[1] < 55% SiO_2 (282)	1.07	—	526	12.7	—	—	—	1.07	34
Composite basalt (250)[3]	1.07	334	461	6.59	—	—	—	0.95	29
Granulites[4]	0.5	—	—	—	—	—	—	2.5	(40)
Hi-calcium granite[5]	3.0	420	440	15	0.72	60	0.13	2.52	110
Oceanic tholeiites[6]	0.09	14	130	0.75	—	—	—	0.12	1.0
Basalt[5]	1.0	330	465	6	0.21	105	0.22	0.83	30
Granodiorite[7]	2.5	500	400	15	0.7	60	0.1	2.5	100
Diorite-granulite	0.6	(100)	500	8	(0.1)	60	(0.1)	1.5	20
Abyssal basalt	0.1	15	130	0.75	(0.1)	100	0.2	0.12	1.0

[1] Composite of specimens analyzed in the Rock Analyses Laboratory, University of Minnesota
[2] Composite of 27 Finnish pre-Cambrian granites
[3] Composite of basalts, analyzed by Turekian and Kulp (1956)
[4] Lambert and Heier (1967) and (1966), Rubidium content based on K/Rb ratios reported orally at AGU meetings
[5] Turekian and Wedepohl (1961)
[6] Engel et al. (1965); U and Pb content, Tatsumoto (1966)
[7] Parentheses indicate very uncertain or arbitrary estimates
[8] Composite of 50 pre-Cambrian from the western United States

pared with the terrestrial abundances based on a chondritic earth. (In the case of Tl, Zn, Cd, Pb, S, and Cl, the terrestrial abundance is based on carbonaceous chondrites.) The results shown here indicate that the refractory elements U and Ba are much more highly enriched than either the alkali metals or some highly volatile elements. Ringwood (1966) has suggested that the under-abundance of Tl, Zn, Cd, and Pb cannot be explained by their chalcophile properties, but that these elements, like Rb, must be depleted in the earth relative to their abundance in meteorites. This is not particularly surprising when one recalls that Tl, Bi, In, Pb, Zn, and Cd are highly depleted in normal chondrites relative to carbonaceous chondrites (Anders, 1965). We may now ask what would be the enrichment of Rb in the crust if the observed Sr isotope composition is explained

Table 3. Crustal abundance and degree of concentration into the crust of some trace elements

	Meteoritic abundance, ppm	MODEL A[8] Abundance (in grams) $\times 10^{-21}$	MODEL A[8] Fraction in crust	MODEL B[9] Abundance (in grams) $\times 10^{-21}$	MODEL B[9] Fraction in crust	MODEL C[10] Abundance (in grams) $\times 10^{-21}$	MODEL C[10] Fraction in crust
U	0.012[1]	0.056	0.77	0.036	0.50	0.033	0.46
Ba	3.6[1]	8.8	0.41	8.7	0.40	6.3	0.30
Sr	11[2]	8.84	0.13	9.0	0.14	10	0.15
Pb	2.0[1]	0.26	0.02	0.22	0.02	0.24	0.02
Zn	50[3]	—	—	2.15	0.007	1.85	0.005
Cd	1.0[4]	—	—	0.004	0.007	0.003	0.005
Tl	0.10[1]	0.009	0.015	0.009	0.015	0.008	0.013
K	860	440	0.085	330	0.064	420	0.0812
Rb	2.7[5]	19	0.116	1.4	0.088	1.35	0.080
"Rb"	0.32[6]	19	0.98	1.4	0.72	1.35	0.67
Cl	300[7]	—	—	34	0.02	3.34	0.02

[1] Reed et al. (1960)
[2] Gast (1963) unpublished data. Pinson et al. (1965)
[3] Nishamura and Sandell (1964)
[4] Schmitt et al. (1963)
[5] Smales et al. (1964)
[6] Depleted by a factor of 8
[7] Reed and Allen (1966)
[8] Crustal abundances of Taylor (1964)
[9] Continental crust (2.09×10^{25} gms) ½ basalt and ½ granodiorite; Oceanic crust, abyssal basalt
[10] Continental crust ½ granodiorite and ½ diorite-granulite; Oceanic crust, abyssal basalt

entirely as the result of Rb depletion, i.e., the first possibly noted above. In this case, the terrestrial abundance of Rb is reduced by a factor of eight. The concentration of Rb in the crust for this assumption required by the various models given in Table 3 is nearly 100% in some cases; i.e., there is almost insufficient Rb in the earth to explain the "observed" crustal abundance. This lack of balance is even more serious than it first appears when it is recalled that mantle-derived volcanic rocks indicate that there are still very significant amounts of Rb in the present mantle of the earth.

Due to the chemical similarity of K and Rb, we may infer that the abundance of potassium in the earth may also differ from that inferred from chondritic meteorites. Birch (1965) and Wasserburg et al. (1964) have noted the apparent systematic differences in the K/U ratio of terrestrial materials and chondritic meteorites. These differences are summarized in Figure 2. A number of recent determinations of these elements are shown in this figure. It is indeed remarkable that two such chemically dissimilar elements show so little variation in their relative abundance in rocks. The constancy of these ratios suggests that the difference between the crustal and perhaps upper mantle K/U ratio and the chondritic K/U ratio is due to a terrestrial depletion in potassium relative to uranium rather than an earth-wide chemical differentiation of these elements. As in the case of the Rb/Sr ratio, this depletion may result from (1) a depletion in potassium alone, (2) an enrichment in uranium, or a combination of (1) and (2). Wasserburg et al. (1964) point out that the first possibility results in serious limitations on the

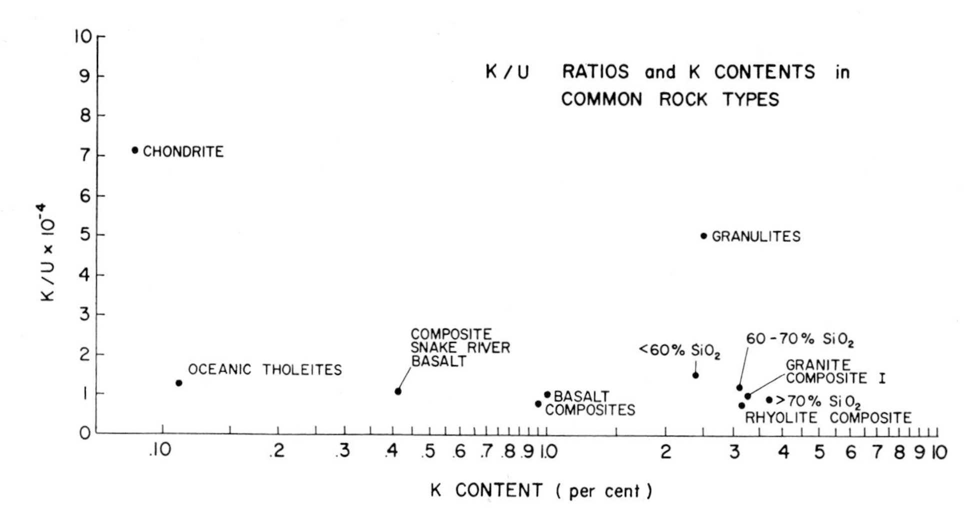

Figure 2. K/U ratios as a function of K content for some common rock types. Uranium contents of composites are new determinations. (Oversby, written communication, 1966.)

terrestrial heat budget and on this basis suggests that the terrestrial K/U ratio is a result of both depletion in K and enrichment in U.

Finally, I should like to note that the atmospheric abundance of Ar^{40} also places significant limits on the terrestrial K abundance. The extent of degassing of the earth has been estimated from the amount of Ar^{40} in the atmosphere (Damon and Kulp, 1958; Turekian, 1959). The estimates made by these authors have assumed that the terrestrial abundance of K is similar to that of chondritic meteorites. The accumulation time used in these calculations was 4.5 b.yr., that is, the age of chondritic meteorites. Since the production rate of Ar^{40} 4.5 b.yr. B.P. was 10 times higher than it is today, most of the terrestrial Ar^{40} was produced early in earth history. As a consequence, calculations of this type are very sensitive to the time when accumulation of Ar^{40} started. If terrestrial xenon and argon have similar histories, i.e., are both the result of outgassing of the earth, the isotope composition of xenon places some useful limits on the accumulation time. The low abundance of Xe^{129} in terrestrial xenon compared to that observed in many chondrites suggests that the separation of the earth from the solar nebula cannot have preceded the formation of these chondrites, i.e., is limited in time by the K-Ar ages of chondrites (cf. Cameron, 1962 and Pepin, 1964). Thus the accumulation time cannot be much greater than 4.6 b.yr. It has also been noted that there is some evidence that terrestrial xenon is enriched in the Xe isotopes resulting from the fission of Pu^{244} (Cameron, 1962; Pepin, 1964). This evidence is much stronger in the case of achondritic meteorites (Rowe and Bogard, 1966; Eberhardt and Geiss, 1966). If we accept this evidence, the earth must have accumulated from the solar nebula after the decay of I^{129} and before the decay of Pu^{244}. This suggests that the accumulation time of terrestrial Ar^{40} is not greater than 4.6 b.yr. and probably not less than 4.4 b.yr. Table 4 shows the extent to which Ar^{40} has been outgassed from the earth for these time limits for several

Table 4. Fraction of Ar^{40} Removed from the Solid Earth

Initial K content	Assumed accumulation time: 4.6 b.yr.	4.4 b.yr.
K 840 ppm	0.094	0.11
K 300 ppm	0.26	0.30
K 120 ppm	0.66	0.75

Decay constants: $\lambda_e = 5.85 \times 10^{-11}$/yr; $\lambda_\beta = 4.72 \times 10^{-10}$/yr

hypothetical terrestrial K contents: a chondritic K content (840 ppm); a K content assuming no terrestrial enrichment in uranium, i.e. $1/7$ of the chondritic K content (120 ppm); and an intermediate K content, $1/3$ of the chondritic abundance (300 ppm). The second of the above three possibilities suggests that depletion of potassium without enrichment in uranium is possible; it implies a very efficient removal of Ar^{40} from the earth. The first possibility is quite inconsistent with the extreme differentiation of U, Th, and Ba noted in Table 3. It seems very unlikely that the degree of concentration of Ar^{40} in the atmosphere should be *less* than the degree to which a refractory element like U or Th is concentrated into the earth's crust. The most likely possibility is the last, which involves both depletion of K and enrichment in U.

The combined evidence summarized here suggests both a depletion in certain volatile elements in the earth and an enrichment in certain refractory elements in the earth relative to carbonaceous chondritic meteorites. The magnitude of this depletion and enrichment is difficult to estimate with confidence. Wasserburg et al. (1964) suggests a U content approximately three times that found in chondrites, i.e., 0.03 ppm. It is seen in Table 3 that a three-fold increase in the Sr and Ba content produces a much more consistent explanation of Sr isotope abundance and crustal Ba and U content. Detailed discussion of the implications of such an enrichment in refractory elements is beyond the scope of this paper. It should be clear, however, that refractory major elements such as Ca and Al should also be enriched in the earth. Thus the mantle may be substantially richer than is usually assumed in these elements, e.g., 5-7% CaO and 7-10% Al_2O_3.

Current opinion leans heavily toward the idea that the planets formed by some condensation process from a primitive solar nebula. Larimer (1967) has calculated the condensation temperatures of a number of elements from a gas with the composition of the solar nebula. These calculations suggest that a condensate enriched in metallic Fe with respect to the oxide-silicate fraction will be accompanied by a silicate fraction enriched in refractory elements, in particular Ca, Al, and Ti. This is qualitatively consistent with the high Fe abundance of the earth relative to meteorites inferred from the uncompressed density of the earth by Urey (1952) and by Anderson and Kovach (1967). Urey noted that the abundance of volatile elements, e.g., Hg and Zn, in the earth is too high to be consistent with the high temperatures usually invoked in the condensation process. This objection is clearly still valid. The observed abundances require both a fraction enriched in refractory elements relative to the solar nebula and a small fraction with volatile elements preserved in ratios appropriate to carbonaceous chondrites.

The Lead Isotope Composition of Upper Mantle Lead

Models of continental and crustal evolution based on Sr and Pb isotope data have been reviewed in a number of recent papers (Patterson and Tatsumoto, 1964; Wasserburg, 1966; Hurley et al., 1962; Cooper and Richards, 1966; Gast, 1967). I will not attempt to review these models in detail here. The object of this discussion is principally to note the significance of lead isotope composition in oceanic volcanic rocks. (The discussion is limited to oceanic rocks because continental volcanic rocks are subject to contamination with lead from crustal rocks, cf. Hamilton, 1966.) The isotope composition of lead in oceanic volcanic rocks has been reported by Gast and Swainbank (1966), Tatsumoto (1966a, b) and Cooper and Richards (1966). Figure 3 summarizes data from several of the above studies as well as reporting some unpublished data obtained

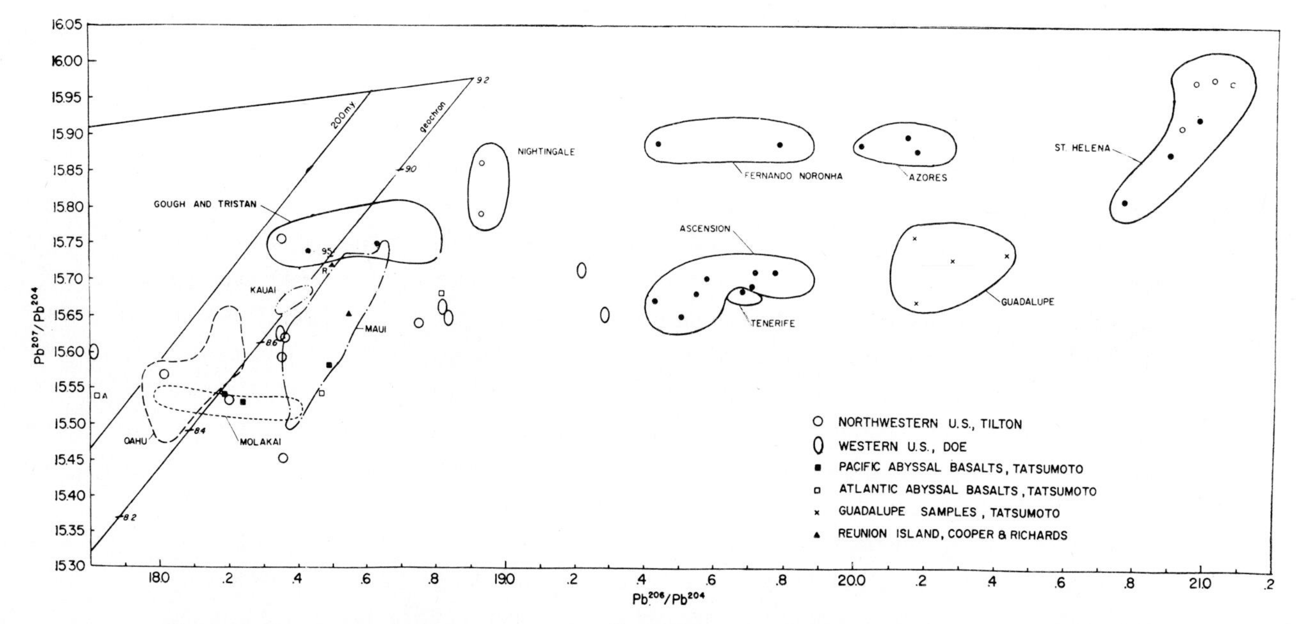

Figure 3. Isotope composition of lead in oceanic volcanic rocks. Data taken from Gast, Tilton, and Hedge (1964), Tatsumoto (1966a and 1966b), Cooper and Richards (1966), Gast (unpublished results, Doe (1967) and Tilton et al. (1964).

by the author. It is quite clear from this figure that the sub-oceanic mantle gives rise to a wide range of lead isotope compositions. The origin of these variations has been discussed by most of the authors mentioned above. Both Cooper and Richards and Tatsumoto prefer to ascribe most of the variations to a mantle-wide differentiation at some time in the past, between 2.7 b.yr. and about 1.2 b.yr. Gast (1967) has suggested that differentiation of the U/Pb ratio can have taken place at almost any time in the past if the cause of this is removal of a volcanic liquid. If we postulate that the observed differences in isotopic composition result from a single fractionation of the U/Pb ratio (μ-value), the time and extent of fractionation for the observed isotope compositions can be compared rather readily. The observed isotope composition will be given in the following equations:

$$[Pb^{206}/Pb^{204}]_{obs} = [Pb^{206}/Pb^{204}]_T + \mu_1\{\exp(\lambda_8 T) - \exp(\lambda_8 \tau_1)\} + \mu_2\{\exp(\lambda_8 \tau_1 - 1)\}$$

$$[Pb^{207}/Pb^{204}]_{obs} = [Pb^{207}/Pb^{204}]_T + \frac{1}{137.8}\left[\mu_1\{\exp(\lambda_5 T) - \exp(\lambda_5 \tau_1)\} + \mu_2\{\exp(\lambda_5 \tau_1) - 1\}\right]$$

where the subscript T designates the initial value for the primeval earth and τ_1 the time when U/Pb ratio was fractionated, and

$\mu_1 = [U^{238}/Pb^{204}]_1$ value during time interval $T > \tau > \tau_1$
$\mu_2 = [U^{238}/Pb^{204}]_2$ value of this ratio during interval $\tau_1 > \tau > 0$.

Figure 4 shows the permitted values of μ_1, μ_2, and τ_1 that will satisfy the above relations for several isotope compositions. It is quite clear that a wide range of both μ_1 and μ_2 values in the mantle are required if the variable lead isotope compositions are due to a single differentiation event. It is also clear that the required differentiation must have taken place at least 500 m.yr. ago in some cases if extreme mantle μ values are to be avoided. The observed oceanic lead isotope compositions thus imply that the present sub-oceanic mantle has been subject to extensive and perhaps frequent differentiation. I should like to suggest that the partial removal of U and Pb with volcanic liquids in the past (i.e., prior volcanic cycles unrelated to the present volcanism) can be the cause of this differentiation in the sub-oceanic mantle. Since there is very little evidence for such old volcanic cycles in oceanic basins, this explanation is more compatible with mantle convection and differential horizontal movement of various parts of the mantle and overlying crust than it is with a static mantle or continents with a chemical complement in the subcontinental mantle.

Even more pertinent to the question of whether or not continental masses are derived from the underlying mantle is a comparison of the isotope composition of lead in oceanic and continental volcanic rocks. MacDonald (1964) and Clark and Ringwood (1964) have shown that the subcontinental mantle must be extensively and substantially depleted in uranium if the continents are entirely derived from the upper mantle underlying them. It is highly unlikely that such depletion in uranium should not be accompanied by very significant fractionations of the U/Pb ratio in this residue. Thus continental evolution by

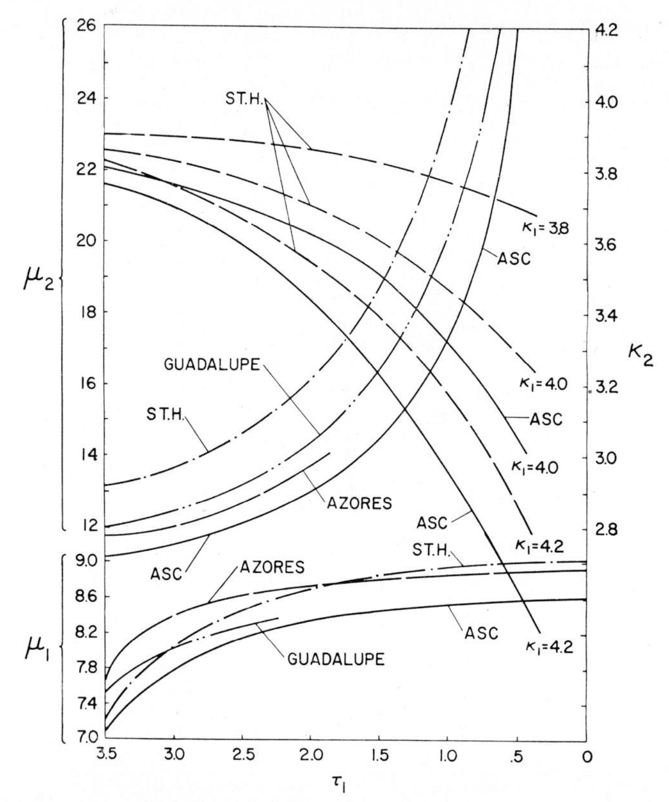

Figure 4. Permitted μ_1, μ_2, κ_1, and κ_2 values for a two-stage interpretation of lead from St. Helena, Guadalupe, Azores, and Ascension $\mu = U^{238}/Pb^{204}$ and $K = Th^{238}/U^{238}$. Where μ_1 and κ_1 are for $\Upsilon > \tau > \tau_1$ and μ_2 and κ_2 are for $\tau_1 > \tau > 0$.

vertical chemical segregation implies significant differences in the lead isotope composition of recent oceanic and continental volcanic rocks, if such rocks are isotopically representative of their upper-mantle source areas. This comparison is somewhat complicated by the possibility that some continental volcanic materials are contaminated with lead from the continental crust. Nevertheless, it is probably significant that published lead isotope compositions from continental volcanic rocks (Doe, 1967; Tilton et al., 1964) are indistinguishable from those shown for oceanic rocks.

Heat Flow

Clark and Ringwood (1964), MacDonald (1964) and Birch (1965) have interpreted the observed continental heat flow and the similarity of oceanic and continental heat flow in terms of an earth where the dominant transport of radioactive heat sources is vertical. Birch concludes from the similarity of oceanic and continental heat flow that differential movement of continental masses with respect to any part of the mantle is unlikely. This conclusion and the models proposed by MacDonald are strongly dependent on the amount of radioactivity concentrated into the mantle. Before using observations on heat flow to restrict the way in which continents evolved, it is necessary to evaluate the degree to which these elements are concentrated into the crust. The suggestion of Lambert and Heier (1966) that the lower crust is depleted in U and Th is very important in this regard. A possible distribution of radioactive elements in the continental crust is shown in Figure 5. At equilibrium the fraction of the total heat flux produced in the continental

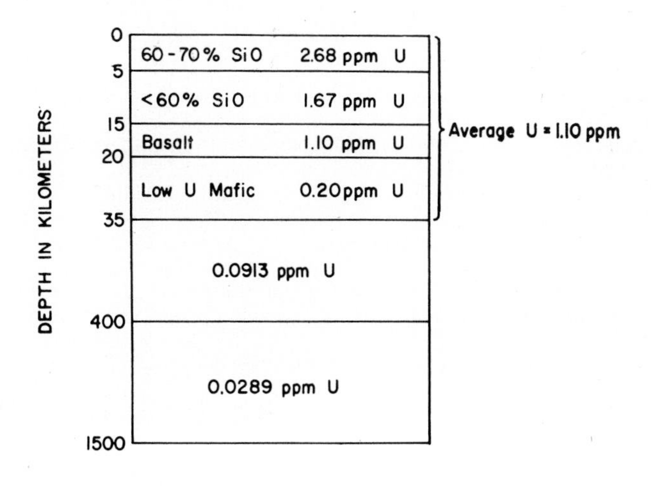

Average Initial Mantle conc. = 0.04 ppm U

Heat Production per sq. cm. of surface area.

	0 - 35 km	= 0.59 μ cal. / sec.
	35 - 400 km	= 0.58 μ cal / sec.
Total	0 - 400 km	= 1.17 μ cal / sec.
	Below 400 km	= 0.58 μ cal / sec.

Figure 5. A possible distribution of uranium in the crust and upper mantle. Heat calculations based on $K/U = 1 \times 10^4$, $Th/U = 4.0$.

crust, according to the present model, may be much less than one-half of the observed heat flow. The heat production rates in subcontinental and sub-oceanic mantle may thus be much more similar than suggested by the heat flow models that are generally accepted.

Summary

I have tried to show that isotopic and chemical observations made on rocks from the earth's surface suggest that the mean composition of the earth differs from that of carbonaceous chondrites—both in the lower abundance of some volatile elements and in the increased abundance of refractory elements. Such differences in the composition of the earth and chondritic meteorites along with the possibility that the lower continental crust has a substantially lower content of large ions than the upper crust, suggest a lesser degree of chemical differentiation for the earth than has heretofore been suggested.

The common occurrence of volcanic rocks in continental areas, and the similarities between these rocks and their oceanic counterparts, argues against widespread and fundamental chemical differences in the sub-oceanic and subcontinental mantle.

Lamont Geological Observatory
Columbia University

REFERENCES

Anders, E. (1965), Chemical fractionations in meteorites, *Meteoritica* 26:17-25 (NASA Contractor report CR-299).

Anderson, D. L., and R. L. Kovach (1967), The composition of the terrestrial planets, *Earth Planet. Sci. Letters* 3:19-24.

Bence, E. A. (1966), The differentiation history of the earth by rubidium-strontium isotopic relationships, Ph.D. Thesis, *M.I.T. Annual Progress Report*, pp. 35-78.

Birch, F. (1958), Differentiation of the mantle, *Geol. Soc. Amer. Bull.* 68:483-86.

Birch, F. (1965), Speculations on the earth's thermal history, *Geol. Soc. Amer. Bull.* 76:133.

Cameron, A. G. W. (1962), The formation of the sun and planets, *Icarus* 1:13.

Clark, S., and A. E. Ringwood (1964), Density distribution and constitution of the mantle, *Rev. Geophys.* 2:35-88.

Cooper, J. A., and J. R. Richards (1966), Lead isotopes and volcanic magmas, *Earth Planet. Sci. Letters* 1:259-69.

Damon, P. E., and J. L. Kulp (1958), Inert gases and the evolution of the atmosphere, *Geochim. Cosmochim. Acta* 13:280-92.

Doe, B. R. (1967), The bearing of lead isotopes on the source of granitic magma, *J. Petrol.* 8:51-83.

Eberhardt, P., and J. Geiss (1966), On the mass spectrum of fission xenon in the Pasamonte meteorite, *Earth Planet. Sci. Letters* 1:99-101.

Engel, A. E. J. (1963), Geologic evolution of North America, *Science* 140:143-152.

Engel, A. E. J., C. G. Engel, and R. G. Havens (1965), Chemical characteristics of oceanic basalts and the upper mantle, *Geol. Soc. Amer. Bull.* 76:719-34.

Fairbairn, H. W., P. M. Hurley and W. H. Pinson (1964), Initial Sr^{87}/Sr^{86} and possible sources of granitic rocks in southern British Columbia, *J. Geophys. Res.* 69:4889-93.

Gast, P. W. (1960), Limitations on the composition of the upper mantle, *J. Geophys. Res.* 65:1287-97.

Gast, P. W. (1962), The isotopic composition of strontium and the age of stone meteorites I., *Geochim. Cosmochim. Acta* 26:927-43.

Gast, P. W. (1967), Isotopic geochemistry of volcanic rocks, in *Basalts, a Treatise on Rocks of Basaltic Composition*, A. Poldervaart, ed., Interscience, N.Y. (in press).

Gast, P. W., and I. G. Swainbank (1966), The isotope composition of lead and relative abundance of U and Pb in oceanic volcanic rocks, *A.G.U. Trans.* 47:194.

Gast, P. W., G. R. Tilton, and C. Hedge (1964), Isotopic composition of lead and strontium from Ascension and Gough Islands, *Science* 145:1181-85.

Goldich, S. R., E. G. Lidiak, C. E. Hedge, and F. G. Walthall (1966), Geochronology of the midcontinent region, U.S., No. 2, northern area, *J. Geophys. Res.* 71:5389-08.

Hamilton, E. I. (1966), The isotopic composition of lead in igneous rocks, 1, The origin of some tertiary granites, *Earth Planet. Sci. Letters* 1:30-37.

Hedge, C. E., and F. G. Walthall (1963), Radiogenic strontium-87 as an index of geologic processes, *Science* 140:1214-17.

Heier, K. S. (1965), Metamorphism and the chemical differentiation of the crust, *Geol. Foren. Stockholm, Forhandl.* 87:249-56.

Hurley, P. M., H. Hughes, G. Faure, H. W. Fairbairn, and W. H. Pinson (1962), Radiogenic strontium-87 model of continent formation, *J. Geophys. Res.* 67:5315-34.

Lambert, I. B., and K. S. Heier (1966), The vertical distribution of thorium and uranium in the continental crust, *A.G.U. Trans.* 47:200.

Lambert, I. B., and K. S. Heier (1967), The vertical distribution of uranium, thorium, and potassium in the continental crust, *Geochim. Cosmochim. Acta* 31:377-90.

Larimer, J. (1967), Chemical fractionations in meteorites, I. Condensation of the elements, *Geochim. Cosmochim. Acta* 31:215.

MacDonald, G. J. F. (1964), Dependence of surface heat flow on the radioactivity of the earth, *J. Geophys. Res.* 69:2933-47.

MacDougall, I., and W. Compston (1965), Strontium isotope composition and potassium-rubidium ratios in some rocks from Reunion and Rodriquez, Indian Ocean, *Nature* 207:252.

McQueen, R. G., J. N. Fritz, and S. P. Marsh (1964), The composition of the earth's interior, *J. Geophys. Res.* 69:2947-65.

Moorbath, S., and G. P. L. Walker (1965), Strontium isotope investigation of igneous rocks from Iceland, *Nature* 207:837.

Moore, J. G. (1966), Base surge at 1965 eruption of Taal volcano, Philippines, *A.G.U. Trans.* 47:194.

Muehlberger, W. R., C. E. Hedge, R. E. Denison, and R. F. Marvin (1966), Geochronology of the midcontinent region, U.S., No. 3, southern area, *J. Geophys. Res.* 71:5409-26.

Murthy, V. R., and W. Compston (1965), Rb-Sr ages of chondrules and carbonaceous chondrites, *J. Geophys. Res.* 70:5297-307.

Nishamura, M., and E. B. Sandell (1964), Zinc in meteorites, *Geochim. Cosmochim. Acta* 28:1055-79.

Patterson, C. (1963), Characteristics of Pb isotope evolution on a continental scale in the earth, in *Isotopic and Cosmic Chemistry*, H. Craig, S. L. Miller, and G. J. Wasserburg, eds., North-Holland Publishing Co.

Patterson, C., and M. Tatsumoto (1964), The significance of lead isotopes in detrital feldspar with respect to chemical differentiation within the earth's mantle, *Geochim. Cosmochim. Acta* 28:1-22.

Pepin, R. O. (1964), Isotopic analyses of xenon, in *The Origin and Evolution of Atmospheres and Oceans*, Peter J. Brancazio and A. G. W. Cameron, eds., John Wiley & Sons, New York.

Pinson, W. H., C. C. Schnetzler, E. Beiser, H. W. Fairbairn, and P. M. Hurley (1965), Rb-Sr age of stony meteorites, *Geochim. Cosmochim. Acta* 29:455-66.

Poldervaart, A. (1955), Chemistry of the earth's crust, pp. 119-144 in *Crust of the Earth, Geol. Soc. Amer. Spec. Pap.* 62.

Reed, G. W., K. Kigoshi, and A. Turkevich (1960), Determinations of concentrations of heavy elements in meteorites by activation analysis, *Geochim. Cosmochim. Acta* 20:122-24.

Reed, G. W., Jr., and R. O. Allen, Jr. (1966), Halogens in chondrites, *Geochim. Cosmochim. Acta* 30:779-800.

Ringwood, A. E. (1966), Chemical evolution of the terrestrial planets, *Geochim. Cosmochim. Acta* 30:41-104.

Ringwood, A. E. (1966), Mineralogy of the mantle, pp. 357-99 in *The Earth's Environment*, P. M. Hurley, ed., MIT Press.

Rowe, M. W., and D. D. Bogard (1966), Xenon anomalies in the Pasamonte meteorite, *J. Geophys. Res.* 71:686.

Schmitt, R. A., R. H. Smith, and D. A. Olehy (1963), Cadmium abundances in meteoritic and terrestrial matter, *Geochim. Cosmochim. Acta* 27:1077-88.

Smales, A. A., T. C. Hughes, D. Mapper, C. A. J. McInnes, and R. K. Webster (1964), The determination of rubidium and caesium in stony meteorites by neutron activation analysis and by mass spectrometry, *Geochim. Cosmochim. Acta* 28:209-33.

Tatsumoto, M. (1966a), Isotopic composition of Pb in volcanic rocks from Hawaii, Iwo Jima and Japan, *J. Geophys. Res.* 71:1721-33.

Tatsumoto, M. (1966b), Genetic relations of oceanic basalts as indicated by lead isotopes, *Science*, 153:1094-1101.

Tatsumoto, M., C. E. Hedge, and A. E. J. Engel (1965), Potassium, rubidium, strontium, thorium, uranium and the ratio of Sr-87 to Sr-86 in oceanic tholeiitic basalt, *Science* 150:886-88.

Taylor, S. R. (1964) Trace element abundances and the chondritic earth model, *Geochim. Cosmochim. Acta* 28:1989-98.

Tilton, G. R., G. L. Davis, S. R. Hart, L. T. Aldrich, R. H. Steiger, and P. W. Gast (1964), Geochronology and isotope geochemistry, pp. 240-56 in *Carnegie Institution of Washington Year Book* 63.

Turekian, K. K. (1959), The terrestrial economy of helium and argon, *Geochim. Cosmochim. Acta* 17:37-43.

Turekian, K. K., and K. H. Wedepohl (1961), Distribution of the elements in some major units of the earth's crust, *Geol. Soc. Amer. Bull.* 72:175-92.

Urey, H. C. (1952), *The Planets*, Yale University Press, New Haven.

Wasserburg, G. J. (1966), Geochronology and isotopic data bearing on development of the continental crust, pp. 357-99 in *The Earth's Environment*, P. M. Hurley, ed., MIT Press.

Wasserburg, G. J., G. J. F. MacDonald, F. Houle, and W. A. Fowler (1964), Relative contributions of uranium, thorium and potassium to heat production in the earth, *Science* 143:465-67.

THE GEOPHYSICAL IMPORTANCE OF HIGH-TEMPERATURE CREEP

DAN P. MC KENZIE

1. Introduction

The deformation of materials in a stress field is in part elastic and in part anelastic. The elastic properties of the earth govern the velocity of seismic waves and are much better known than the anelastic. The purpose of this paper is to examine the laboratory experiments on metals and ceramics at high temperature and under constant stress in order to understand how the rocks of the mantle may deform anelastically. When a stress is applied to material under such conditions the immediate response is principally elastic and is followed by transient anelastic deformation, called transient creep. After a sufficiently long time, which depends on both the temperature and the material, the deformation rate becomes constant. Such steady-state creep governs the deformation in most high-temperature creep experiments. In the earth steady-state creep probably controls post-glacial uplift and tectonics in the lower crust and mantle. If convection is the cause of continental drift it may also be governed by such creep. The geophysical importance of experiments on pure materials at high temperatures is that they may be used to set a lower limit on the deformation rate under a given stress within the earth. In a polyphase solid other creep mechanisms can increase, but not decrease, this rate.

At present the principal technique used to study the lower crust and mantle is seismology. The periods of the elastic waves used for this purpose are less than an hour, and therefore their propagation is probably not affected by steady-state creep. Various phenomena, such as earth tides and the Chandler wobble, have time scales longer than seismic waves. Unfortunately the anelastic response at these frequencies cannot at present be separated from other factors which affect the amplitudes. It is therefore unlikely that the anelastic behavior of the earth between frequencies of 1 and 0 cycles/hour will be determined experimentally in the near future.

Several authors (Jeffreys, 1959; Orowan, 1965) have attempted to use the results of laboratory experiments on creep to understand how the earth's mantle might deform. At low temperatures most crystalline materials show a rather sudden transition from elastic to plastic flow as the shear load is increased. Both authors use this idea of a yield stress to interpret the deformation of the mantle. It is clearly dangerous to extrapolate the creep behavior of rocks and metals deformed rapidly at low pressure and temperature to determine the anelastic properties of the mantle, where the pressure is large and the creep rate small. Though such extrapolations cannot be avoided at present, the theory of creep processes suggests that temperature, not pressure, determines the creep mechanism. Thus the deformation of dense mantle materials is likely to be similar to that of low-porosity ceramics at high temperature and at atmospheric pressure.

The mantle is probably a complicated mixture of phases and may well be partially melted in its upper parts. Such a material can deform in many more ways than those discussed here. The idealized creep theory may then be used to place a lower limit on the strain rate as a function of stress (Nabarro, 1948).

2. Creep Mechanisms

A perfect crystal can yield only at shear stresses comparable with the shear modulus μ. Since real solids fracture at stresses of 10^{-2} to $10^{-5}\mu$ they must do so by mechanisms involving imperfections. Monatomic crystals contain point defects, in the form of interstitial atoms, impurities, and vacancies, and line defects such as screw and edge dislocations. An edge dislocation occurs in a crystal where a plane of atoms ends, and a screw occurs where two sides of a cut are displaced by one atomic unit (Read, 1953). The planes of atoms around a screw dislocation form a spiral. More complicated stacking faults are possible, especially in compounds. Several excellent books (Cottrell, 1953; Read, 1953; Friedel, 1964; Weertman and Weertman, 1964, Hull, 1965) have been written on the behavior of dislocations in a stress field, but they are mainly concerned with nearly perfect single crystals of metals at low temperatures. Under these conditions creep takes place by the movement and interaction of edge and screw dislocations. Since crystals can only grow where dislocations intersect their surfaces, no crystal is ever free from them. If both edge and screw dislocations were present in their equilibrium concentration there would be extremely few of them because their free energy is so large. Crystals might then behave more like perfect solids. The reason why the equilibrium concentration is probably never achieved is because in general dislocations can only be created and destroyed at a grain boundary. The only exception to this rule is a Frank-Read source (Read, 1953) which can create or destroy two dislocations of opposite sign in the same slip plane. This process is more effective at increasing than at decreasing the dislocation density.

In the absence of a shearing stress, dislocations in high-melting-point materials are not mobile at room temperature. They become mobile as the temperature is increased and tend to diffuse toward grain boundaries to reduce their concentration toward the equilibrium value. This process is known as annealing. If a shearing force is applied to a crystal, the dislocations experience a force; if they move, creep will take place. The shear stress required to cause movement is called the yield stress. Since some dislocations can move more easily than others, the yield point is not sharp, and limited creep can take place at lower stresses. However, under small loads the force on a moving dislocation is not sufficient to drive it through the elastic stress field around a fixed dislocation. If two edge dislocations or an edge and a screw actually intersect, the total length of the dislocations is increased. The energy required to do this must be obtained from the stress field. It is even more difficult for two screw dislocations to cross because a vacancy or an interstitial atom is produced (Read, 1953). Grain boundaries are effective at preventing the motion of dislocations because they behave as a dense two-dimensional network of interlocking dislocations. These ideas have been used by Mott (1953) to produce a theory of exhaustion or transient creep.

At temperatures greater than perhaps one third of the melting temperature, the mobility of point defects also becomes important. Unlike dislocations, mobile interstitials and vacancies at these temperatures are present in their equilibrium concentration. Point defects affect creep in two ways; they may carry material from places of high stress to those of low stress, or they may remove or add material to the points at which dislocations are pinned. The temperature dependence of both processes is governed by the activation energy for self-diffusion, but the stress dependence of the creep rate in the two cases is entirely different.

Diffusion or Nabarro-Herring creep is governed by the diffusion of point defects from one grain boundary to another. If the grain boundaries are represented by an array of dislocations, these act as sources and sinks of point defects, but they do not move in response to the applied load. In dense rocks and metals the energy required to form a vacancy is very much less than that required to produce an interstitial atom; therefore the equilibrium concentration of interstitials may be neglected. Vacancies can then be produced only along grain boundaries and dislocations. In the absence of an applied load the equilibrium concentration of vacancies will be constant. If a shear stress is suddenly applied the equilibrium concentration will be changed because pressure gradients are set up across the individual crystals. Therefore a transient vacancy current is generated which establishes a steady concentration gradient. The transient creep produced by this current has been observed (Folweiler, 1961). However, creep must continue even in the steady state because the pressure gradient causes a vacancy concentration gradient, and diffusion down the gradient produces creep. The change in equilibrium concentration is always sufficiently small to be considered as a small perturbation to the equilibrium concentration. Since the applied load merely perturbs an existing situation, the material has no finite strength, and at high temperatures there is nothing corresponding to a yield stress. Detailed analysis of the steady-state creep problem (Herring, 1950; see Appendix) gives

$$\dot{\epsilon} = \frac{C D V_a}{k T a^2} \sigma \tag{2.1}$$

where k is the Boltzman constant, T the absolute temperature, $\dot{\varepsilon}$ the strain rate, σ the stress, D the diffusion coefficient for vacancies, V_a the activation volume, and a the mean grain radius. The constant C is a numerical factor which depends on the geometry and boundary conditions. If all shear stresses on crystal boundaries are absent, C is about 10. Equation (2.1) shows that the strain rate depends linearly on stress, thus the deformation may be described by a viscosity

$$\eta = \frac{k T a^2}{C D V_a} \tag{2.2}$$

Since

$$D = D_0 \exp\left[- \frac{E + p V_a}{k T} \right] \tag{2.3}$$

where E is the activation energy for self-diffusion, p the pressure and D_0 is a constant, (2.2) becomes

$$\eta = \frac{k\,T\,a^2}{C\,D_0\,V_a} \exp\left[\frac{E + p\,V_a}{k\,T}\right] \tag{2.4}$$

The equations of motion are then easily derived and are the equations of fluid dynamics with a temperature-dependent η.

Diffusion can also take place along the grain boundaries rather than through the body of the crystal. Coble (1963) has demonstrated that the resulting creep obeys an equation similar to (2.1) but with a viscosity proportional to the cube of the grain size. The mechanism is important at low temperatures and small grain sizes, but is probably too slow to dominate creep within the mantle.

Another creep mechanism which is often mentioned because it gives the viscous equations of motion is grain boundary sliding. This mechanism is only important if one of the principal stresses is negative. If this is not the case, a polycrystalline solid cannot deform in this way because voids cannot form in the material. Schumacher (1941) pointed out that such a solid can only remain pore-free if the shapes of the crystals change as the solid deforms. In fact Herring (1950) assumed the grain boundaries to be inviscid and still obtained a diffusion-limited creep rate. It seems probable that throughout the mantle the hydrostatic pressure is greater than the shearing stress. Thus all principal stresses are positive and grain boundary sliding is not important. However, laboratory creep experiments at atmospheric pressure often have one negative principal stress, and void formation may indeed explain some of the rapid creep rates that have been observed (Warshaw and Norton, 1962; Passmore et al., 1966).

In diffusion creep, dislocations merely act as sources and sinks of vacancies. However, if the stress is sufficiently large to cause the dislocations themselves to move, the nature of the creep changes. At low temperatures obstacles to dislocation movement, such as stationary dislocations, grain boundaries, and impurities, limit the creep to a transient response. At high temperatures vacancy diffusion can remove the obstacles or move atoms to produce dislocation climb. Either process permits creep to continue and prevents exhaustion; thus transient creep becomes less important. Weertman (1957) has derived expressions for the creep rate when dislocation climb is the rate-limiting process. In his model, dislocations are produced at Frank-Read sources in different glide planes. They produce creep when they climb and annihilate each other. He obtained

$$\dot{\epsilon} = \text{const.}\ \frac{D}{M^{1/2}}\left(\frac{\sigma}{\mu}\right)^{4.5} \tag{2.5}$$

where M is the density of Frank-Read sources. The empirical relationship for high-temperature creep is (Sherby, 1962)

$$\dot{\epsilon} = \text{const.}\ D\,a^2\left(\frac{\sigma}{\mu}\right)^{5} \tag{2.6}$$

Thus there is good agreement between eqs. (2.5) and (2.6) except in the grain-size dependence, since it is unlikely that $M \propto 1/a^4$. The principal temperature dependence is through D, and therefore both creep regimes have the same activation energy, that for self-diffusion. A comparison of eqs. (2.1) and (2.6) suggests that dislocation climb will dominate over diffusion creep at a critical stress σ_c

$$\sigma_c \propto 1/a \tag{2.7}$$

The experiments clearly show that diffusion creep controls the deformation to higher stresses as the grain size is reduced.

The theories of Herring and Weertman were put forward to explain creep in metals. Though the general ideas apply equally well to ionic solids, there are several slight differences caused by the periodic nature of the electric charge distribution. In an ionic solid, diffusion must preserve charge neutrality. If this constraint were not satisfied the electric field produced would cause rapid diffusion to restore neutrality. Thus diffusion is rate-limited by the movement of the slower ion. The periodic nature of the charge distribution also limits the slip planes in which dislocations may move (Fig. 1). Slip on

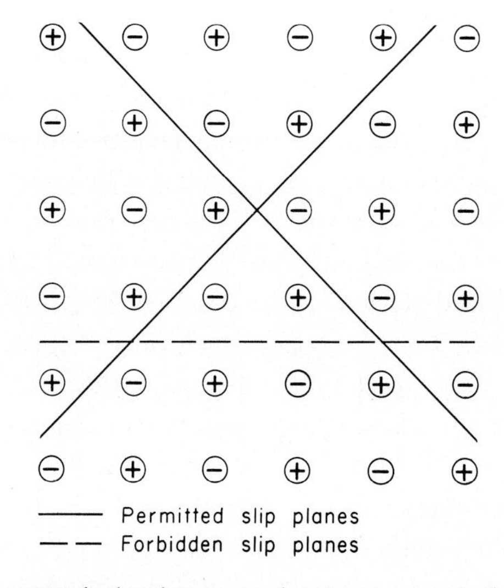

Figure 1. In ionic crystals the charges on the atoms prevent slip in certain planes.

planes which would cause similarly charged atoms to become nearest neighbors is not possible. There is no such restriction in metals. In a polycrystalline material, creep by dislocation movement can take place only if there are five active slip directions. This condition is known as von Mises' criterion and is often not satisfied by ionic solids at low temperatures. Such solids must fracture or creep by the formation of voids when the elastic strength is exceeded. Even at high temperatures dislocation creep is difficult, and diffusion dominates at higher stresses than in metals. One reason why diffusion creep remained unobserved is that it only controls creep in metals at stresses $\lesssim 10^{-5}\,\mu$ and at temperatures close to the melting point. It is therefore much easier to detect in ionic solids.

The two creep mechanisms discussed in this section are not the only ones which may operate in the mantle. Twinning, in particular, may dominate both at high stresses. However, it is clear from the experiments on ceramics that diffusion and dislocation climb together explain the general creep behavior, and therefore form a realistic model for steady-state creep in the mantle.

3. High-Temperature Creep Experiments

Few creep experiments have been performed on rocks and minerals. Those most often cited (Lomnitz, 1956; Griggs et al., 1960; Misra and Murrell, 1964) were carried out at low temperature and high shear stress and therefore may apply to the crust but not to the mantle as a whole. The high-temperature creep of ceramics at low stresses is more relevant to the mechanical properties of the earth's deep interior. A large number of experiments have recently been carried out on these materials because of their engineering interest. They are used in nuclear piles and turbines because of the thermodynamic advantages of operating at as high a temperature as possible. This work is rarely discussed in the geophysical literature, and the principal purpose of this section is to review those experiments which are relevant to the earth.

If the theories of creep discussed above are correct, the behavior of all materials should be the same if they are compared at the same values of $\sigma/\mu = \pi$ and T/E. Though the discussion of dislocation movements shows that metals do not behave in the same way as ceramics, each group does in outline obey such an "equation of state." The shear modulus has been determined for most materials of interest, but the same is not true for E. Especially in the case of polyatomic crystals it is not clear whether E for the anions or cations should be used, even if they are known. However, there is an approximate theory of melting (Sherby, 1962) which relates E to the melting temperature T_M. Since only an outline of the creep regimes is required $T/T_M = \tau$ is used instead of T/E.

In pure metals diffusion creep is difficult to observe because dislocations move easily on the many active slip planes and therefore dominate creep at all except the lowest stresses. Despite the experimental difficulties Greenough (1952) observed diffusion creep in silver at $\pi = 4 \times 10^{-7}$ and $\tau = 0.96$. He also succeeded in demonstrating that the creep rate in polycrystals was at least a factor of ten faster than in single crystals. Diffusion creep has been measured in gold at $\pi \leqq 2 \times 10^{-6}$, $\tau = 0.98$ by Buttner et al. (1952); and in copper at $\pi \lesssim 2 \times 10^{-6}$, $0.91 < \tau < 0.98$ by Pranatis and Pound (1955). When $\pi \gtrsim 2 \times 10^{-6}$, dislocation climb dominates in metals and creep obeys eq. (2.6). Sherby (1962) reviews the extensive work which has been carried out in the high-stress nonlinear creep regime in metals. Bonding in copper, silver, and gold is principally metallic and is produced by the delocalization of the outer electron from each atom. In contrast to covalent bonds, metallic bonds are not strongly directional and therefore only weakly oppose dislocation movement. Metals like magnesium, aluminum, and zirconium have more than one free electron from each atom, and therefore the bonding is partly covalent. For this reason diffusion creep dominates to higher stresses, and has been observed in magnesium at $0.6 < \tau < 0.75$, $10^{-5} < \pi < 10^{-4}$ (Jones, 1965), in aluminum at $\tau = 0.99$, $\pi < 5 \times 10^{-6}$ (Harper and Dorn, 1957), and in zirconium at $\pi < 6 \times 10^{-4}$, $\tau = 0.40$ (Jones, 1966). The experiments on aluminum are puzzling because the creep rate is a thousand times more rapid than eq. (2.1) predicts but in other respects resembles diffusion creep. Squires et al. (1963) have also observed creep in magnesium-zirconium alloys, where only the magnesium diffuses to form pure magnesium zones on the tension boundaries. The difference in composition is clearly visible and strikingly supports the diffusion mechanism (Fig. 2).

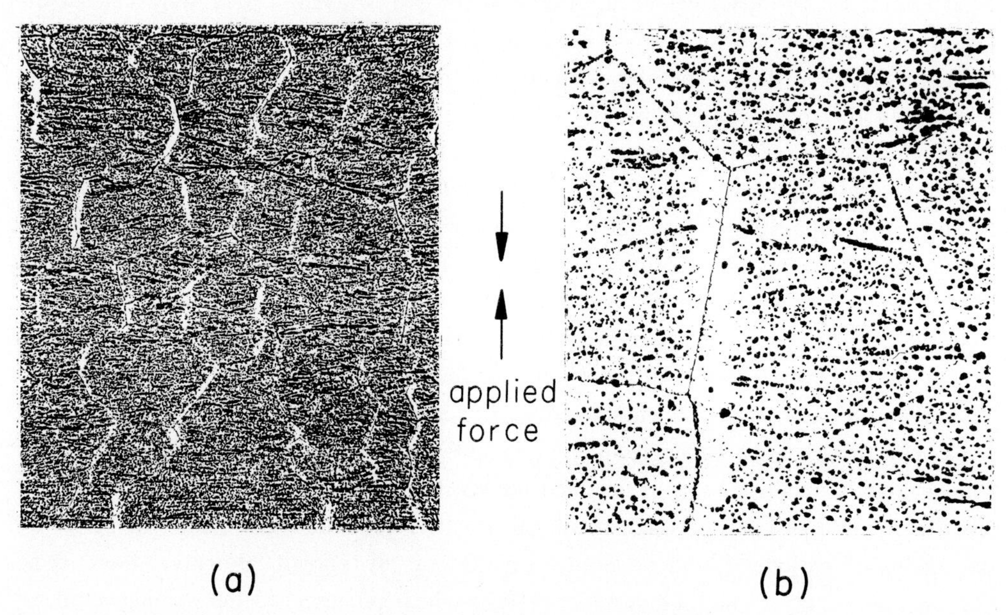

Figure 2. (a) Magnesium-zirconium alloy after deformation by diffusion creep. The magnesium diffuses to the tension boundaries, but the dark streaks of zirconium do not move and therefore are absent where the grains have grown. (b) Shows the grain boundary growth in detail.

Dislocations in ceramics are even less mobile than those in polyvalent metals. Most experiments have used BeO, MgO, Al_2O_3, or UO_2 at $0.4 < \tau < 0.9$ and $10^{-5} < \pi < 10^{-3}$. Except in single crystals, dislocation climb only dominates when $\pi \gtrsim 3 \times 10^{-4}$. The value of τ corresponding to the experimental temperature may be obtained from the melting point; that of MgO, BeO, and Al_2O_3 is given by Birch et al. (1942), that of UO_2 by Hausner (1965). The shear modulus for polycrystalline MgO and Al_2O_3 is quoted by Anderson and Nafe (1965), for BeO by Bentle (1966), and for UO_2 by Wachtman et al. (1965).

Most creep experiments are carried out in a three- or four-point bending apparatus inside a furnace. The specimen is a beam supported at two points with a load applied at one or two others. The deflection as a function of time and certain assumptions about the stress within the specimen then determine the creep rate as a function of stress. Most experiments have been carried out between 1000°C and 1900°C. Below these temperatures polycrystalline ceramics fracture and do not creep. The combined results from these four oxides cover the region $0.4 < \tau < 0.9$ because their melting temperatures range from about 2000°C to 2800°C.

The creep behavior at the highest τ values is obtained from alumina. Folweiler (1961) showed that creep was by diffusion when the grain size was between 7 and 34μ, with an activation energy of 130 kcals/mole. He used stresses of $10^{-5} < \pi < 10^{-3}$. Even at the highest stresses his results obeyed eq. (2.1) rather than (2.6). However, Warshaw and Norton (1962) obtained dislocation dominated creep when $\pi \gtrless 2 \times 10^{-5}$ with grain sizes of 50-100μ. The creep rate in both experiments was a thousand times too rapid for Nabarro-Herring creep, which Coble and Guerard (1963) suggest was due to grain separation. Warshaw and Norton's results are especially doubtful because the yield point in single crystals is $\pi \gtrless 8 \times 10^{-5}$ (Wachtman and Maxwell, 1957; Kronberg, 1962). The activation energy for O^{2-} self-diffusion is 152 kcals/mole, but only 110 kcals/mole if diffusion is along grain boundaries (Oishi and Kingery, 1960). The aluminum ion has an activation energy of 114 kcals/mole (Paladino and Kingery, 1962) but does not diffuse preferentially along grain boundaries. Thus if the grains are small oxygen will diffuse along the grain boundaries and creep will be rate-limited by body diffusion of Al^{3+}. Paladino and Coble (1963) believe that creep will become rate-limited by O^{2-} at larger grain sizes (20-40μ).

The experiments on BeO are more consistent. Barmore and Vandervoort (1965) used grain sizes from 8.5 to 40μ, $\pi \leq 1.9 \times 10^{-4}$, and observed diffusion creep only, with an activation energy of 99 kcals/mole. The same authors (1963) obtained some dislocation movement at $\pi = 2.4 \times 10^{-4}$ with a grain size of 10μ. These results were extended to higher stresses $\pi \sim 10^{-3}$ by Fryxell and Chandler (1964), who observed a change from diffusion to dislocation creep when $\pi \simeq 2.5 \times 10^{-4}$. Austerman (1964) believes that creep is rate-limited by Be^{2+} body diffusion, though the activation energies do not agree very well (De Bruin and Watson, 1964).

Creep in single crystals of UO_2 has a yield stress $\pi \sim 10^{-4}$, then approximately obeys eq. (2.6) with an activation energy of 120 kcals/mole (Armstrong et al., 1966). In polycrystalline specimens with grain sizes 6-40μ, diffusion creep controls the deformation when $\pi \lesssim 6 \times 10^{-4}$, at greater stresses dislocation creep takes over (Armstrong et al., 1962; Armstrong and Irvine, 1964). The activation energy is 90 kcals/mole, which agrees with that of 88 kcals/mole for U^{4+} self-diffusion obtained by Auskern and Belle (1961), though not with that of 73 kcals/mole obtained by Yajima et al. (1966).

Diffusion creep has been observed in polycrystalline magnesia in the range $5 \times 10^{-5} < \pi < 2.5 \times 10^{-4}$ and $0.48 < \tau < 0.60$ by Vasilos et al. (1964) and Passmore et al. (1966). Passmore et al. believe that the activation energy increases as the grain size decreases, and that the limiting process changes from O^{2-} extrinsic diffusion to perhaps intrinsic Mg^{2+}. Their results are surprising because a high activation energy process would be expected to dominate at all grain sizes if it rate-limits the creep for the smallest. It will be interesting to see if further work supports their interpretation. Experiments at high temperatures, $0.4 < \tau < 0.6$, using single crystals (Cummerow, 1963; Rothwell and Neiman, 1965) show a yield point $\pi \simeq 10^{-4}$ beyond which creep takes place according to a law similar to eq. (2.6), but with a stress exponent between 3 and 7. It is probable that the creep rate is governed by dislocation climb.

These results are all plotted together in Figure 3 and it is clear that there are two creep regimes. The position of the boundary between them depends on the grain size, but in general occurs at a higher value of π in ceramics than it does with metals. The diffusion rate is governed by body diffusion of the metal cation in all the experiments so far carried out. This is a somewhat unexpected result because the oxygen ion is bigger than the cation and might therefore be expected to diffuse more slowly. The reason why this is not the case is because O^{2-} ions diffuse along dislocations and grain boundaries at a lower activation energy than that for body diffusion of cations. There is no evidence that the intrinsic diffusion of cations is enhanced in this way. Thus theoretically it is expected that oxygen ion body diffusion will rate-limit creep at grain sizes larger than any of those used so far. It is important to test this hypothesis, since there may be enough dislocations in crystals to permit oxygen ions to continue to move along grain boundaries even in large crystals. It is also important to geophysics to re-examine creep in magnesia, even though body diffusion was the rate-limiting process in all the experiments which have been carried out.

4. Geophysical Applications

The most obvious and important result of the experiments is that finite strength is not observed in polycrystalline materials at high temperature. The rock of the earth's mantle probably has a larger grain size than the ceramics discussed in Section 3, but this cannot affect the conclusion that the rock will creep under any load, however small. Provided the stress is sufficiently small (Fig. 3), the much abused viscous constitutive equation applies and, in general, may always be used to put a lower limit on the creep rate. Indeed when Nabarro (1948) originally suggested the diffusion creep mechanism he remarked "if the . . . argument could be extended to include crystals held at their melting points under external pressures comparable with their internal pressures this estimate could be used as a lower limit to the viscosity of the lithosphere of the earth." The various viscous solutions to geophysical problems (Haskell 1935, 1936; Crittenden 1963) are therefore more relevant to the mantle than has often been believed. It is likely that similar solutions can be used to examine the existence and consequences of convection in the mantle. The absence of finite strength makes the existence of such convection more likely. Figure 4 is a summary of the experiments using ceramics. A great advantage of using τ and π as the dimensionless variables is that the creep regime boundaries are then approximately independent of pressure. The reason for this is because the pressure only affects the movement of dislocations through the shear modulus, and the activation volume for vacancy formation controls both the melting temperature and ionic diffusion. Thus if the values of μ and T_M at the relevant pressure are used to obtain τ and π, the regime boundaries in Figure 4 still apply. This argument does not hold for low-temperature creep, and it also breaks down if the activation volume of the dislocation is greater than that of the vacancy. In both cases the boundaries in Figure 4 probably move to higher π values with increasing pressure. The points corresponding to two harmonics of the external gravity field are also plotted. The stress fields required to support these harmonics were obtained from Kaula's (1963) estimates for an elastic mantle. Since the shear modulus within the mantle is accurately known (Bullard, 1957), the values of π corresponding to these points are probably well determined. The same

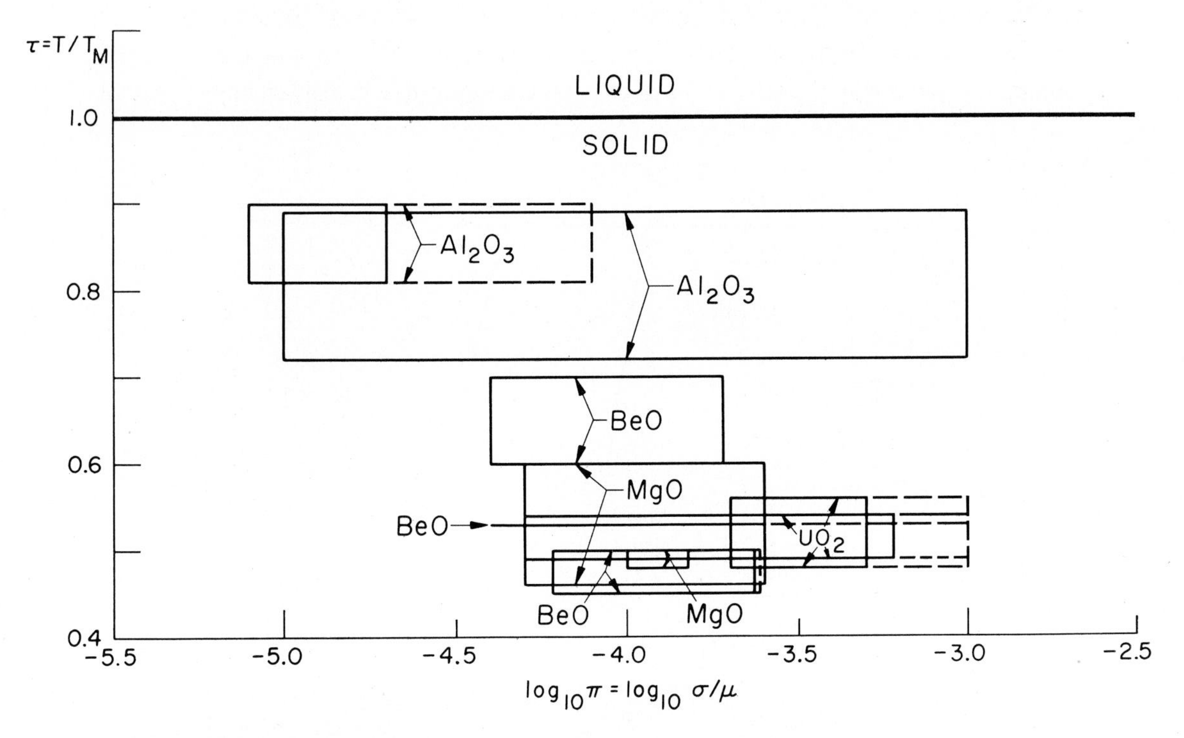

Figure 3. High-temperature creep in ceramics. Solid lines surround regions in which diffusion creep takes place, broken lines outline regions of nonlinear creep.

is not true for the values of τ, which are derived from Uffen's (1952) estimates of the melting temperature and the temperature distribution obtained from the electrical conductivity (McKenzie, 1967). Figure 4 suggests that the decay of the nonhydrostatic gravity field will be controlled by diffusion creep, and not dislocation climb.

The theory of diffusion creep expresses the viscosity in terms of the other parameters in eq. (2.4); if their values were known the viscosity η of the mantle could be estimated directly. The value of the viscosity obtained from experiments on metals is $\sim 10^{13}$ poise, that from ceramics $10^{15} \rightarrow 10^{17}$ poise. Unfortunately only poor estimates exist for both the temperature and the activation energy within the mantle. For this reason the values obtained by Zharkov (1963) and Gordon (1965) are of little use in understanding the long-term mechanical properties of the mantle. At present such attempts only demonstrate that diffusion creep can account for the observed rates of mantle deformation (McKenzie, 1967).

The forces responsible for earthquakes must also produce creep. Though at present the stress required to cause an earthquake in the mantle is rather uncertain, Press and

Brace (1966) estimate that the stress release in the 1964 Alaskan earthquake was about 100 bars. This shearing stress is used to obtain the value of π in Figure 4, that of τ is for the low-velocity layer. Though other earthquakes have different values of τ and π, most estimates for those below 40 km are in the region of creep by dislocation climb.

At shallow depths, rocks can fail by brittle fracture, and therefore the origin of earthquakes in this region is part of the general problem of the origin of stresses in the crust and upper mantle. However, at high pressure rocks yield by creep and not by brittle failure, thus the origin of deep earthquakes is obscure even if sufficient energy is present

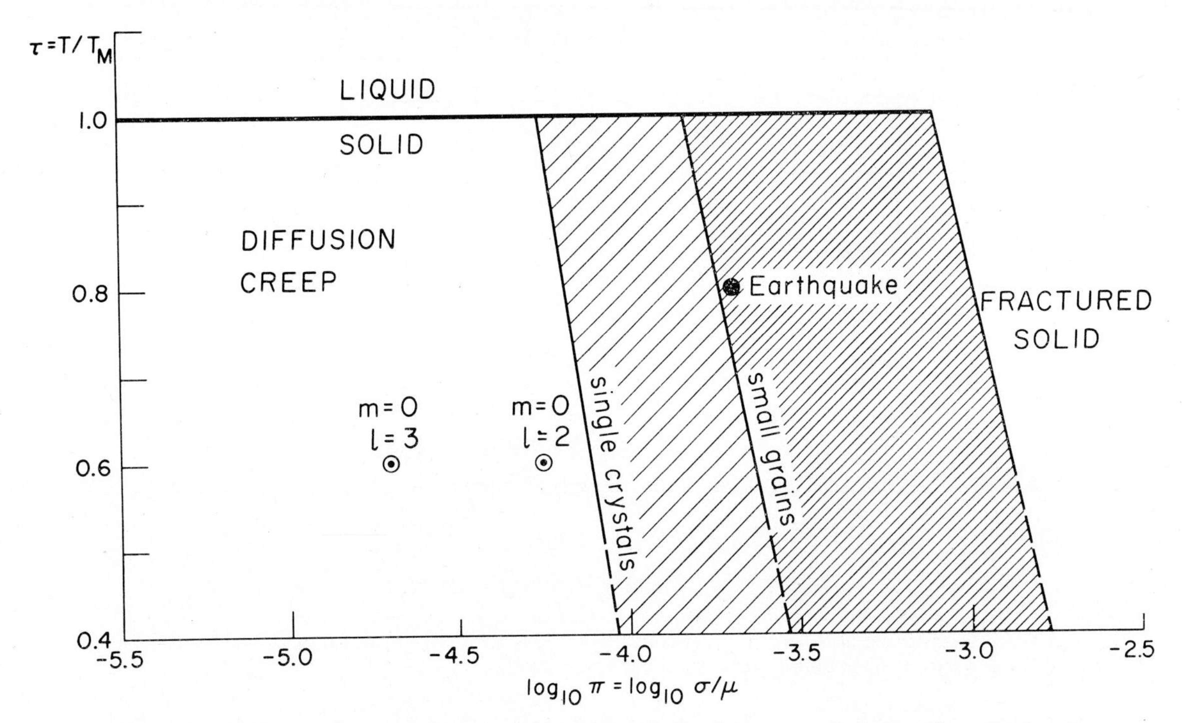

Figure 4. Sketch of creep regimes for ceramics. Creep is by dislocation climb for all grain sizes in the densely hachured region, and by diffusion only for small grain sizes in lightly hachured region. Points marked ⊙ are obtained from the harmonics of the external gravity field.

in the stress field. There is some experimental evidence that the plasticity of rocks at low temperature and high hydrostatic pressure may be due to the production of a large number of mobile dislocations as the pressure is applied. Bullen et al. (1964) found that iron remained ductile after an applied hydrostatic pressure of 10 kbar had been released. If the specimen was annealed at atmospheric pressure and then recycled it showed a yield point. This result suggests that the free dislocations may be fixed by annealing at high pressure, and that iron would then show brittle failure even at high pressure. If rocks behave in the same way their plasticity in high-pressure experiments is caused by the experimental conditions, and the results cannot be applied to the earth.

Though the pressure may or may not affect brittle fracture, the temperature certainly does. Thus the problem of the origin of deep earthquakes remains. The explanation of their existence may be an instability in the dislocation creep regime, an idea which is supported by the depth distribution of earthquakes. Sykes (1966) showed that there were fewer earthquakes in the Kermadec-Tonga region at depths between 200 and 400 km than in the regions above and below. In this region the temperature is closer to the melting temperature than anywhere else within the mantle. The resulting high creep rates may release stress continuously rather than as earthquakes.

Though diffusion creep in the lower mantle probably governs the decay of the low-order harmonics of the gravity field, the occurrence of earthquakes above 700 km may well demonstrate that the upper mantle principally creeps by dislocation climb. The equations of motion for such flow are nonlinear and must be solved numerically in even the simplest cases. Viscous solutions to upper mantle problems will then be of qualitative interest only.

An interesting consequence of creep by dislocation climb is the orientation of polycrystalline grains. Mineral orientation is common both in geology and in the laboratory (Griggs et al., 1960) and may be used to examine the orientation of the stress tensor. Hess (1964) has suggested similar orientation may produce regional seismic anisotropy in the upper mantle. Attempts have been made to use equilibrium thermodynamics to relate the grain orientation to the stress field (Kamb, 1961; McLellan, 1966). However, if creep by dislocation movement is responsible, such theories are not likely to apply. At low stresses diffusion creep will dominate, and vacancies, unlike dislocations, move isotropically (Austerman and Wagner, 1966). Thus orientation at low stresses may agree with equilibrium theories, but the alignment will probably be poor. This difference between the consequences of the two creep mechanisms may allow their relative importance within the earth to be determined from measurements of seismic anisotropy.

This paper is principally concerned with steady-state creep. It is obviously necessary to examine transient creep and low-frequency acoustic attenuation in the same manner. Such a survey would perhaps determine the origin and importance of seismic attenuation.

Appendix: Diffusion Creep

The argument by which Herring (1950) derived the effective viscosity of a creeping solid is extremely condensed. Also his final equation contains a misprint which has caused it to be quoted incorrectly (Gordon, 1965; McKenzie, 1967). The purpose of this appendix is to expand Herring's discussion, especially his physical arguments. The rather tedious algebra is omitted.

The concentration of vacancies, n, is a solid at a pressure p_0 is

$$n = n_0 \exp\left[-\frac{H + p_0 V_a}{kT}\right] \tag{A.1}$$

n_0 is a constant, H the activation energy, and V_a is the activation volume. If a shearing force is applied, it produces a small change p_1 in the pressure field. At high temperatures solids fracture before $p_1 V_a$ becomes comparable with kT. Thus the change in the vacancy concentration δn may be obtained by expanding (A.1)

$$\delta n = -\frac{p_1 V_a}{k T} n \qquad (A.2)$$

The vacancy flux is $-D_v \nabla \delta n$ where D_v is the self-diffusion coefficient for vacancies. The volume transport of material $\mathbf{j}$ is in the opposite direction to the vacancy flux, and therefore is

$$\mathbf{j} = D_v V_v \nabla \delta n \qquad (A.3)$$

where V_v is the volume occupied by each atom in a perfect crystal.

The diffusion coefficient D is defined in terms of the bulk diffusion of the solid. In this problem such transport is by vacancy diffusion alone:

$$D = \frac{n}{N} D_v \qquad (A.4)$$

$N = 1/V_v$ and is the number of atoms per cubic centimeter. Thus (A.3) becomes

$$\begin{aligned} \mathbf{j} &= D \frac{N}{n} V_v \nabla \delta n \\ &= -\frac{D V_a}{k T} \nabla p_1 \end{aligned} \qquad (A.5)$$

Equation (A.5) is essentially the solution of the creep problem. However, the equation is more useful in the form of an effective viscosity, even though the viscosity so obtained depends on the geometry of the crystals and on the boundary conditions. A useful and probably realistic model is a sphere strained in simple shear. This strain may be produced by normal stresses applied to its surface. Tangential stresses on its surface are taken to be zero, since in most solids they are relaxed by grain boundary diffusion. The solution of the viscous problem with the same boundary conditions can then be obtained from the correspondence principle. The solution of the elastic equations with the correct boundary conditions gives p_1, and hence $\mathbf{j} \cdot \mathbf{a}_r$ on the sphere's surface ($\mathbf{a}_r$ is the unit vector along the radius). The effective viscosity is then obtained by requiring the surface deformation rates to be the same.

The diffusion model used here is a small sphere set in a large block of material. The origin of the coordinates is chosen to be the center of the sphere. If the shear stress in the block in the absence of the sphere is σ, and is in the xz (and zx) direction, then the displacement $\mathbf{u}$ is

$$\mathbf{u} = \frac{\sigma}{\mu} (z, 0, 0) \qquad (A.6)$$

in Cartesian coordinates. Though $\mathbf{u}$ will be affected by the sphere it is probably a good approximation to assume that the displacement at the sphere's surface is given by (A.6). The strain boundary condition is that the normal displacement should be continuous at the sphere's surface:

$$[\mathbf{u} \cdot \mathbf{a}_r]_{r=a} = \frac{\sigma}{2\mu} a \sin 2\theta \cos \phi \tag{A.7}$$

The equivalent normal velocity $\mathbf{v}$ in the viscous problem may be obtained from the correspondence principle:

$$[\mathbf{v} \cdot \mathbf{a}_r]_{r=a} = \frac{\sigma}{2\eta} a \sin 2\theta \cos \phi \tag{A.8}$$

(A.8) and (A.6) then give:

$$-\frac{D V_a}{k T} [\nabla p_1 \cdot \mathbf{a}_r]_{r=a} = \frac{\sigma}{2\eta} a \sin 2\theta \cos \phi \tag{A.9}$$

p_1 must be obtained from the elastic equations for an incompressible solid:

$$\begin{aligned} \mu \nabla^2 \mathbf{u} &= \nabla p_1 \\ \nabla \cdot \mathbf{u} &= 0 \end{aligned} \tag{A.10}$$

The displacement field within the sphere and p_1 are obtained from the boundary conditions by standard methods. The solution for the pressure field is

$$p_1 = -\frac{21\sigma}{10a^2} r^2 \sin 2\theta \cos \phi \tag{A.11}$$

Substitution into (A.9) then gives

$$\eta = \frac{k T a^2}{8.4 D V_a} \tag{A.12}$$

In this notation Herring's equation (11) should read

$$\eta = \frac{k T a^2}{10 D V_a} \tag{A.13}$$

Thus the two treatments agree well. The small difference in the numerical factor is produced by the boundary conditions.

ACKNOWLEDGMENTS

I would like to thank R. L. Squires for supplying the photographs in Figure 2.

Department of Geodesy and Geophysics
University of Cambridge

DISCUSSION

DR. HURLEY: Some of the heavy elements do not occur as solid phases within the mantle, but as thin fluid layers between grains. The thickness of the layer is controlled by the temperature. What do you estimate the effect of the fluid intergranular material will be on the creep rates?

DR. MC KENZIE: Your remarks do not suggest the fluid can form large pores filled with liquid. Thus its principal effect will be to relax shearing stresses at the grain boundaries. Since Herring's theory requires the transverse stress to vanish between grains I believe the fluid layer will improve the agreement between theory and experiment.

DR. PRESS: It now seems probable that parts of the upper mantle are partially melted. Do you expect this to affect the creep rates?

DR. MC KENZIE: Partial melting can greatly increase the creep rates, if liquid is permitted to flow round the grains. The problem is then similar to creep experiments at atmospheric pressure when voids are allowed to form, except that the voids will be filled with liquid. However, if the fraction of liquid present is increased, creep will continue to be rate-limited by diffusion in the solid phase until the grains disaggregate. These remarks are not valid if the liquid and solid are in equilibrium, when no arguments of this kind can be used because the solid can melt to fill a potential void.

REFERENCES

Anderson, O. L., and J. E. Nafe (1965), The bulk modulus—volume relationship for oxide compounds and related geophysical problems, *J. Geophys. Res.* 70:3951-63.

Armstrong, W. M., W. R. Irvine, and R. H. Martinson (1962), Creep deformation in stoichiometric uranium dioxide, *J. Nucl. Mater.* 7:133-41.

Armstrong, W. M., and W. R. Irvine (1964), Creep of urania base solid solutions, *J. Nucl. Mater.* 12:261-70.

Armstrong, W. M., A. R. Causey, and W. R. Sturrock (1966), Creep of single-crystal UO_2, *J. Nucl. Mater.* 19:42-49.

Auskern, A. B., and J. Belle (1961), Uranium ion self diffusion in UO_2, *J. Nucl. Mater.* 3:311-19.

Austerman, S. B. (1964), Self diffusion in BeO, *J. Nucl. Mater.* 14:248-57.

Austerman, S. B., and J. W. Wagner (1966), Cation diffusion in single-crystal and polycrystalline BeO, *J. Amer. Ceram. Soc.* 49:94-99.

Barmore, W. L., and R. R. Vandervoort (1965), High temperature plastic deformation of polycrystalline beryllium oxide, *J. Amer. Ceram. Soc.* 48:499-505.

Bentle, G. G. (1966), Elastic constants of single-crystal BeO at room temperature, *J. Amer. Ceram. Soc.* 49:125-28.

Birch, F., J. F. Schairer, and H. C. Spicer (1942), *Handbook of Physical Constants, Geol. Soc. Amer. Spec. Pap.* 36.

Bullard, E. C. (1957), The density within the earth, *Koninkl. Ned. Geol-Mijnb. Genoot. Verhandel. Geol. Ser.* 18:23-41.

Bullen, F. P., F. Henderson, M. M. Hutchison, and H. L. Wain (1964), The effect of hydrostatic pressure on yielding in iron, *Phil. Mag.* 9:285-97.

Buttner, F. H., E. R. Funk, and H. Udin (1952), Viscous creep of gold wires near the melting point, *Trans. AIMME* 194:401-408.

Coble, R. L. (1963), A model for boundary diffusion controlled creep in polycrystalline materials, *J. Appl. Phys.* 34:1679-82.

Coble, R. L., and Y. H. Guerard (1963), Creep in polycrystalline aluminum oxide, *J. Amer. Ceram. Soc.* 46:353-54.

Cottrell, A. H. (1953), *Dislocation and Plastic Flow in Crystals*, Oxford.

Crittenden, M. D., Jr. (1963), Effective viscosity of the earth derived from isostatic loading of pleistocene Lake Bonneville, *J. Geophys. Res.* 68:5517-30.

Cummerow, R. L. (1963), High temperature steady state creep rate in single crystal MgO, *J. Appl. Phys.* 34:1724-29.

De Bruin, H. J., and G. M. Watson (1964), Self diffusion of beryllium in unirradiated beryllium oxide, *J. Nucl. Mater.* 14:239-47.

Folweiler, R. C. (1961), Creep behaviour in pore free polycrystalline Al_2O_3, *J. Appl. Phys.* 32:773-78.

Friedel, J. (1964), *Dislocations*, Pergamon.

Fryxell, R. E., and B. A. Chandler (1964), Creep, strength, expansion, and elastic moduli of sintered BeO as a function of grain size, porosity, and grain orientation, *J. Amer. Ceram. Soc.* 47:283-91.

Gordon, R. B. (1965), Diffusion creep in the earth's mantle, *J. Geophys. Res.* 70:2413-17.
Greenough, A. P. (1952), The deformation of silver at high temperature, *Phil. Mag.* 43:1075-82.
Griggs, D. T., F. J. Turner, and H. C. Heard (1960), Deformation of rocks at 500 to 800° C., Ch. 4 in *Geol. Soc. Amer. Mem.* 79.
Harper, J., and J. E. Dorn (1957), Viscous creep in aluminum near its melting temperature, *Acta Met.* 5:654-65.
Haskell, N. A. (1935), The motion of a viscous fluid under a surface load 1, *Physics* 6:265-69.
Haskell, N. A. (1936), The motion of a viscous fluid under a surface load 2, *Physics* 7:56-61.
Hausner, H. (1965), Determination of the melting point of uranium dioxide, *J. Nucl. Mater.* 15:179-83.
Herring, C. (1950), Diffusional viscosity of a polycrystalline solid, *J. Appl. Phys.* 21:437-45.
Hess, H. H. (1964), Seismic anisotropy of the uppermost mantle under the oceans, *Nature* 203:629-31.
Hull, D. (1965), *Introduction to Dislocations*, Pergamon.
Jeffreys, H. (1959), *The Earth*, Cambridge.
Jones, R. B. (1965), Diffusion creep in polycrystalline magnesium, *Nature* 207:70.
Jones, R. B. (1966), Diffusion creep in zirconium alloys, *J. Nucl. Mater.* 19:204-207.
Kamb, W. B. (1961), The thermodynamic theory of non-hydrostatic stressed solids, *J. Geophys. Res.* 66:259-71.
Kaula, W. M. (1963), Elastic models of the mantle corresponding to variations in the external gravity field, *J. Geophys. Res.* 68:4967-78.
Kronberg, M. L. (1962), Dynamical flow properties of single crystals of sapphire, *J. Amer. Ceram. Soc.* 45:274-79.
Lomnitz, C. (1956), Creep measurements in igneous rocks, *J. Geol.* 64:473-79.
McKenzie, D. P. (1967), The viscosity of the mantle, *Geophys. J.* 14:297-305.
McLellan, A. G. (1966), A thermodynamic theory of systems under nonhydrostatic stresses, *J. Geophys. Res.* 71:4341-47.
Misra, A. K., and S.A.F. Murrell (1964), An experimental study of the effect of temperature and stress on the creep of rocks, *Geophys. J.* 9:509-35.
Mott, N. F. (1953), A theory of work hardening of metals II: Flow without slip lines, recovery and creep, *Phil. Mag.* 44 (7th Series):742-65.
Nabarro, F. R. N. (1948), Deformation of crystals by motion of single ions, *Rept. Conf. Strength Solids, Bristol*, pp. 75-90.
Oishi, Y., and W. D. Kingery (1960), Self diffusion of oxygen in single-crystal and polycrystalline aluminum oxide, *J. Chem. Phys.* 33:480-86.
Orowan, E. (1965), Convection in a non newtonian mantle, continental drift, and mountain building, *Phil. Trans., A*, 258:284-313.
Paladino, A. E., and W. D. Kingery (1962), Aluminum-ion diffusion in aluminum oxide, *J. Chem. Phys.* 37:957-62.
Paladino, A. E., and R. L. Coble (1963), Effect of grain boundaries on diffusion controlled processes in aluminum oxide, *J. Amer. Ceram. Soc.* 46:133-36.
Passmore, E. M., R. H. Duff, and T. Vasilos (1966), Creep of dense polycrystalline magnesium oxide, *J. Amer. Ceram. Soc.* 49:594-600.
Pranatis, A. L., and G. M. Pound (1955), Viscous flow of copper at high temperatures, *Trans. AIMME* 203:664-68.
Press, F., and W. F. Brace (1966), Earthquake prediction, *Science* 152:1575-84.
Read, W. T. (1953), *Dislocations in Crystals*, McGraw-Hill.
Rothwell, W. S., and A. S. Neiman (1965), Creep in vacuum of MgO single crystals and the electric field effect, *J. Appl. Phys.* 36:2309-16.
Schumacher, E. E. (1941), Discussion of a paper by A. A. Smith, Jr., 'Creep and recrystallization in lead,' *Trans. AIMME* 143:171-78.
Sherby, O. D. (1962), Factors affecting the high temperature strength of polycrystalline solids, *Acta Met.* 10:135-47.
Squires, R. L., R. T. Weiner, and M. Philips (1963), Grain boundary denuded zones in a magnesium–½ wt.% zirconium alloy, *J. Nucl. Mater.* 8:77-80.
Sykes, L. (1966), The seismicity and deep structure of island arcs, *J. Geophys. Res.* 71:2981-3006.
Uffen, R. J. (1952), A method of estimating the melting point gradient in the earth's mantle, *Trans. Amer. Geophys. Union* 33:893-96.
Vandervoort, R. R., and W. L. Barmore (1963), Compressive creep of polycrystalline BeO, *J. Amer. Ceram. Soc.* 46:180-84.
Vasilos, T., J. B. Mitchell, and R. M. Spriggs (1964), Creep in polycrystalline magnesia, *J. Amer. Ceram. Soc.* 47:203-204.
Warshaw, S. I., and F. H. Norton (1962), Deformation behavior of polycrystalline Al_2O_3, *J. Amer. Ceram. Soc.* 45:479-86.
Wachtman, J. B., Jr., and L. H. Maxwell (1957), Plastic deformation of ceramic—oxide single crystals II, *J. Amer. Ceram. Soc.* 40:377-85.

Wachtman, J. B., M. L. Wheat, H. J. Anderson, and J. L. Bates (1965), Elastic constants of single crystal UO_2 at 25° C., *J. Nucl. Mater.* 16:39-41.

Weertman, J. (1957), Steady-state creep through dislocation climb, *J. Appl. Phys.* 28:362-64.

Weertman, J., and J. R. Weertman (1964), *Elementary Dislocation Theory*, Macmillan.

Yajima, S., H. Furuya, and T. Hirai (1966), Lattice and grain boundary diffusion of uranium in UO_2, *J. Nucl. Mater.* 20:162-70.

Zharkov, V. N. (1963), The viscosity of the interior of the earth, in Magnitskii, ed., *Theoretical Seismology and Physics of the Earth's Interior*, Jerusalem.

VISCOSITY OF THE EARTH'S MANTLE

ROBERT K. MC CONNELL, JR.

In examining the literature on viscosity of the mantle one becomes aware of quite a strong contradiction between the apparent viscosities predicted from examination of long-wavelength deformations, such as the flattening of the geoid, and those from the short-wavelength deformation, such as isostatic rebound of Fennoscandia.

This contradiction may be illustrated with reference to Figure 1, which compares the relaxation time τ as a function of wave number for two models, a layer of unit viscosity and thickness h over a very-high-viscosity halfspace, and the same layer over a very-low-

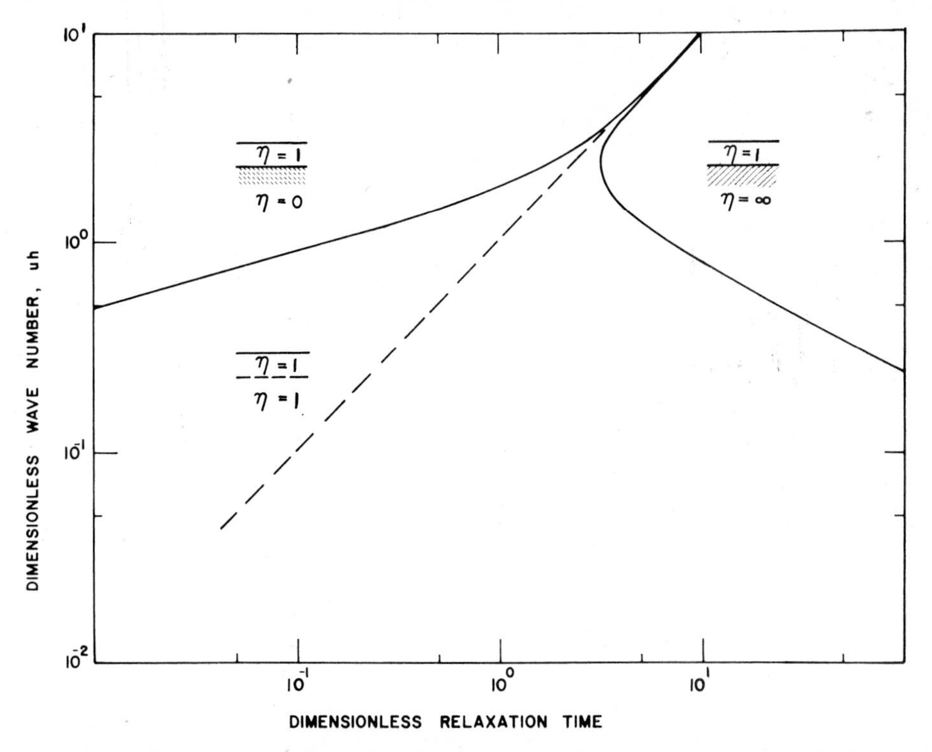

Figure 1. Comparison of relaxation time as a function of wave number for a layer of unit viscosity over a halfspace with very-high-viscosity halfspace, very-low-viscosity, and equal viscosity.

viscosity halfspace with the curve from an unlayered halfspace of unit viscosity. The derivation of such relaxation time curves is discussed by McConnell (1965). For the unlayered halfspace the relaxation time increases uniformly with increasing wave numbers (decreasing wavelength). The model with high viscosity at depth is similar to the uniform one at wave numbers where the product $uh >> 1$, but when $uh << 1$ this has much greater relaxation times, which decrease rapidly with decreasing wave numbers as a result of the influence of the high-viscosity material at depth. As long as one is dealing with approximately Newtonian viscosities, the slope of the small-wave number end of this curve represents a bound to possible rates of change of relaxation times at a given wave number no matter how great the viscosity of the underlying layers. The model with the very-low-viscosity material underneath is similar to the other two for $uh >> 1$, but as uh becomes much less than 1, the times decrease sharply. For small wave numbers the slopes of the two limiting models represent two extremes such that if any one point on an experimentally determined curve can be established, permissible relaxation time curves cannot lie in the shaded region beyond the two lines with these slopes drawn through the point (Fig. 2). Although under some circumstances the presence of an elastic layer may flatten the first curve a little over a short wavelength range, it should not be enough to significantly affect the following arguments.

Relaxation times for postglacial rebound of Fennoscandia allow one to estimate short- and intermediate-wavelength relaxation times through which the bounding curves may be drawn. Any interpretation of long-wavelength data which implies relaxation times which lie outside these bounds suggests that either the Newtonian viscosity model does not hold; the data is being interpreted wrongly; or the cause of the long-wavelength relaxation is not related to the short-wavelength phenomena.

As the derivation of relaxation time spectra from beach level data has been discussed elsewhere (McConnell, 1963, 1965, 1968), we need not go into it here except to say that these spectra provide our best estimates for wave numbers of the order of 5×10^{-8} cm^{-1} (wavelengths of several hundred kilometers). Figure 3 shows a relaxation time spectrum determined from Fennoscandian shorelines alone, upon which is superimposed the forbidden region for Newtonian viscosity in a layered halfspace. This spectrum, which was derived assuming that where no tilting has taken place since the melting of the glaciers not much deformation has occurred either, indicates a relaxation time which first increases with increasing wavelength and then bends over and approaches a constant value. The bend results directly from the assumption that there has been no deformation over the Baltic shield beyond Leningrad since 9800 years B.P.

If this relaxation time spectrum is correct, it presents several inconsistencies. One is that for low wave numbers the presently remaining unisostatically compensated deformation predicted using the calculated relaxation times is quite a bit less than that predicted from gravity anomalies. Similarly the present predicted upwarping velocity is much lower than actually observed (McConnell, 1968). Both of these discrepancies suggest that the low-wave-number portion of the curve probably has a longer relaxation time than computed from the beach level data. If this is true, it means that the assumption of no deformation near the outer boundary of the deformed region is incorrect. Fortunately, it also suggests why no tilting has been observed: at long wavelengths, which have a long relaxation time, the rate of tilting and of deformation would be very small even though a great deal of deformation remains uncompensated. Thus we can conclude that the long-wavelength portion of the curve computed from the beach level data alone is probably not reliable.

The piece of evidence most often used to estimate the relaxation time spectrum at very long wavelengths is the difference between the observed second-order harmonic of the earth's external gravity field and the theoretical value for a rotating fluid spheroid. If this "nonhydrostatic bulge" is assumed to be an unrelaxed artifact of a period when the earth was rotating faster, as Munk and MacDonald (1960) and McKenzie (1966) believe, relaxation times of the order of 10^7 years or 3×10^{14} seconds are required. As these values fall well outside the region which is compatible with the Fennoscandian data (Fig. 3), if the simple model of Newtonian viscosity is correct, the bulge cannot

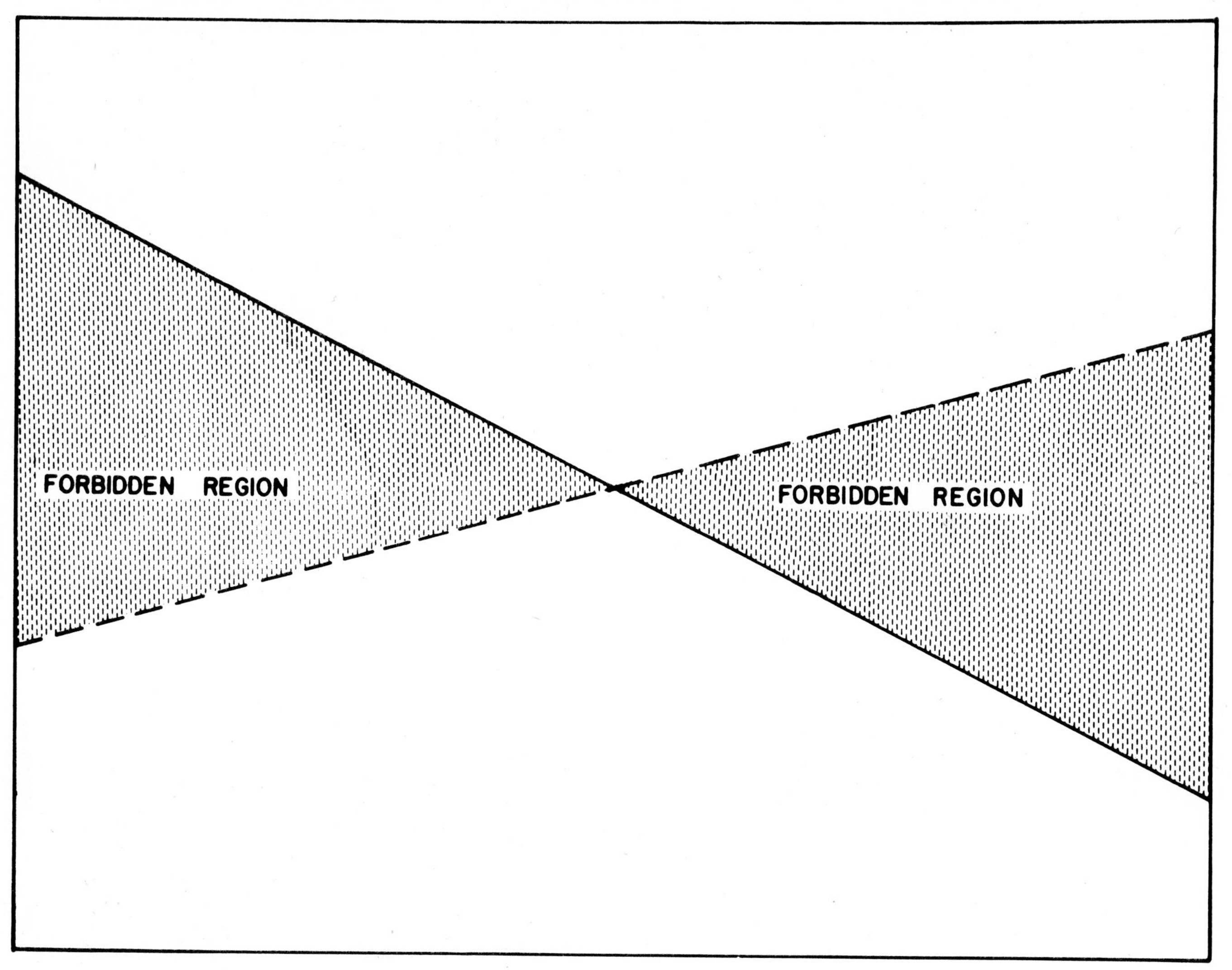

Figure 2. Permitted region for relaxation time spectrum for halfspace made up of layers of uniform density and Newtonian viscosity after one point on curve has been established.

be due to lack of adjustment of the upper mantle. It can of course, as pointed out by McKenzie (1966), be attributed to lack of equilibrium at the core mantle boundary; this would not be inconsistent with the Fennoscandian data, as relaxation at different interfaces cannot be compared by means of the limiting line.

The exact position of the theoretical curve at very low wave numbers (long wavelengths) is somewhat uncertain, perhaps by as much as a factor of two, as a result of the neglect of the sphericity of the earth in our calculations. However, since at the moment we are concerned with order-of-magnitude effects, the layered viscous halfspace model with overlying elastic layer would seem to be sufficiently accurate.

McKenzie (1966) has estimated the ability of the Pleistocene ice sheets to generate the nonequilibrium gravitational anomaly of the earth and has concluded that the magnitude corresponds closely with observed $P_2{}^0$ component of the earth's external gravity field. However, he has interpreted Farrand's (1962) observation that only $1/10$ of the equilibrium isostatic adjustment under the North American ice sheet remains to mean that $9/10$ has been accomplished; and he assumed the same relaxation time for surface deformations

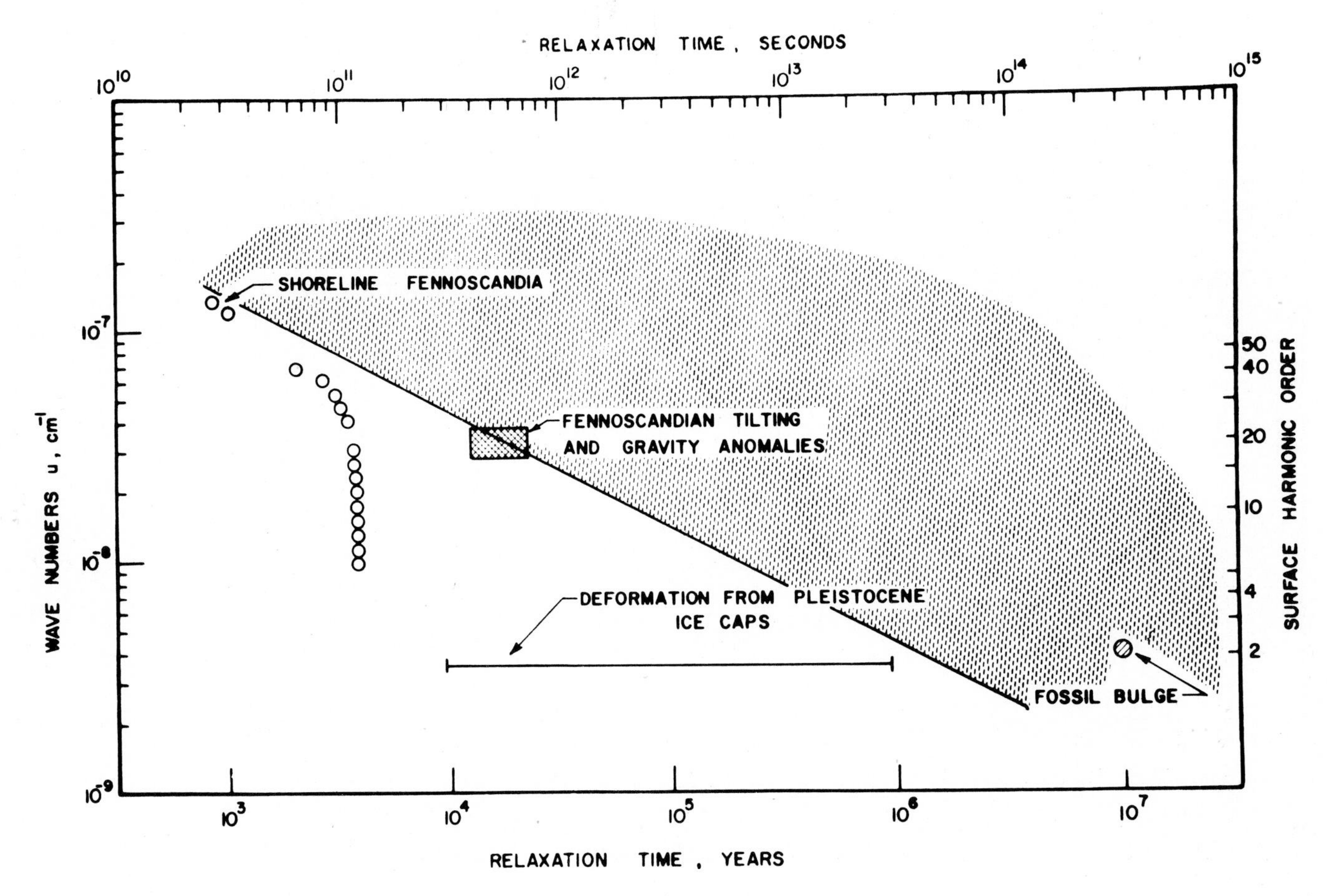

Figure 3. Comparison of relaxation times computed from several sources with permitted region for layered viscous halfspace.

of the entire earth. Using this assumption, he concludes that any flattening from the ice sheets should have disappeared and that the remaining portion must be attributed to slowing down of the earth. If we accept McKenzie's argument for the order of magnitude of the glacial flattening, but recognize both the uncertainty in the relaxation time interpretations from North America and the danger of applying relaxation times measured at one wavelength to a different wavelength, we would seem justified in using his calculation as evidence for a Pleistocene cause and then estimating the range of possible values for the relaxation time at the very long wavelengths from other considerations.

The total duration of the Pleistocene is estimated to be about 1.5×10^6 years and the duration of the last glaciation, which ended on the order of 10^4 years ago, to have been about 6×10^4 years (Ericson, et al., 1964). If these figures are correct, the permitted relaxation times must lie in the range of 10^4 years to about 10^6 years, i.e., 3×10^{11} seconds to 3×10^{13} seconds. Although more detailed considerations of the amplitude and duration of the individual ice ages which further restrict the permissible range will be discussed elsewhere, on the basis of the data presented here one cannot tell which of the times is most appropriate. (As of September 1967 it appears that 10^{13} seconds may be the best value.)

If the viscous model is correct and the short-wavelength relaxation times estimated from the Fennoscandian data are also correct then the entire range of relaxation times which would be estimated from the assumption of flattening of the earth by Pleistocene ice sheets lies within the permitted region (Fig. 3).

We now consider those models whose relaxation times would be compatible with a Pleistocene cause of the earth's equatorial bulge. We may start with model 62-12 described by McConnell (1963, 1965) designed to fit the original Fennoscandian data. This model has a surface elastic layer 120 km thick with an average rigidity of 6.5×10^{11} dynes/cm^2, which has been able to maintain elastic stresses over periods of the order of 10^4 to 10^5 years. Immediately below the elastic layer is a region of relatively low viscosity extending to depths of about 400 km, which is underlain in turn by a region extending to 1,200 km over which the viscosity increases by several orders of magnitude. We note that at low wave numbers this model predicts relaxation times so short that no appreciable fraction of the deformation imposed by the ice caps could remain.

The problem now arises of modifying the model in such a way that it will predict long-wavelength relaxation times which fall within the expected range for Pleistocene glaciation. We may start with model 62-12 and first try to bring about the desired changes by increasing viscosities below 400 km. It is clear from Figure 4 that only by increasing the viscosity below 400 km to values of the order of 10^{23} poise or greater can relaxation times which fall within the appropriate range for a Pleistocene cause be predicted. The models which seem to be most compatible with the observations are shaded in Figure 4b. Even though they are consistent with the very-long-wavelength data, it is interesting to note that they all predict relaxation times around $2\text{-}3 \times 10^{11}$ seconds, comparable to those estimated by Haskell (1937) and Heiskanen and Vening Meinesz (1958) for the

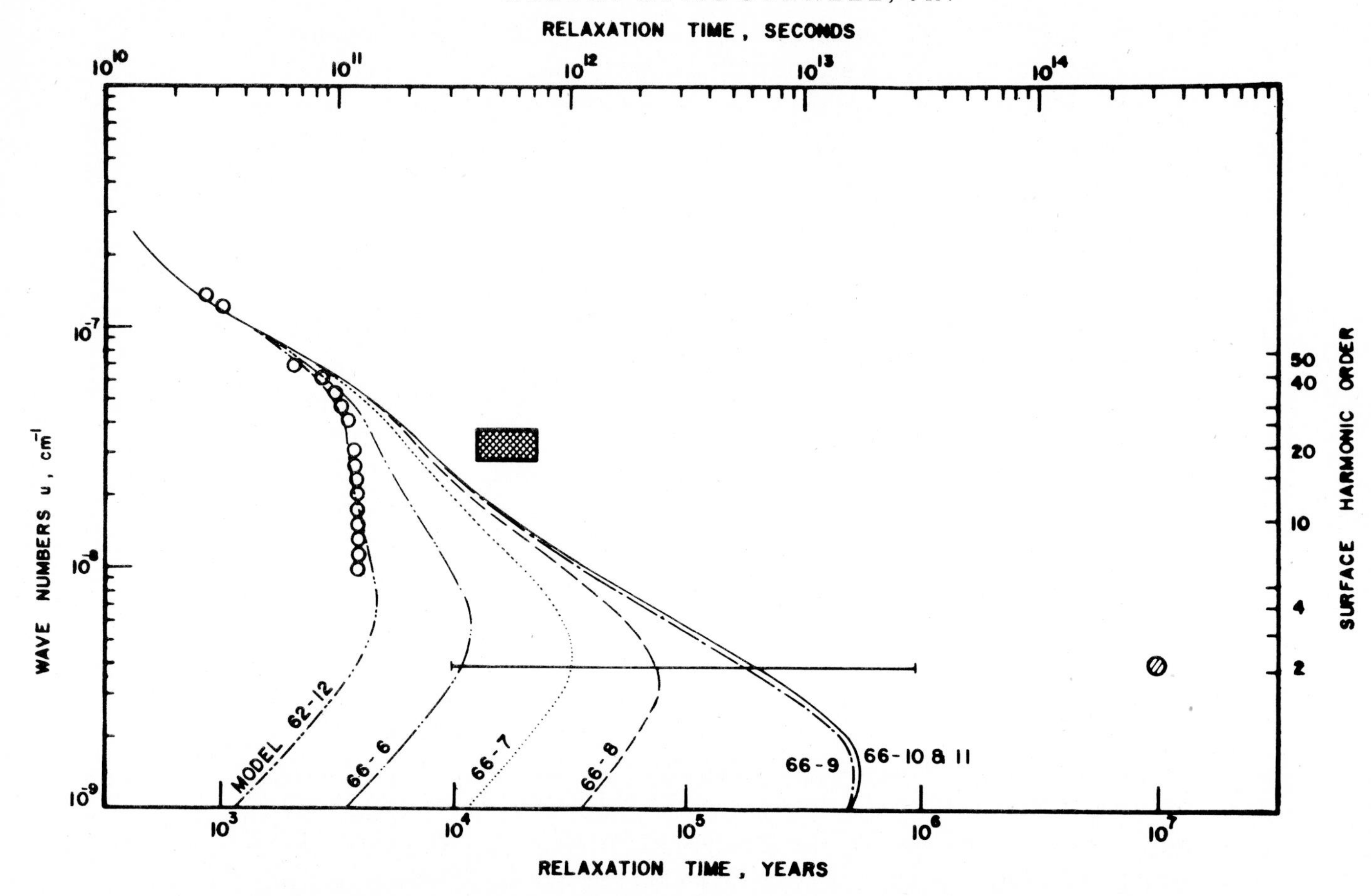

Figure 4. Comparison of theoretical relaxation time spectra for a number of models with viscosity increases below 400 km with various data.

relaxation time appropriate to the bulk of the Fennoscandian data, using present uplift rates and past beach levels, but shorter than those estimated by McConnell (1967) from present upwarping rates and with remaining depression. Figure 5 shows the effect of a viscosity increase which starts around 200 rather than 400 km. In this case the times for the Fennoscandian data at values of wave numbers around 2×10^{-8} are considerably greater, and once more the shaded areas illustrate those models which seem to be most compatible with a Pleistocene cause of the long-wavelength deformation. This model suggests a viscosity which increases from somewhere around 10^{21} poise between 100 and 200 km to 10^{23} poise around 400 km and 10^{24} somewhat below that.

Figure 6 shows the effect of increasing the relative viscosity of the layer between 200 and 400 km by a factor of 100 over model 62-12. A comparison of the extent of the shaded areas for the last two sets of models shows that in fact the likely viscosity ranges do not differ greatly.

We may therefore conclude that although relaxation time spectra for models fitting the assumed data points are not very sensitive to the details of the viscosity gradient

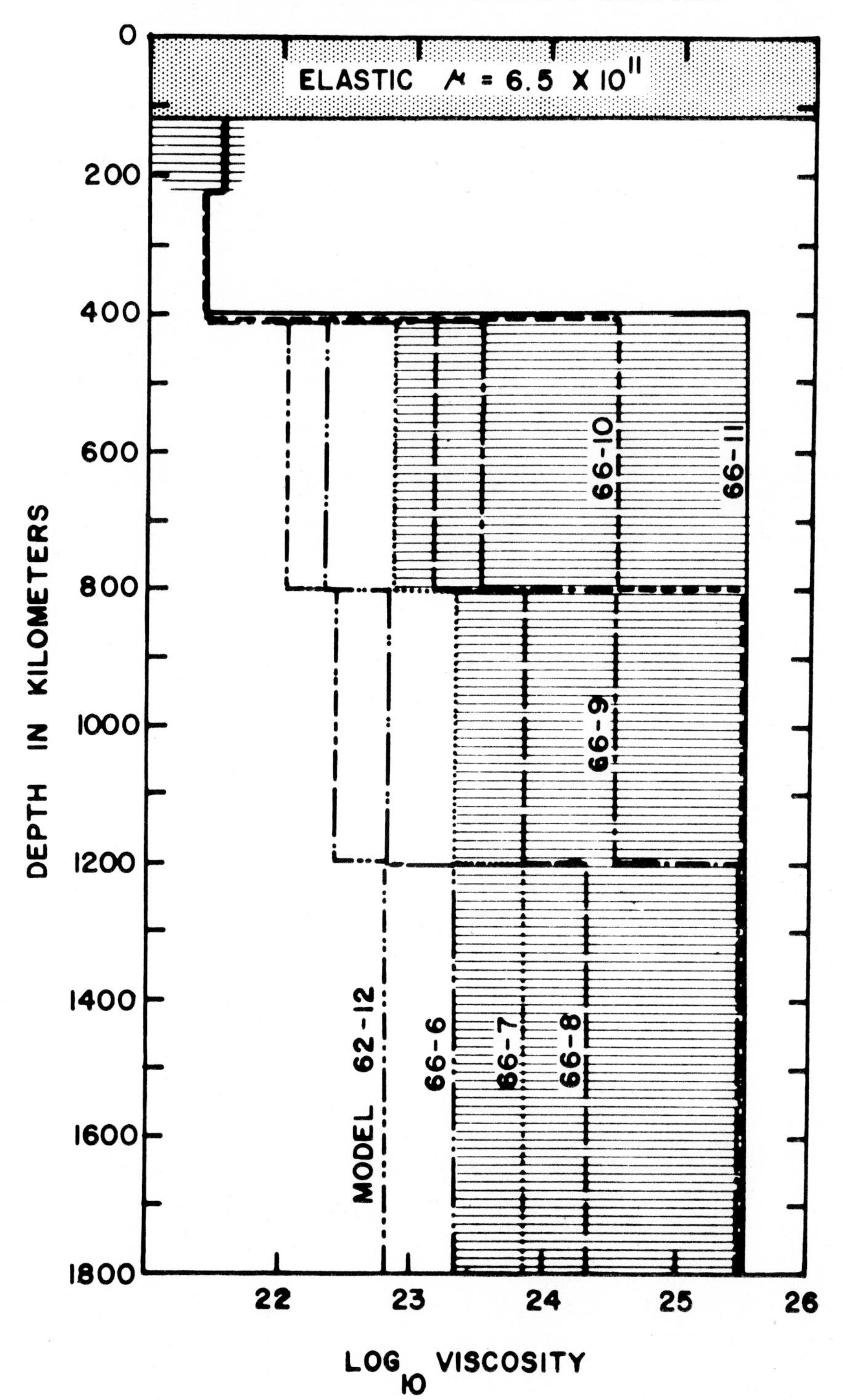

Figure 4 (*continued*)

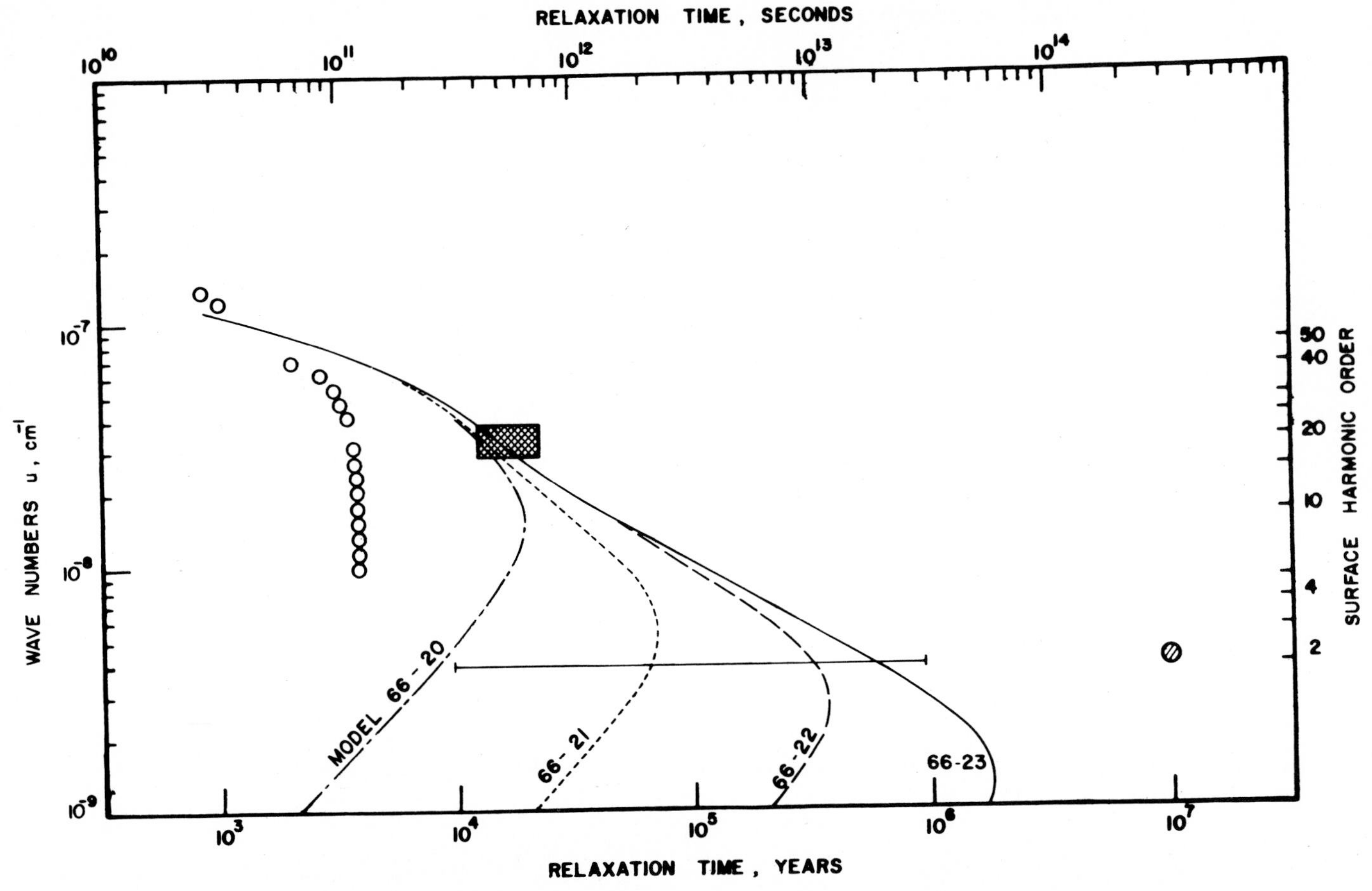

Figure 5. Relaxation time spectra for a number of models with viscosity increases below 200 km.

between 200 and 400 km, a steeper gradient in this region than previously estimated would seem to be required. If the strong viscosity contrasts implied by the data are correct, the case for crustal drift being driven by lower mantle convection would seem to be so tenuous that one must look for the cause of the drifts in the upper mantle itself.

Arthur D. Little Incorporated

DISCUSSION

DR. WASSERBURG: What would happen if you made the elastic layer very thin?

DR. MC CONNELL: The large-wave-number end of the curve in Figure 6a would move out—toward longer relaxation times. If you make the surface layer so it isn't really elastic at all, you have to treat it as a viscous layer, and you are forced to put the low-viscosity region very shallow—maybe 40 or 50 km down.

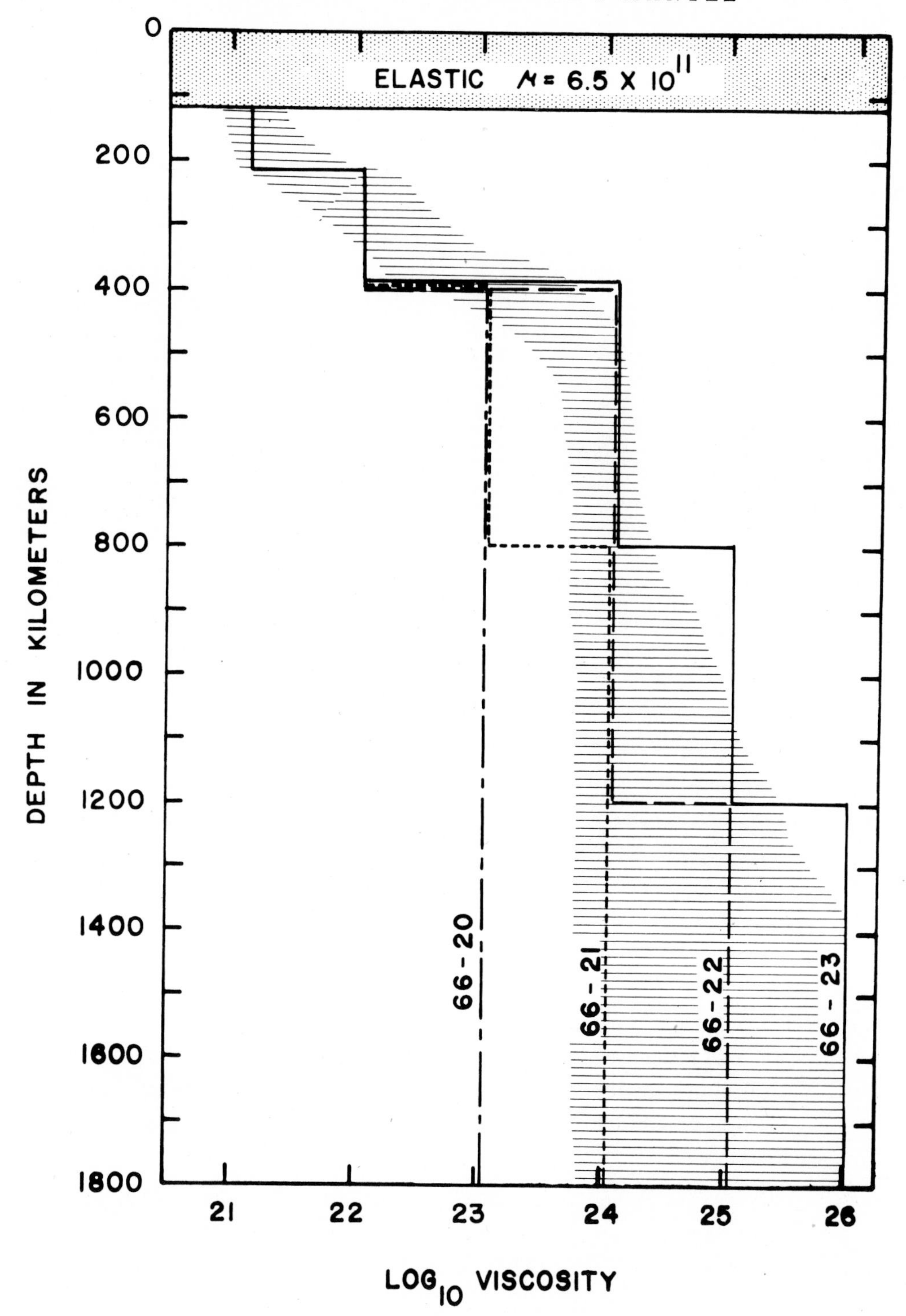

Figure 5 (*continued*)

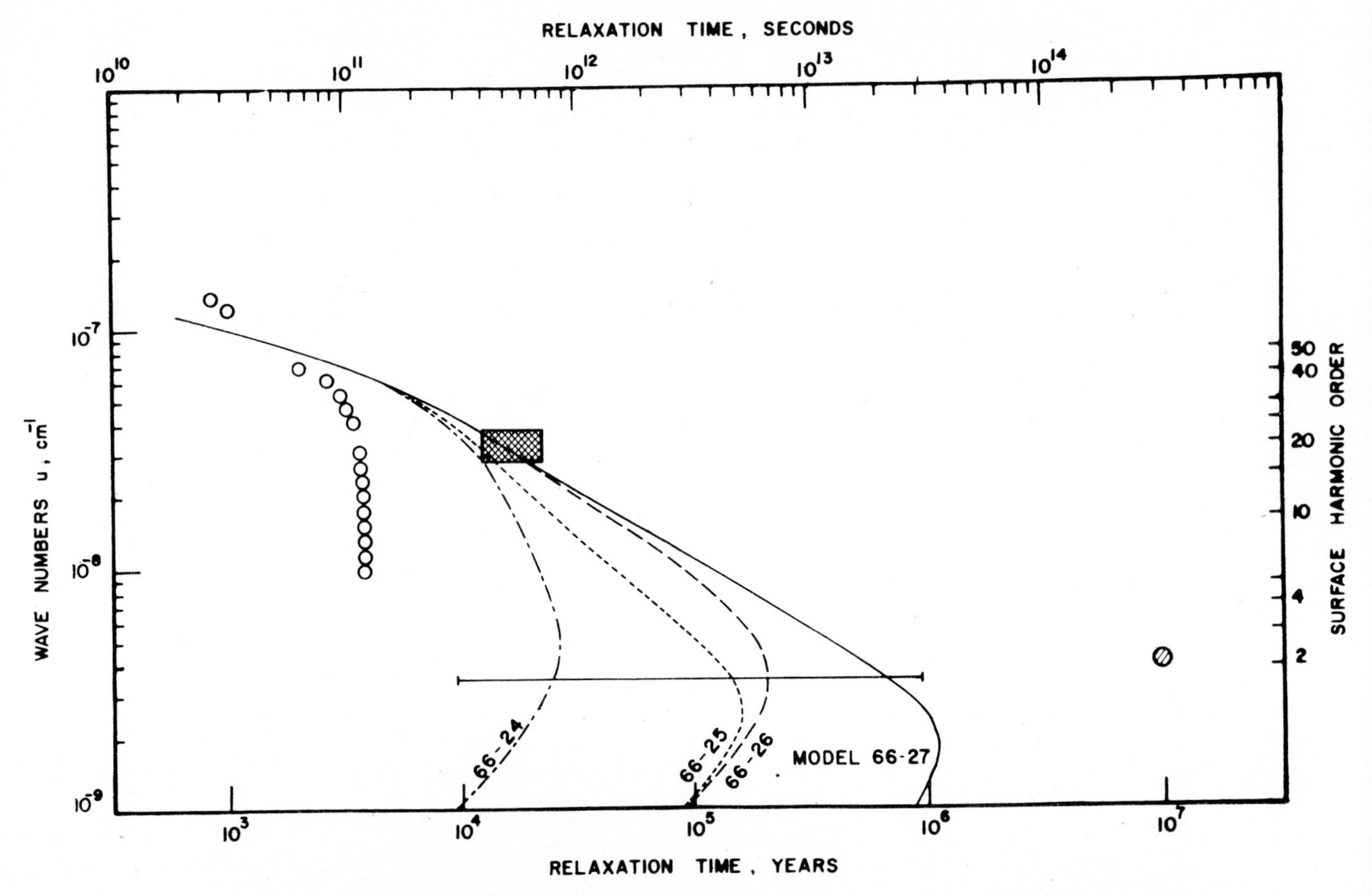

Figure 6. Relaxation time spectra for models with more abrupt viscosity increase below 200 km.

DR. MC KENZIE: What sort of relaxation time do you get for a radius of about 1,000 km, that is, for Fennoscandia as a whole?

DR. MC CONNELL: It looks like about 3×10^4 years.

DR. MC KENZIE: What I am after is that there is another method. You know vaguely how much ice load is involved—say 3 km. So the crust would deform about a kilometer down. When you melt the ice, the immediate elastic uplift is about half of that. So you have about 500 meters to uplift. What is the actual height of the highest raised beaches above sea level now?

DR. MC CONNELL: The highest reported are at least 250 to 300 meters at the center of the deformation region.

DR. MC KENZIE: That will give you the stress. We have already established the scale. The result should be the order of the relaxation time. That is the other method.

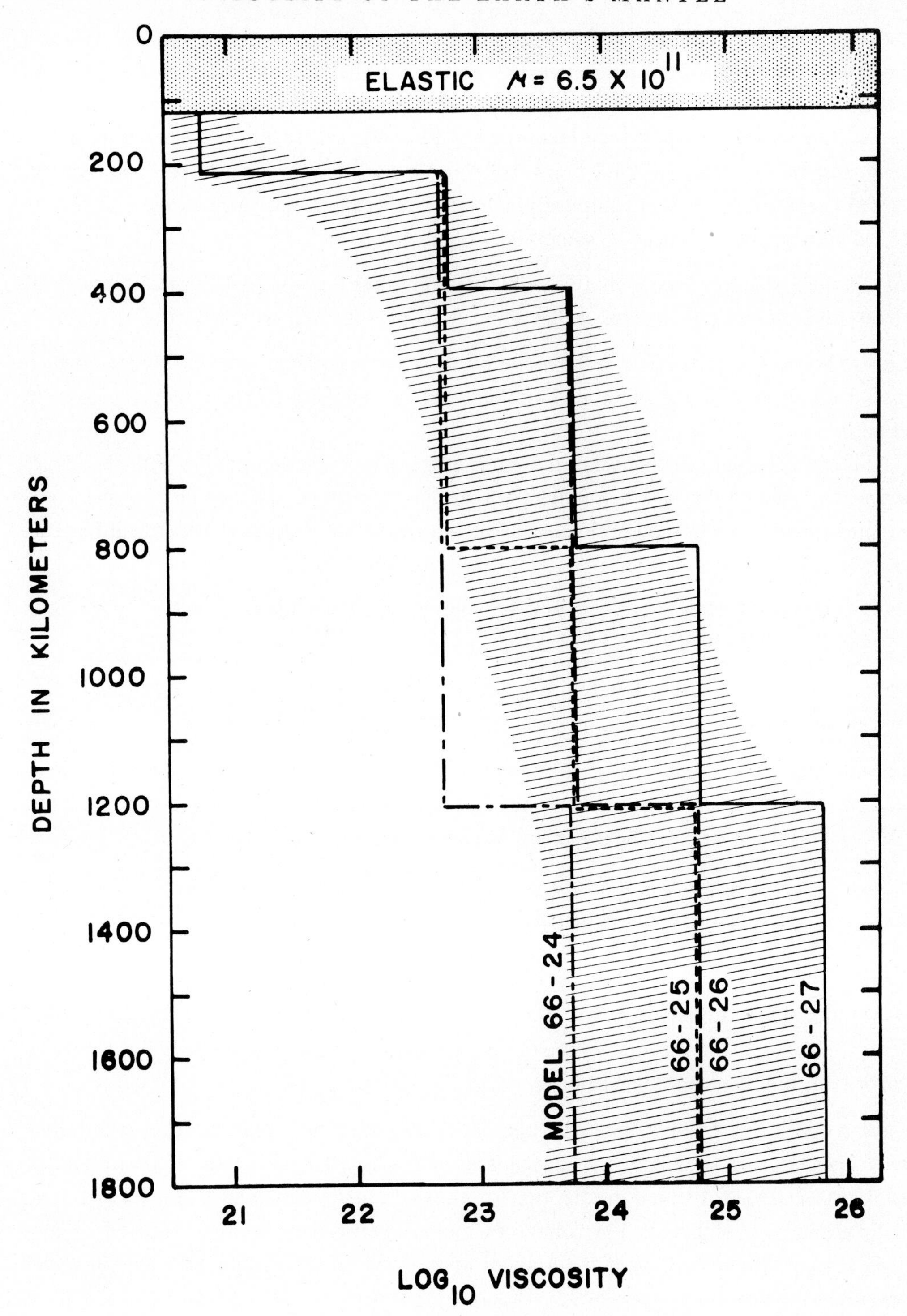

Figure 6 (*continued*)

DR. MC CONNELL: This is similar to the approach used here in the sense that you are assuming how much deformation is left and using the original deformation and the velocities. It might be interesting to try this. In any event, these are very crude calculations. Numbers which differ by a factor of less than two are within the uncertainty in the ice loads. In addition you don't know how long it took for the area to feel the loading. If you knew whether it was completely unloaded between glacial stages, then you probably would be able to calculate something meaningful.

DR. ANDERSON: Do I take it that you are assuming that the non-hydrostatic ellipticity of the earth is due to the ice caps rather than the changing rotation rate?

DR. MC CONNELL: Yes. That is the assumption. You can see the point in Figure 3 which would correspond to the other assumption, that of changing rotation rate due to tidal friction.

It could be we are really seeing the latter effect. But to be compatible with the short-wavelength data from Fennoscandia the relaxation must be taking place at the core-mantle boundary or in the lower mantle. Such a relaxation process at depth will have its own line on the wave number–relaxation time diagram.

DR. ANDERSON: It doesn't really belong on your curves, since it isn't in fact a surface load.

DR. GAST: You have a low-viscosity zone here under a shield area. Is that consistent with the seismic observations of the difference between low-velocity zones between shield areas and ocean areas?

DR. ANDERSON: The data are pretty sparse. There is a suggestion that there is a low-velocity zone for shear waves under shields, although it is possibly not very pronounced for compressional waves. The shear low-velocity zone is actually more pertinent to these calculations.

DR. GAST: Then is there any contradiction?

DR. ANDERSON: No.

DR. PRESS: I am troubled by one thing. The Russians have reported vertical motion comparable to Fennoscandia in regions that have never been glaciated, but with the same rates and the same dimensions.

DR. MC CONNELL: I think there are other ways of getting deformation. This is only one way: the glaciers loaded it, and we are measuring the way this particular region responded to the vertical load. I would predict that as soon as you start relaxing the glaciated area, with a long wavelength, you will have to move material beyond the glaciated region. It is not unreasonable to predict tilting or uplift at distances comparable to this wavelength away from the margin.

REFERENCES

Ericson, D. B., M. Ewing, and G. Wollin (1964), The pleistocene epoch in deep-sea sediments, *Science* 146:723-32.

Farrand, W. R. (1962), Postglacial uplift in North America, *Amer. J. Sci.* 260:181-99.

Haskell, N. A. (1937), The viscosity of the asthenosphere, *Amer. J. Sci.* 33:22-28.

Heiskanen, W. A., and F. A. Vening Meinesz (1958), *The Earth and Its Gravity Field*, McGraw-Hill Book Company, Inc., New York.

McConnell, R. K., Jr. (1963), The viscoelastic response of a layered earth to the removal of the Fennoscandian ice sheet, Ph.D. Thesis, University of Toronto.

McConnell, R. K., Jr. (1965), Isostatic adjustment in a layered earth, *J. Geophys. Res.* 70:5171-88.

McConnell, R. K., Jr. (1968), Viscosity of the mantle from relaxation time spectra of isostatic adjustment, *J. Geophys. Res.* (in press).

McKenzie, D. P. (1966), The viscosity of the lower mantle, *J. Geophys. Res.* 71:3995-4010.

Munk, W. H., and G. J. F. MacDonald (1960), *The Rotation of the Earth*, Cambridge University Press.

2. EVIDENCE FROM THE OCEAN BASINS

THE PALEOMAGNETISM OF OCEANIC CORES*

N. D. OPDYKE

One of the outstanding contributions of paleomagnetism to the study of the earth has been to establish firmly that the earth's magnetic field has not always been of only one polarity but has changed polarity many times in the past, a phenomenon generally known as reversal of the earth's field.

The reality of field reversal has become increasingly obvious, although minerals and rocks with self-reversing properties are known and well documented (Nagata, 1952). In a large number of field tests, rocks reheated or baked by extrusives or intrusives were found to have the same polarity as the rock mass which caused the reheating. A series of related studies by Cox et al. (1964), McDougall and Tarling (1963), Doell et al. (1966), McDougall and Wensink (1966), in which K-Ar dating and paleomagnetism have been done on the same rocks, has shown that rocks of a single age possessed directions of one polarity. These times of single polarity were arranged in a system of polarity epochs which contained short intervals of opposite polarity, called "events."

Paleomagnetic study of oceanic cores has been proceeding sporadically since 1938, but it was not until 1964 that the first paper describing reversals in a short core from the Pacific Ocean was published (Harrison and Funnel, 1964). There followed papers by Lin'kova (1965), who worked on gravity cores from the Arctic basin, and by Harrison (1966). The paleomagnetic study of long piston cores began at Lamont Geological Observatory in 1965 and picked up momentum in February 1966 with the observation made by Glass and Foster that cores from high latitudes around Antarctica were strongly magnetized and stable. As a result of this study a paper was published (Opdyke et al., 1966), describing work on piston cores from the Antarctic.

The techniques used in the study of the Antarctic cores is the same as that used for all cores studied paleomagnetically at Lamont. Since the cores were unoriented and only the direction down is known, it is only the determination of the inclination that is of any stratigraphic value. Specimens are cut from the core at 10-cm intervals using a band saw. These specimens are rough cubes approximately 2 cm on a slide in which the direction toward the bottom of the cores is noted. The specimens are then partially demagnetized in alternating fields of from 50 to 150 oersteds to remove any viscous remanent magnetization (VRM) which the specimen may have acquired. The specimens are then measured on a 5-cps spinner magnetometer described by Foster (1966).

Figure 1 shows a plot of inclination in piston core V16-134 from Antarctic waters after magnetic cleaning. The change in inclination observed in this core is very sharp, and the interpretation is unambiguous. This core gives a sequence of normally and reversely magnetized sections which is the exact duplication, from top to bottom of the

* Lamont Geological Observatory Contribution No. 1194

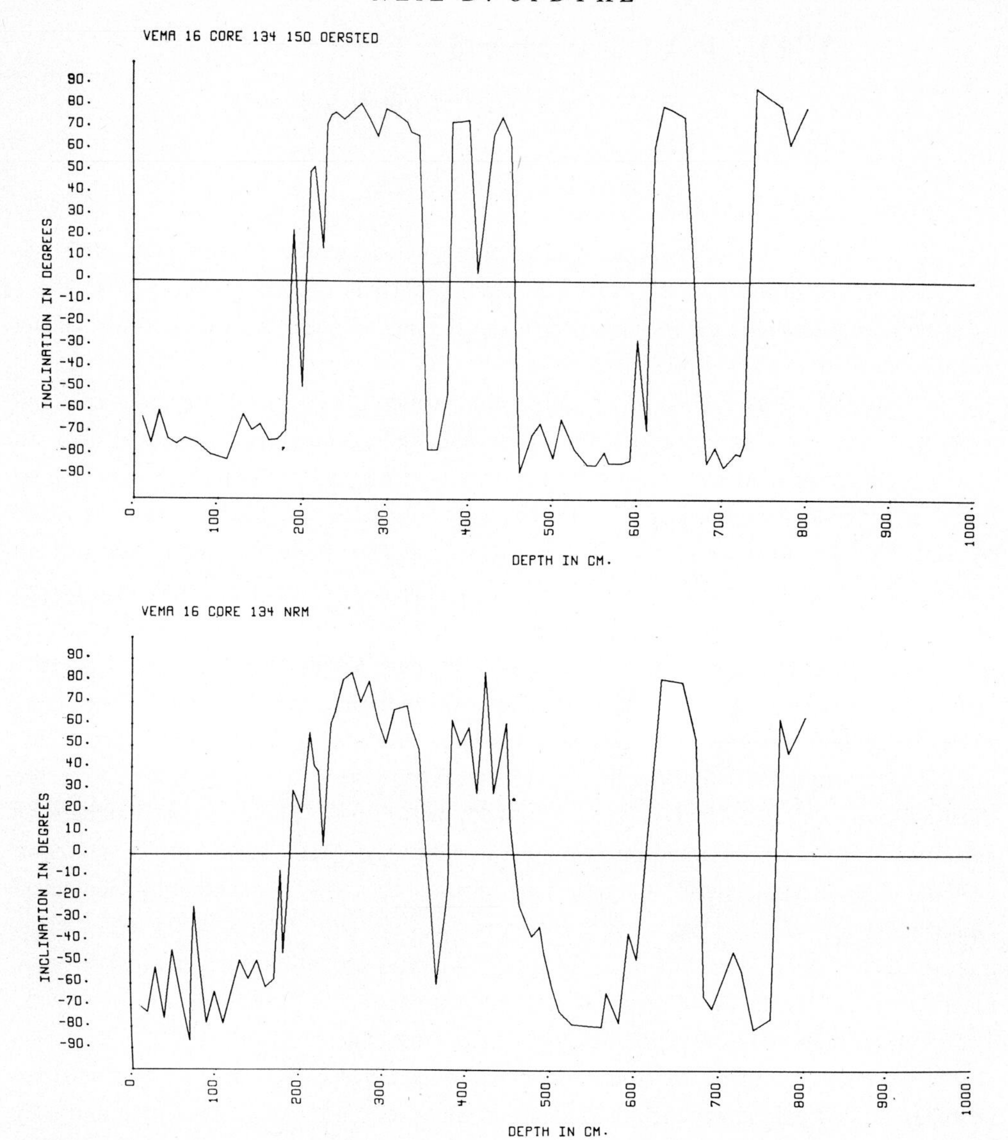

Figure 1. Inclination in core V16-134, top, after cleaning by 150 oersted AC field; bottom, before cleaning.

core, of the magnetic stratigraphy worked out on lava flows by Cox et al. (1964) using K-Ar dating. This detailed correspondence can only be due to a single controlling mechanism, namely the change of polarity of the earth's magnetic field. There can no longer be any reasonable doubt as to reality of this phenomenon.

The cores chosen for this study had previously been used by Hays (1965), in a study of radiolarian zones in Antarctic cores. These zones are based on the appearance and disappearance of radiolarian species. It can be seen from Figure 2 that where the radiolarian zones are abbreviated the paleomagnetic zones are short and where the zone boundaries occur far down in the core the magnetic stratigraphy also indicates a high

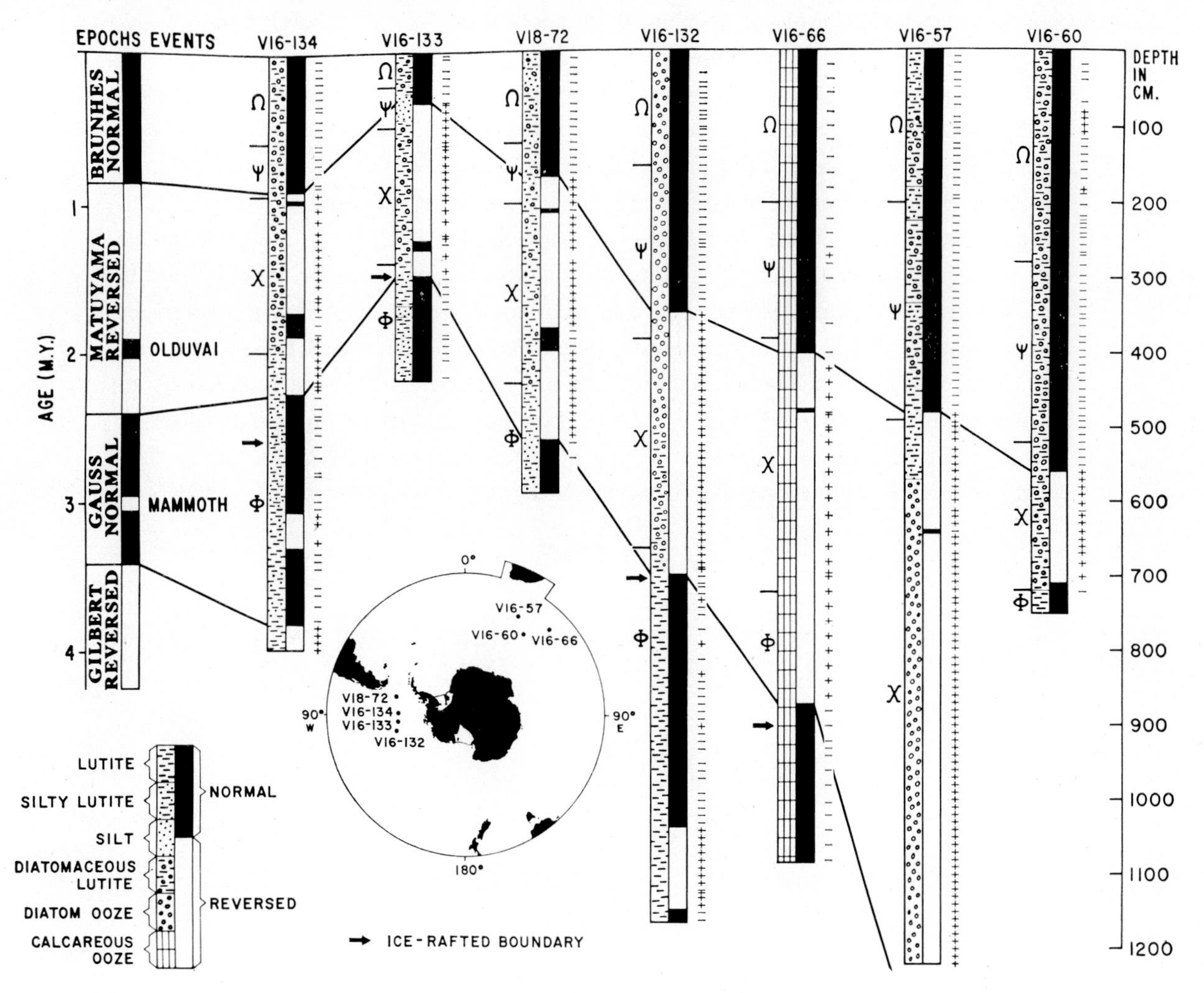

Figure 2. Stratigraphy of seven Antarctic cores.

rate of deposition. It is clear that both the radiolarian zones and the magnetic stratigraphy have the same time dependence.

It can also be seen in Figure 2 that the change of polarity between the Brunhes normal polarity series and the Matuyama polarity series falls very close in these cores to the Ψ-X boundary based on the extinction of radiolarian species. The X-Φ boundary is also seen to fall very close to the base of the Olduvai event. These observations seem at first glance to support the theory put forward by Uffen (1963) that reversals of polarity might cause extinctions due to an increase in radiation. Many questions remain to be answered regarding this particular problem, and the question must remain open.

Using the dates from the reversals that have previously been determined from the work on lavas by Cox et al. (1964), it is possible to estimate the age of anything that occurs in the cores if the rate of sedimentation is constant. In the Antarctic cores it is possible to determine the age of the point in the cores where ice-rafted debris first appears. In the cores studied so far this takes place at from 2.5 to 2.7 m.yr. B.P.

In the studies of the Antarctic cores and in a subsequent study of cores from the North Pacific Ocean by Ninkovich et al. (1966), an attempt was made to see what could be learned about the behavior of the magnetic field during reversals of magnetization. In these studies the intensity of magnetization in each specimen was normalized by weight. It was found that in general the intensity of magnetization declined at the point in the core where reversals occurred. In an attempt to study this phenomenon more closely in the cores from the North Pacific, they were continuously sampled across the reversals. It was found in all cases that the intensity of magnetization decreased before and after the actual change in inclination occurred. In several cases intermediate directions of magnetization were observed (Fig. 3). These data were interpreted by Ninkovich et al. (1966) to indicate that during reversal of the earth's field intensity of the dipole field begins to decrease until the actual change in inclination takes place; it then appears to

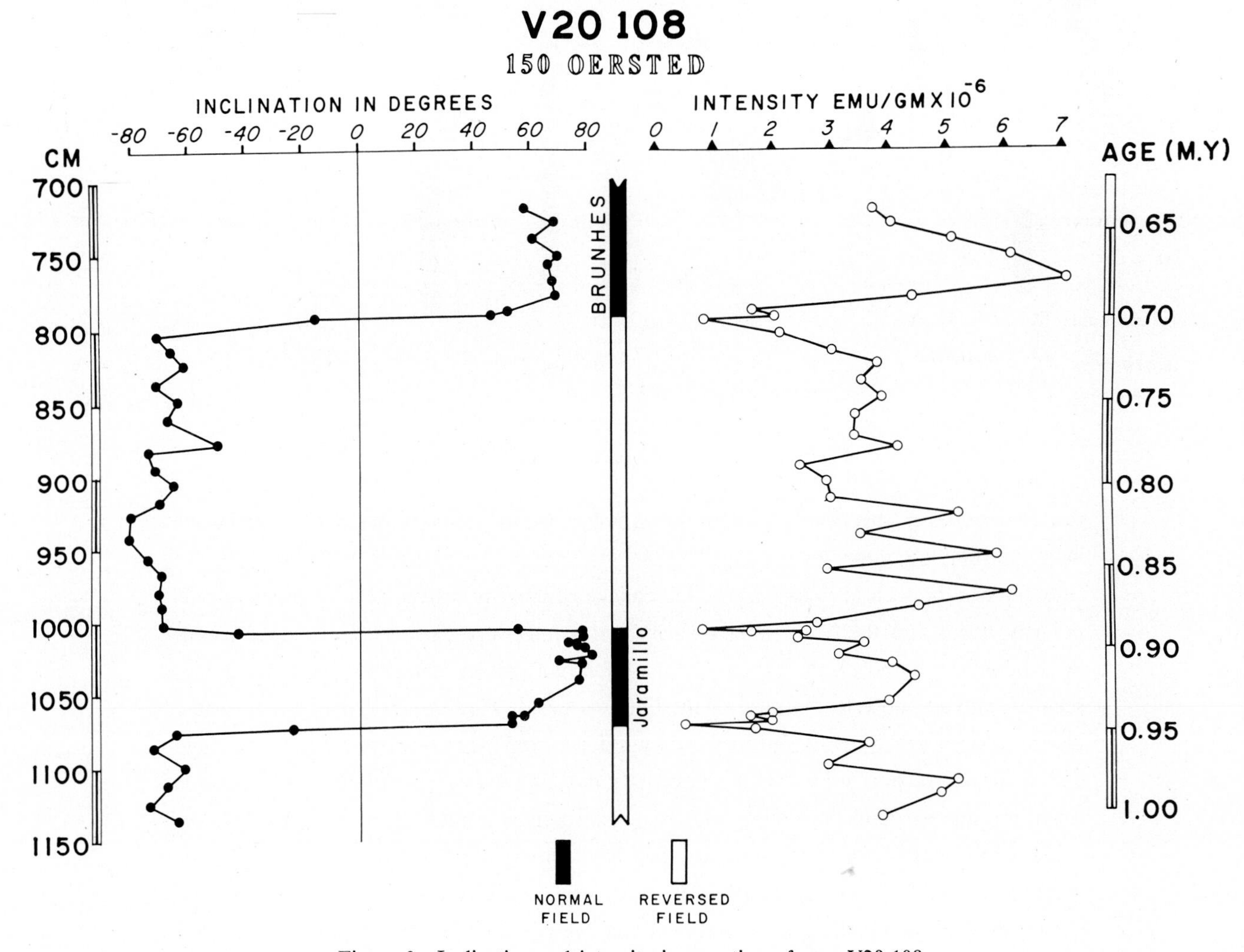

Figure 3. Inclination and intensity in a section of core V20-108.

grow in the opposite direction. The period of time represented by each specimen is long (2,000 to 10,000 years) compared to the secular variation; therefore, it seems probable that each specimen retains a record of the dipole field freed of secular variation. Support for this hypothesis has recently been reported by Harrison (1966), who shows that the Koenigsberger ratio decreases across the reversal, which would tend to indicate that the magnetizing field had become less efficient, although the amount of material that was available for magnetization remained substantially the same. An estimate of the amount of time involved in this decrease of intensity can be made, and it would appear that the dipole field takes about 20,000 years to decay and build in the opposite direction. The actual change in direction probably takes less than 2000 years to occur, since the time taken for reversal cannot be resolved in these cores even with continuous sampling. This estimate seems to be in substantial agreement with that of Harrison (1966), who gave 1500 years as the length of time taken for reversals to occur.

Table 1

Core	Lat.	Long.	I_n	I_r	I_d
V16-57	45° 14′ S	29° 29′ E	65°	63°	63°
V16-134	61° 54′ S	91° 15′ W	71°	71°	75°
V20-107	43° 24′ N	178° 52′ W	57°	64°	62°
V20-109	47° 19′ N	179° 39′ W	62.8°	61.6°	65°

I_n is average inclination of normally magnetized specimens.
I_r is average inclination of reversely magnetized specimens.
I_d is the expected inclination of the axial dipole field.

Table 1 gives the mean inclinations of cores that have been studied and published within the last year (Opdyke et al., 1966; Ninkovich et al., 1966). Several things immediately become apparent. There is no real difference in inclination between the normal and reversed sections of the cores. The inclination in the cores is closer to the theoretical axial dipole field than to the value of the inclination at the sites of the present field. This would indicate that the rate of deposition is slow enough to average out the secular variation.

Because the dates of the reversals at the magnetic epoch boundaries have been reasonably well established, it is possible to plot the reversal against its depth of occurrences in the core. If the K-Ar dates for the reversals are correct and the rate of sedimentation is taken as constant, then the plot of the points will be a straight line (Fig. 4). Such a plot has been made for cores in the Antarctic and North Pacific studies. This technique immediately gives the time of occurrence of magnetic events in the core and also an estimate of the length of time which is spanned by the events. From a variety of cores, both published and unpublished, it is possible to estimate the duration of time represented by the Jaramillo event and the Olduvai event. The Jaramillo event would appear to be about 50,000 years in length beginning about 0.95 m.yr. B.P. and terminating about 0.90 m.yr. B.P. The best estimate for the length of time involved in the Olduvai event would be about 150,000 years, extending roughly from 1.95 m.yr. B.P. to

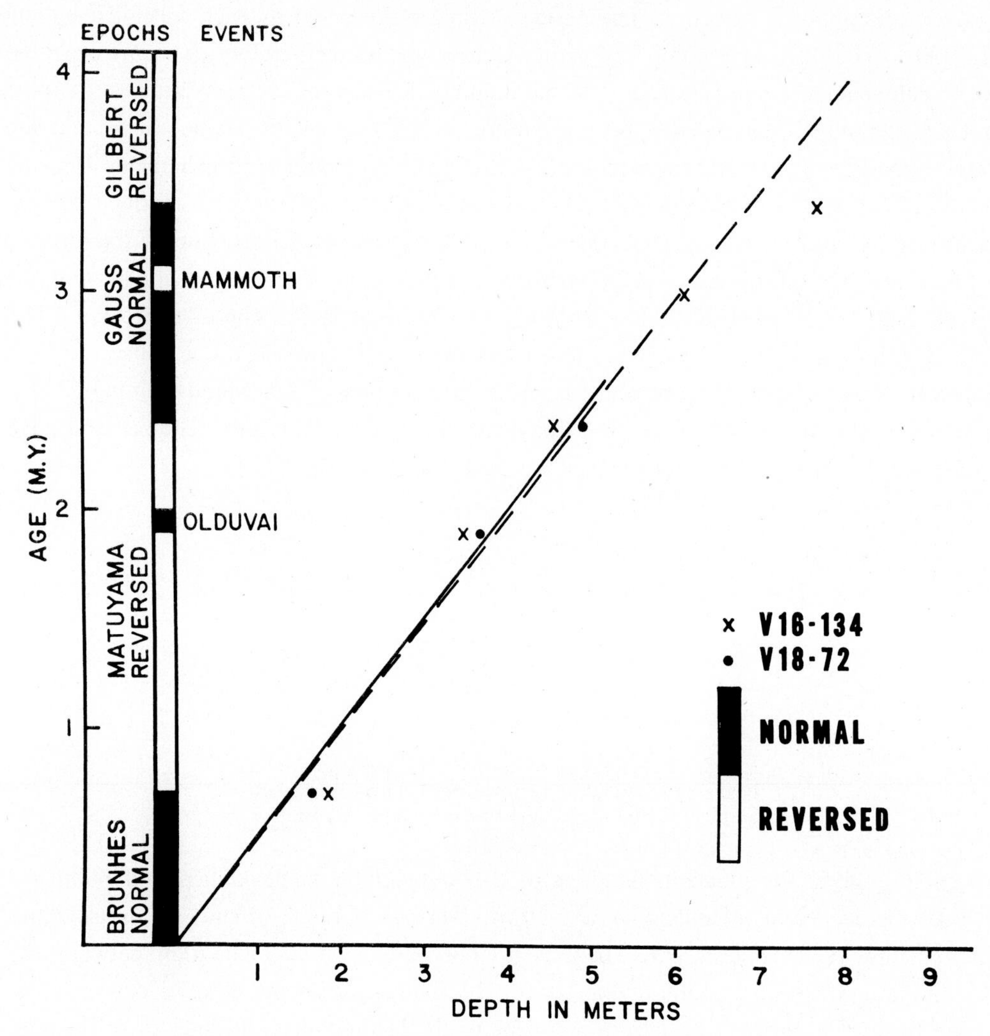

Figure 4. Depth of observed reversals in inclination as a function of age, based on the magnetic reversal time scale.

1.80 m.yr. B.P. Ninkovich et al. (1966) have presented evidence to show that there may be a small section of reversed polarity near the top of the Olduvai event (Fig. 5). At the present time there is no valid reason to doubt the reality of this split in the Olduvai event; however, confirmation of this event within an event must be substantiated by further work. If it is substantiated, then events with a duration of the order of 10,000 years will be a possibility, indicating a very unstable condition for the polarity of the earth's dipole field.

Unfortunately not all piston cores and not all lithologies are equally amenable to paleomagnetic study. The two lithologies with which we have had the most difficulty are so-called "red clays" and Globigerina oozes. The former because they seem to be slightly unstable and the latter because they tend to be very weakly magnetized. Dickson and Foster (1966) have studied very intensively a single red clay core from the central

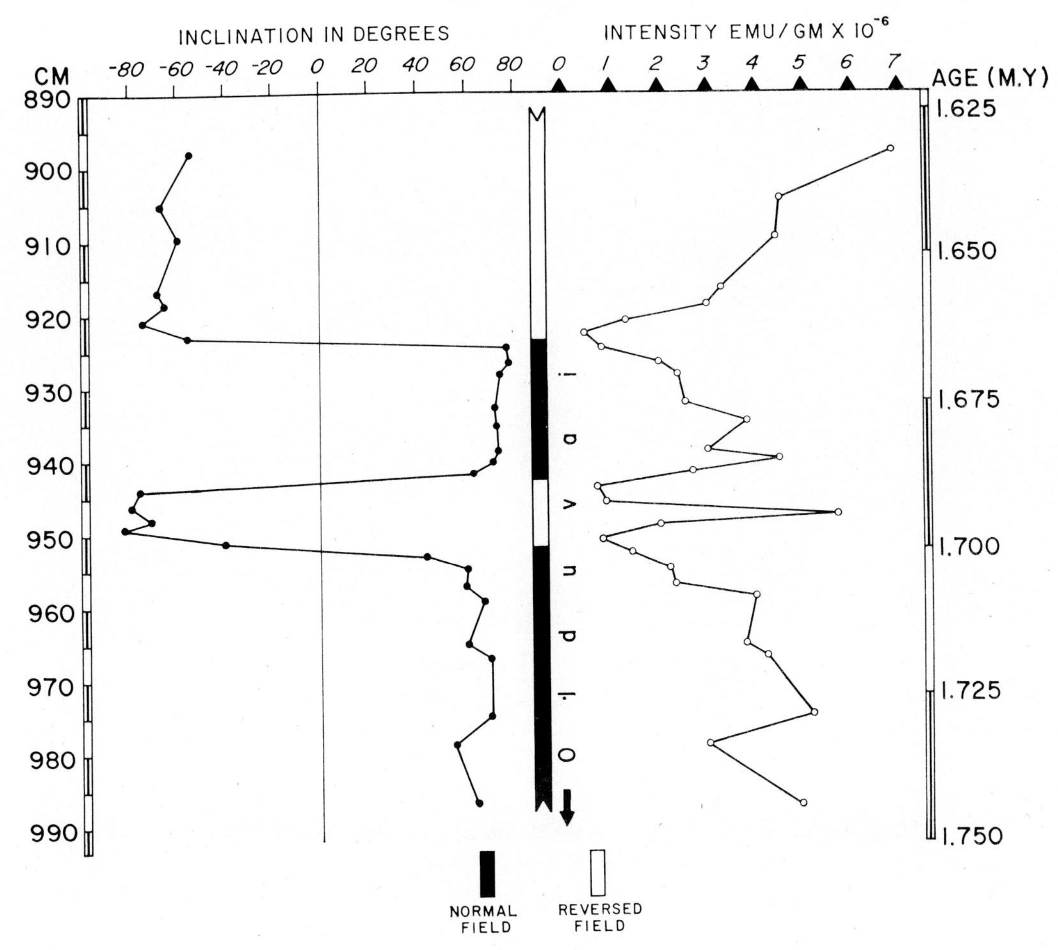

Figure 5. Inclination and intensity in a section of core V20-109.

Pacific V21-65. The core was continuously sampled every two centimeters along its length. The data were very noisy, and in order to enable the authors to interpret the core more easily, a 5-point running mean was calculated and plotted (Fig. 6). The stratigraphy then became much more easily identifiable. Alternating field demagnetization curves were run on material from the core and these show that the NRM (natural remanent magnetization) is only partially stable. Curie-point determinations on specimens from the core show that the magnetization is due to hematite and minor magnetite. The magnetic mineral in the Antarctic and North Pacific cores, on the other hand, is most probably magnetite.

The magnetic stratigraphy of the V21-65 extends well into the Gilbert reversed polarity epoch. Other cores studied at Lamont extend to greater ages. Three cores from the Antarctic which have been studied by Hays and Opdyke (1966) give magnetic stratigraphy which extends back to an estimated age of 5 m.yr. B.P. These authors recog-

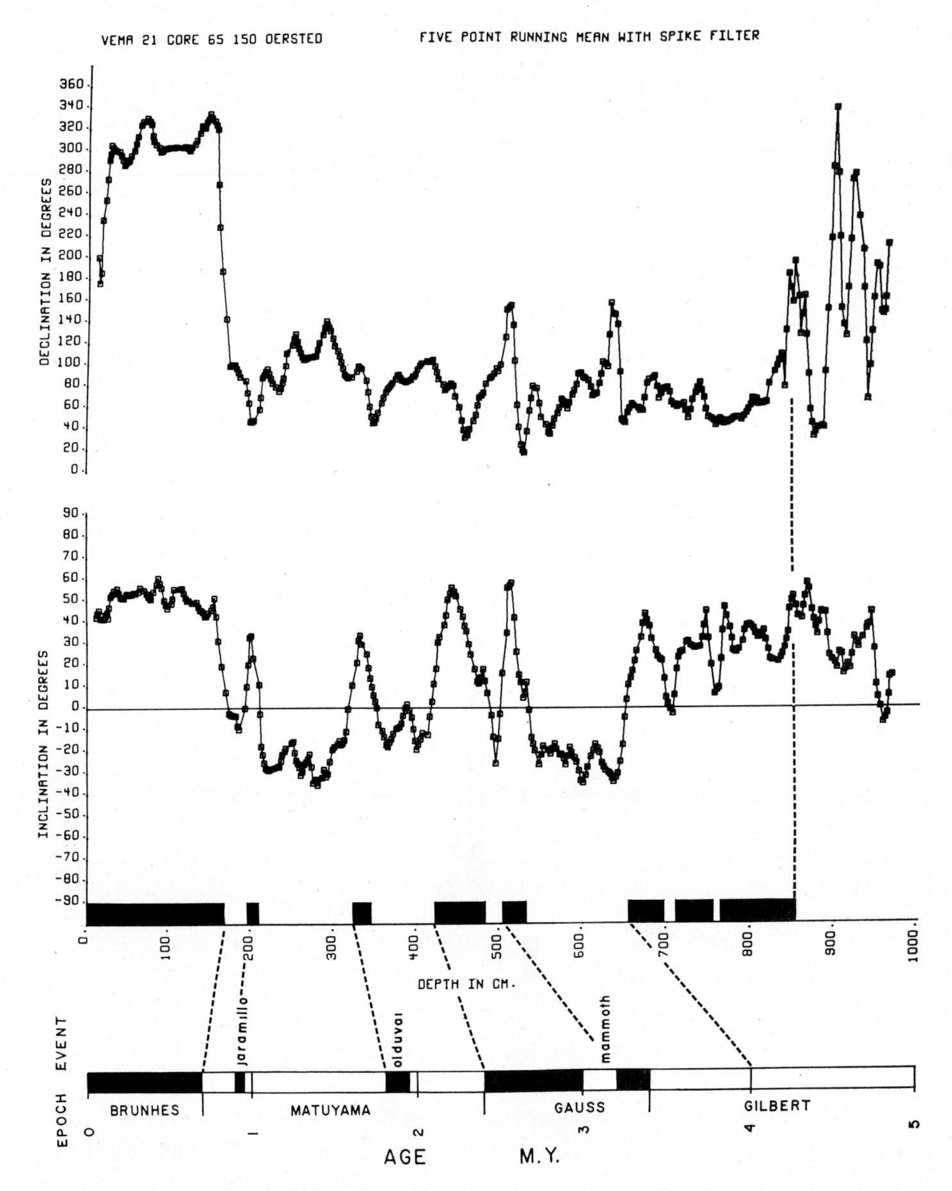

Figure 6. Inclination and declination in core V21-65 after smoothing the data.

nize three previously known events which occur in the Gilbert reversed polarity series. These three events, which have been designated events No. a, b, and c in order of increasing age, are estimated to lie between 3.8 and 4.5 m.yr. B.P. In these cores the Gilbert reversed polarity series has been defined as the beginning of the reversed section which precedes these three events. The beginning of the Gilbert reversed polarity series as defined has a probable age of about 5 m.yr. B.P.

Figure 7 shows the magnetic stratigraphy of V21-148, a core from the western Pacific, at latitude 42°.05′N and longitude 160°36′E. This core contains a record of the magnetic stratigraphy through the three events in the Gilbert reversed polarity epoch to the pre-

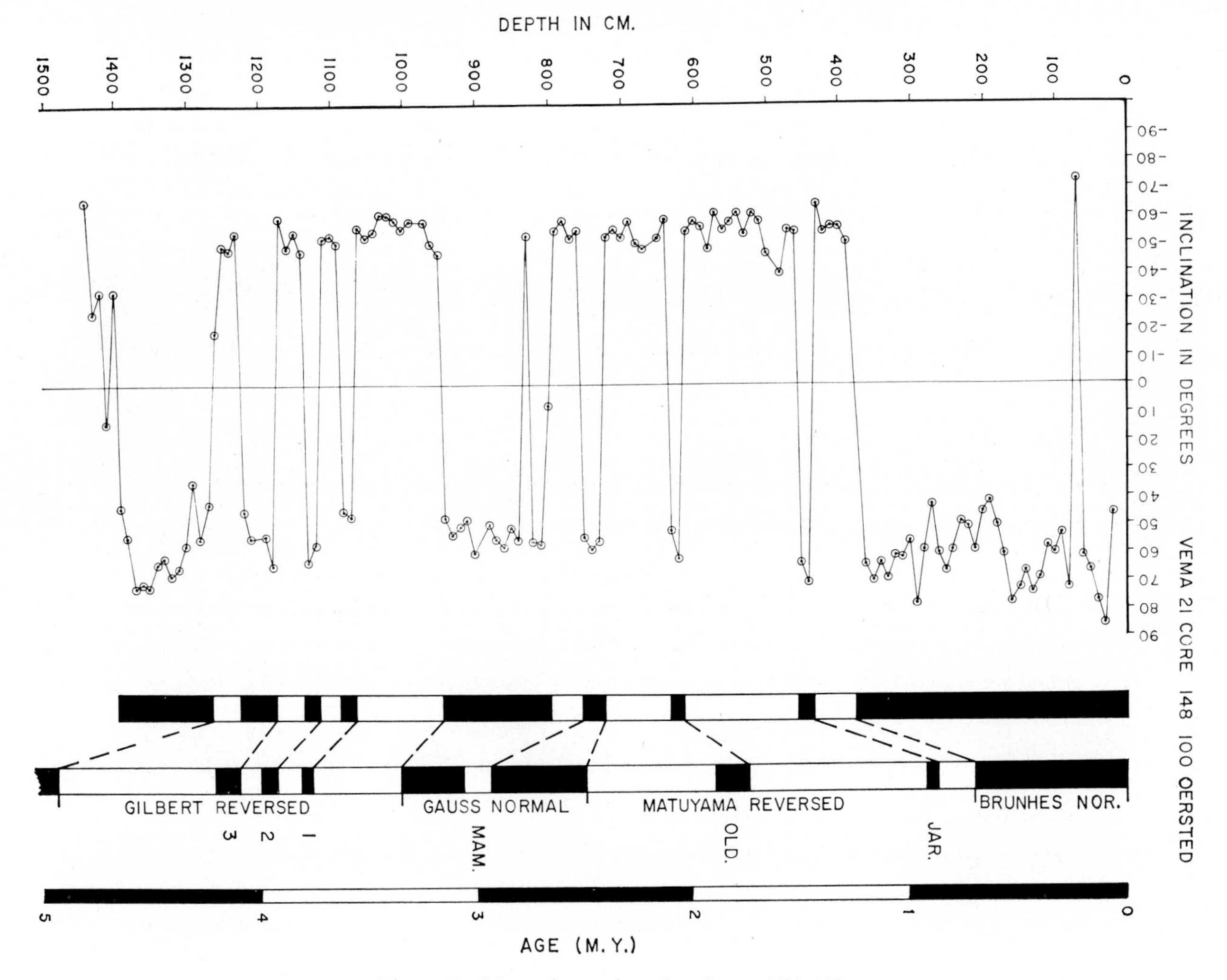

Figure 7. Magnetic stratigraphy of core V21-148.

ceding polarity epoch. This core penetrates to this level because a hiatus is present which represents most of the time represented by the upper Gauss normal polarity series.

In conclusion it is clear that the development of the paleomagnetic study of cores will (1) be of great assistance to stratigraphic studies in the Plio-Pleistocene of the ocean sediments; (2) enable marine geologists to determine rapidly rates of sedimentation at sea in low-sedimentation-rate areas; (3) help in the study of the history of the earth's magnetic field.

Acknowledgments

I wish to express my profound gratitude to Dr. Ewing and the chief scientists on Lamont vessels who have carried on the coring program at sea. Without the efforts of these men, the present summary could not have been written. I would also like to express my gratitude to Dr. Heirtzler, Mr. Foster, Mr. Glass, Dr. Ninkovich, and Mr. Dickson, who have given valuable help and advice since the inception of this program at Lamont. Paleomagnetic research at Lamont is supported by the National Science Foundation under Grants GP-4004 and GA-824. The work was also partially supported by Office of Naval Research N00014-67A-0108-0004. The coring program at Lamont is supported by National Science Foundation GA-558 (B). This support is gratefully acknowledged.

Lamont Geological Observatory
Columbia University

DISCUSSION

DR. MENARD: I believe that cores such as V16-134 are in areas where part of the time you had ice rafting and part of the time you didn't.

DR. OPDYKE: That is right.

DR. WASSERBURG: There seems to be an apparent correlation between intensity and polarity. Is this true or just apparent from looking at the data?

DR. OPDYKE: Generally speaking, there is not any correlation. The intensity seems to wander all over. If we knew more about what was causing the intensity variation, we would be much better off. The only thing we have been able to determine with certainty is that we seem to have an almost one-to-one correlation between reversals and the drop in intensity.

DR. HIDE: What is the shortest period of time in which the field kept the same polarity?

DR. OPDYKE: That would be about ten thousand years—the very short interval in the Olduvai. I must caution you that the shortest absolutely substantiated such interval is the Jaramillo event, with something of the order of fifty thousand years. I wouldn't like to say that the ten thousand year interval is absolutely real until I find it at least two more times.

DR. HALES: Is the magnetic transition paralleled exactly by the paleontologic, or is there any kind of time delay?

DR. OPDYKE: The cores in the Antarctic do show this change in paleontology just at the reversal boundary. It doesn't always fall before and it doesn't always fall after. It looks like the change in fauna occurs sometimes just afterward and sometimes just before the last reversal of the field.

DR. HALES: You mean the change in the fauna controls the magnetic field?

DR. OPDYKE: I am not saying anything of the sort.

DR. IRVING: How widespread is the ash layer?

DR. OPDYKE: It extends over at least 600 miles, and probably originated in the Aleutian arc. It has been dated by Dr. Dymond at 1.4 m.yrs. Incidentally, the sedimentation rates that we obtained have been more or less what people in geochemistry have been getting.

DR. BROECKER: I ran only the cores that you looked at paleomagnetically. We get the same sedimentation rate you get within 30 or 50%. Of course we can only date back 400,000 years. Your first point is at 700,000.

DR. OPDYKE: There are no alarming discrepancies.

MR. THOMSEN: What can you say about the erratic behavior of your data at the top of the cores?

DR. OPDYKE: It is quite a common feature. Commonly we find the first specimen reversed, which is nonsense. If you look at the top of the core and have any experience with deep sea cores, you can see, when you open the core tray, that you are dealing with a core which was taken some years ago. A top layer of clay which may have been two inches thick frequently is found to have lost its water, which may have amounted to a large percentage of the layer. You can expect disorientation of the magnetic particles during drying out, I believe.

MR. THOMSEN: In the act of taking the core?

DR. OPDYKE: In the act of taking the core, or in the act of the core drying out.

Dr. Irving performed an experiment one time where he settled particles in a column in field-free space, then put a field on them, tapped gently, and the particles all lined up. Is that right?

DR. IRVING: Something like that.

DR. OPDYKE: If there is a lot of water present, there is no reason to believe particles can't realign.

DR. BULLARD: The top of a core sometimes runs out, too.

DR. OPDYKE: Yes, and sometimes people pick it up and stuff it back in.

DR. IRVING: Reversed! (*Laughter*) Have you done any formal statistics on whether the reversals are exact?

DR. OPDYKE: No. I don't think it is necessary. For instance, in Vema 16-57, the upper normal sequence has an inclination of 65° and the reversed an inclination of 63°. That is a difference which I do not believe to be real. For another, Vema 16-134, both normal and reversed give 71°.

DR. MUNK: What is the smallest thickness core you can now use to get a sample?

DR. OPDYKE: It depends on the intensity. We can measure specimens down to a cubic centimeter, as long as the intensity is high enough, and our magnetometer is sensitive enough, and the core is homogeneous.

DR. BULLARD: A centimeter corresponds to how long?

DR. OPDYKE: Depending on the rate of sedimentation, a centimeter means from one thousand to ten thousand years.

DR. BULLARD: If it was ten thousand years it would have time to average for variation.

DR. OPDYKE: Yes. I think I am looking at the dipole field averaged to the position of the rotational pole.

DR. BULLARD: Is there any danger, that if you get down to a millimeter, you might get the secular variation?

DR. OPDYKE: Yes. Of course we want to get cores with higher rates of sedimentation. The difficulty we have run into is that cores with higher rates of sedimentation seem to be more weakly magnetized. Only in areas with high volcanic activity do we seem to find high-intensity cores.

REFERENCES

Cox, A., R. R. Doell, and G. B. Dalrymple (1964), Reversals of the earth's magnetic field, *Science* 144:1537.

Dickson, G. O., and J. H. Foster (1966), The magnetic stratigraphy of a deep-sea core from the North Pacific Ocean, *Earth Planet. Sci. Letters* 1:458-62.

Doell, R. R., G. B. Dalrymple, and Allen Cox (1966), Geomagnetic polarity epochs: Sierra Nevada, 3, *J. Geophys. Res.* 71:531-41.

Foster, J. H. (1966), A paleomagnetic spinner magnetometer using a flux gate gradiometer, *Earth Planet. Sci. Letters* 1:463.

Harrison, C.G.A. (1966), The paleomagnetism of deep-sea sediments, *J. Geophys. Res.* 71:3033-44.

Harrison, C.G.A., and B. M. Funnell (1964), Relationship of paleomagnetic reversals and micropaleontology in two late Cenozoic cores from the Pacific Ocean, *Nature* 204:566.

Harrison, C.G.A., and B.L.K. Somayajulu (1966), Behavior of the earth's magnetic field during a reversal, *Nature* 212:1193-95.

Hays, J. D. (1965), Radiolaria and late Tertiary and Quaternary history of the Antarctic seas. *Amer. Geophys. Union Antarctic Res. Ser.* No. 5, p. 125.

Hays, J., and N. Opdyke (1966), Five million years of magnetic and radiolarian stratigraphy (Abstr.), *Geol. Soc. Amer. Program 1966 Ann. Meetings.*

Lin'kova, T. L. (1965), Some results of paleomagnetic study of Arctic Ocean floor sediments (Trans. 1966 by E. R. Hope), pp. 279-81 in *The Present and Past of the Geomagnetic Field*, 279-81, Moscow "Nauka" Press.

McDougall, I., and H. Wensink (1966), Paleomagnetism and geochronology of the Pliocene-Pleistocene lavas in Iceland. *Earth Planet. Sci. Letters* 1:232-36.

McDougall, I., and D. H. Tarling (1963), Dating of polarity zones in the Hawaiian Islands, *Nature* 200:54-56.

Nagata, T. (1952), Reverse thermo-remanent magnetization, *Nature* 169:704.

Ninkovich, D., N. D. Opdyke, B. C. Heezen, and J. H. Foster (1966), Paleomagnetic stratigraphy, rates of deposition and tephrachronology in North Pacific deep-sea sediment, *Earth Planet. Sci. Letters* 1:476-92.

Opdyke, N. D., B. Glass, J. D. Hays, and J. Foster (1966), Paleomagnetic study of Antarctic deep-sea cores, *Science* 154:349-57.

Uffen, R. J. (1963), Influence of the earth's core on the origin and evolution of life, *Nature* 198:143.

MAGNETIC ANOMALIES ASSOCIATED WITH MID-OCEAN RIDGES

F. J. VINE

Introduction

In 1962 Hess revived and expanded the concept of ocean floor spreading in order to explain various geological and geophysical findings pertaining to the ocean basins (Hess, 1962, 1965). In so doing he formulated what is for most earth scientists the only plausible mechanism to account for continental drift.

Hess postulated that the oceanic crust is a surface expression of a slowly convecting mantle, that it is formed by hydration and partial fusion of the mantle at the crest of a mid-ocean ridge and is partially resorbed into the mantle beneath the trench systems. Thus spreading by continuous generation of oceanic crust at ridge crests is thought to produce relative movements of the order of a few centimeters per year between crustal blocks separated by active ridge crests, faults, and/or trench systems. This rate of spreading implies that the ocean basins are ephemeral features compared with the age of the earth and have life spans of a few hundred million years.

In 1963, Vine and Matthews and, apparently simultaneously, Morley and Larochelle (1964) suggested that if the oceanic crust has been formed in this way then its remanent (permanent) magnetization might record changes in the intensity and polarity of the earth's magnetic field as spreading occurs, and that the resulting magnetization contrasts should be revealed in the short-wavelength disturbances or "anomalies" in the earth's magnetic field observed over the ocean basins. As new crust is formed at the crest of a mid-ocean ridge and cools through the Curie temperature it will acquire a remanent magnetization parallel to the ambient direction of the earth's magnetic field. If the earth's field reverses its polarity intermittently then strips of crust of alternate polarity will be produced paralleling and, in the simplest case, distributed symmetrically about, the axis of the ridge.

At the time this suggestion was made none of its three premises was generally accepted, and there was little or no evidence to support it. However, since then the importance of remanent magnetization in interpreting oceanic magnetic anomalies has become more firmly established and widely held (Ade-Hall, 1964; Vogt and Ostenso, 1966; Opdyke and Hekinian, 1967), the likelihood of reversals of the earth's magnetic field in the past has been strikingly confirmed by independent field tests (Cox, Dalrymple, and Doell, 1967; Cox et al., this volume; Opdyke, this volume), and new evidence has come to light which supports the hypothesis. It has been possible therefore to provide virtual proof of the third premise, ocean floor spreading (Vine, 1966).

A Crustal Model

In invoking ocean floor spreading and the Vine-Matthews hypothesis it is clearly necessary to consider the composition and bulk magnetic properties of the oceanic crust. In attempting to simulate oceanic magnetic anomalies all authors have been obliged to assume high magnetization contrasts extending up into, or entirely within, Layer 2 (e.g., Mason, 1958; Heirtzler and LePichon, 1965; Vogt and Ostenso, 1966). The only satisfactory explanation of this is that Layer 2 consists largely, if not entirely, of volcanic extrusives and intrusives of basaltic chemistry, modified locally by metamorphic and metasomatic effects, i.e., the hard rock material commonly dredged from oceanic areas (e.g., Cann and Vine, 1966; Muir and Tilley, 1964, 1966; Melson and Van Andel, 1966). Layer 3 presents more of a problem; likely rock types, capable of satisfying the observed seismic velocities, are gabbro (chemically equivalent to basalt), and serpentinite (hydrated peridotite) derived possibly by hydration and depletion of mantle material. Serpentinite has several attractions as regards ocean floor spreading as discussed by Hess (1962, 1965). In either case the remanent magnetization of Layer 3 is probably not as significant as that of Layer 2. This has been discussed previously by Vine and Matthews (1963), Vine and Wilson (1965) and Cann and Vine (1966). Thus the bulk of the remanent magnetization of the oceanic crust is thought to reside in the basaltic layer (Layer 2), and the intensities of magnetization assumed are in good agreement with measurements on dredged rock samples (e.g., Opdyke and Hekinian, 1967). Such measurements also confirm the assumption that susceptibility contrasts within the oceanic crust are of very minor importance in that the induced magnetization is typically an order of magnitude less than the remanent magnetization and is uniformly directed. Thus for simplicity, in demonstrating the efficacy of the Vine-Matthews hypothesis, remanent magnetization contrasts have been assumed, and confined entirely to Layer 2 (Pitman and Heirtzler, 1966; Vine, 1966), although magnetic sources can potentially exist to the depth of the Curie point isotherm within the upper mantle (i.e., 600°-700°C).

Figure 1, A, is an attempt to summarize and portray this model assuming a literal interpretation of the Vine-Matthews hypothesis, a rate of spreading of 3 cm/yr per spreading limb as deduced for the Juan de Fuca Ridge, southwest of Vancouver Island (Vine, 1966), and the reversal time scale derived below. Figure 1, B, shows a portion of the summary map of magnetic anomalies observed over the Juan de Fuca Ridge (Raff and Mason, 1961). Black areas on this map correspond to areas in which the earth's magnetic field is enhanced, and white areas to those in which it is reduced. As can be seen from the observed and simulated profiles (Fig. 1, C and D), these areas of positive and negative anomaly approximate to predicted "blocks" of normally and reversely magnetized crust, respectively, for this latitude and orientation of the ridge crest. This and the other simulated profiles in this paper have been computed by a technique analogous to that given by Talwani and Heirtzler (1964).

It will be noticed that the simulated anomaly profile contains details which are not present on the observed profile. This is thought to be due to a number of effects:

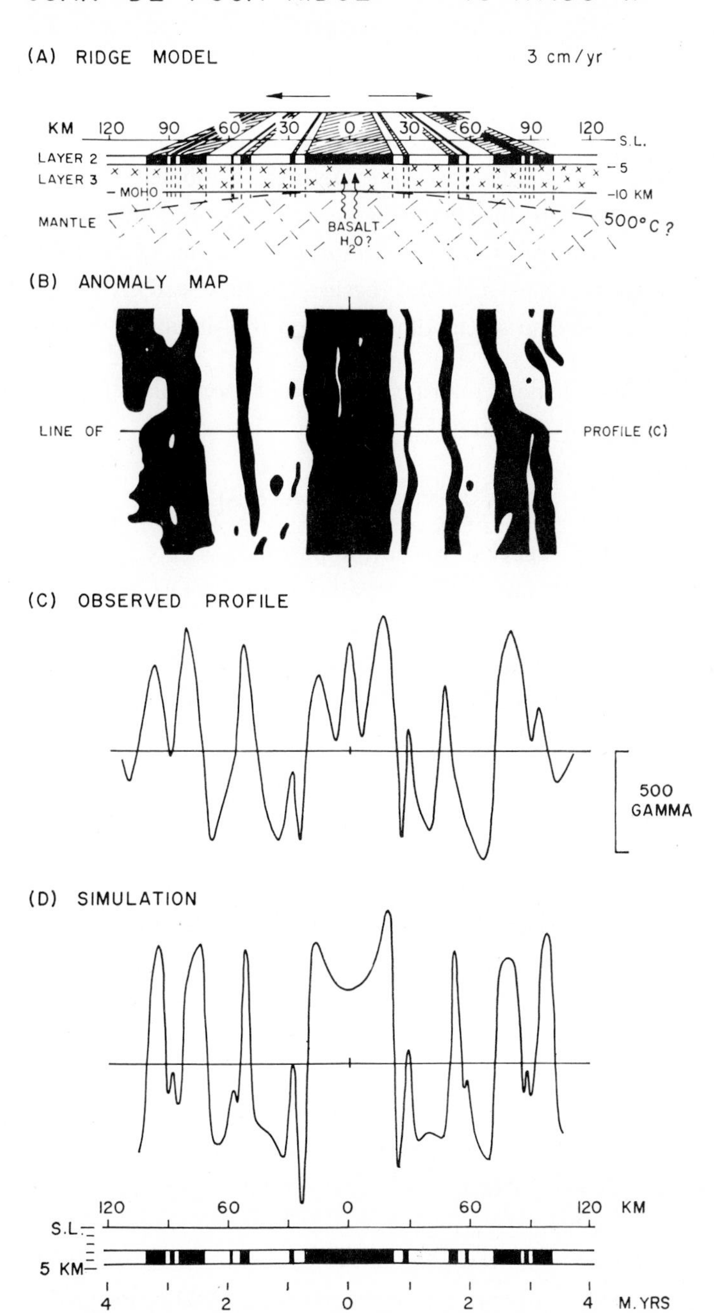

Figure 1. (A) A schematic representation of the crustal model discussed in the text, applied to the Juan de Fuca Ridge, southwest of Vancouver Island (see Fig. 4). Shaded material in layer 2, normally magnetized; unshaded, reversely magnetized.

(B) Part of the summary map of magnetic anomalies recorded over the Juan de Fuca Ridge (Raff and Mason, 1961). Black, areas of positive anomalies; white, areas of negative anomalies.

(C) A total-field magnetic anomaly profile along the line indicated in (B).

(D) A computed profile assuming the model and reversal time scale discussed in the text. Intensity and dip of the earth's magnetic field taken as 54,000 gamma and + 66°; magnetic bearing of profile 087°. (1 gamma = 10^{-5} oersted). (S.L. = sea level.)

Note: Throughout, observed and computed profiles have been drawn in the same proportion: 10 km horizontally is equivalent to 100 gamma vertically. Normal or reverse magnetization is with respect to an axial dipole vector, and the effective susceptibility assumed is ± 0.01 except for the central block at a ridge crest (+ 0.02).

1. The observed profile is, unfortunately, taken from a contoured map of magnetic anomalies (Raff and Mason, 1961), which introduces a certain, and unknown, amount of smoothing.

2. In making the simulation, magnetic sources beneath Layer 2 have been ignored. Consideration of these would also tend to smooth the computed profile.

3. The division of the crust into homogeneously magnetized "blocks" of normally and reversely magnetized material is almost certainly unrealistic, in fact, ungeologic. Some contamination of blocks almost certainly occurs, as suggested by Vine and Matthews (1963), and this too will tend to smooth details on the simulated profile.

An attempt to derive a more realistic model, and mechanism of formation, for the oceanic crust will be discussed later.

The Reversal Time Scale

The four main polarity "epochs" within the past 4 m.yr. are now thought to be well defined by potassium-argon dating of terrestrial lava flows (Cox et al., this volume). They would appear to be bounded by reversals at 0.7, 2.4, and 3.32 m.yr. B.P. Within these epochs there are events of opposite polarity and much shorter duration. There is considerable difficulty in determining the length of these events by the K-Ar method because of the resolution limits of the technique. It seems probable that the work on deep-sea sediment cores is more suited to defining the lengths of these events, although clearly less capable of defining their absolute age because of variations in sedimentation rate. The magnetic anomalies associated with certain mid-ocean ridge crests may also help to define the reversal time scale (Vine, 1966). On the basis of a profile across the East Pacific Rise it was suggested that the Mammoth event at approximately 3 m.yr. B.P. was split. This has now been independently confirmed by McDougall and Chamalaun (1966), working on basalt samples from Hawaii (who named the additional event the Kaena), and by Dickson and Foster (1966), working on deep-sea cores. A further implication of the magnetic profiles observed over ridge crests in the Pacific, where the rate of spreading is typically 3 cm/yr or higher, is that the Olduvai event at approximately 2 m.yr. B.P. is in fact considerably younger than is commonly quoted. However, the anomalies do not preclude, in fact to some extent they suggest, a short event of say 30,000 years duration around 1.95 m.yr. B.P.

Clearly there is a limit to the resolving power of this method, and for the fastest rates of spreading determined thus far it is probably about 20,000 years, perhaps a little less. It is quite possible, therefore, that if shorter events exist they might be discernible by means of K-Ar dating or measurements on sediment cores but have no expression in magnetic anomaly profiles recorded at or above sea level. The resolution of oceanic magnetic anomalies is obviously intimately related to the mechanism of formation of the oceanic crust, as discussed later. The anomalies recorded at sea level would appear to set considerable restraints on this mechanism, and should it be subsequently found that the resolution is greater at depth the restrictions will be even more severe.

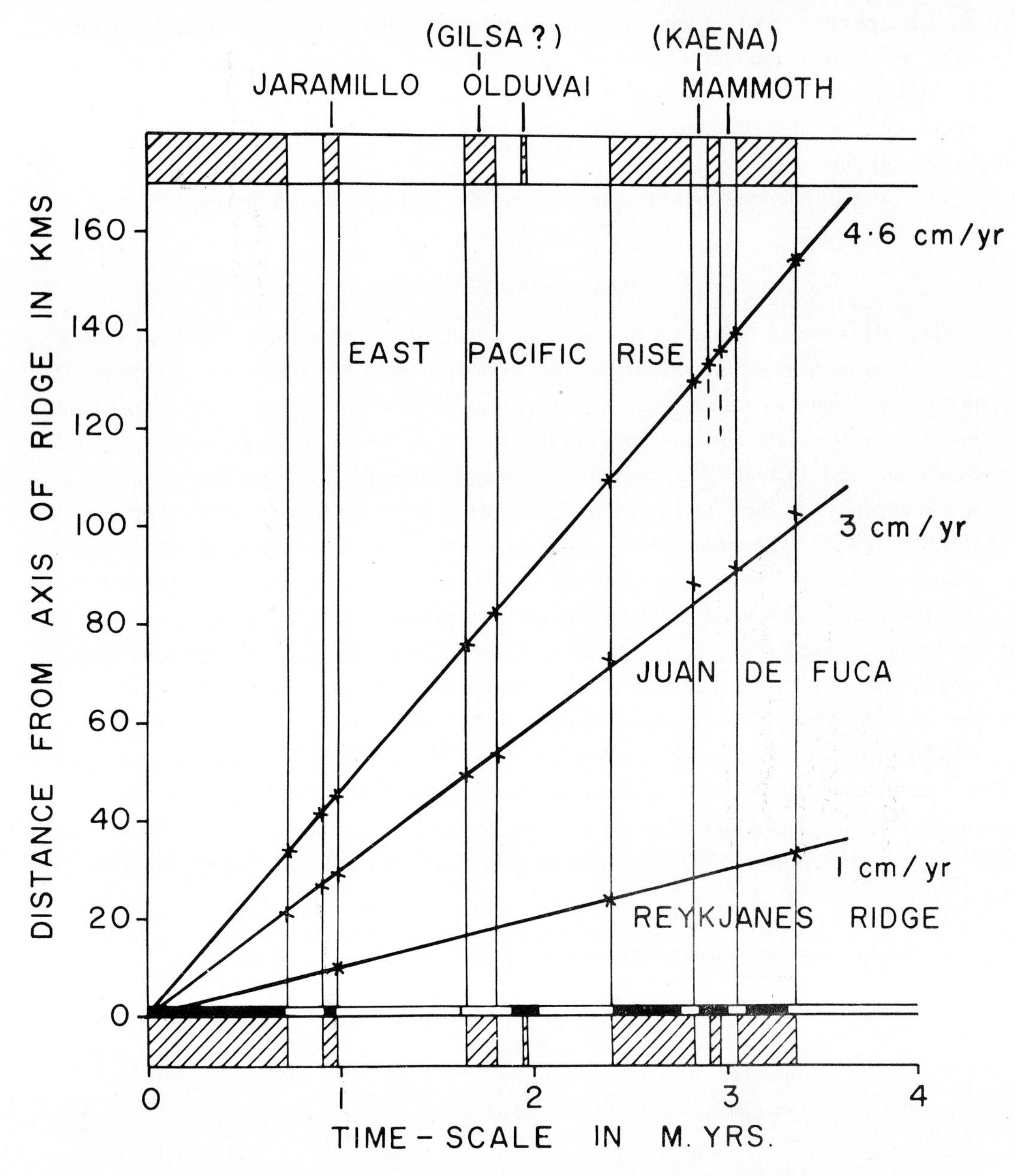

Figure. 2. Inferred normal-reverse boundaries in the oceanic crust plotted against a suggested reversal time scale. (Shaded intervals, periods of normal polarity.) The thin black-and-white time scale indicated immediately above the abscissa is that suggested by Cox et al. (this volume).

The reversal time scale derived in Figure 2 is an attempt to satisfy as much of the relevant data as possible. Thus the main reversal boundaries approximate to those defined by K-Ar dating, and the lengths of events are consistent with those deduced from deep-sea sediment cores where available. It is also felt that the positioning of the main Olduvai (or Gilsa) event is more consistent with the results from deep-sea sediments (e.g., Ninkovich et al., 1966). It is interesting to note that this interval (1.64 to 1.80 m.yr.

B.P.) is a pronounced data gap in the plot of polarities reliably dated by the K-Ar method (Cox, et al., this volume). Clearly the slight revisions suggested are details and are of no consequence, for example, for rates of spreading of 2 cm/yr or less; however, in the Pacific, where rates of spreading are typically 3 cm/yr or higher, such revisions are consistently indicated.

This revised time scale and the rates of spreading deduced from it are used throughout this paper.

Ocean Floor Mapping

Since the reversal time scale has only been defined by independent techniques back to 3.5 m.yr. B.P., it is only strictly possible to deduce rates of spreading at the crests of mid-ocean ridges, as for example in Figure 2 (Vine, 1966). However, it is of great interest to extrapolate these rates beyond this central region and hence extend the reversal time scale back in time. The most suitable magnetic profile available for this purpose would appear to be the *Eltanin* 19 crossing of the East Pacific Rise at 51°S (Pitman and Heirtzler, 1966). This profile exhibits a remarkable degree of symmetry about the topographic axis of the Rise, even in small details (Fig. 3). It is something of a relief to learn that there are in fact some slight deviations from perfect symmetry. Thus if the rate of spreading deduced from the center of this profile (Fig. 2) is assumed for its whole length, the reversal time scale may be extended back to 11 m.yr. B.P. by assuming that the magnetic anomaly gradients continue to define boundaries between essentially normally and essentially reversely magnetized crust (Fig. 3). This is exactly analogous to earlier extrapolations of this type but assumes a slightly different rate of spreading based on the revised reversal time scale for the past 3.5 m.yr. presented above (cf. Pitman and Heirtzler, 1966; Vine, 1966). The time scale deduced here is listed digitally in Table 1.

Table 1. Intervals for which the earth's magnetic field was of normal polarity during the past 11 million years.

Times in millions of years before present		
0.00-0.72	4.26-4.46	7.37-7.41
0.90-0.98	4.59-4.65	7.78-8.13
1.64-1.80	4.72-4.93	8.22-8.31
1.93-1.96	5.50-5.76	8.50-8.57
2.39-2.82	5.85-6.16	8.63-9.79
2.90-2.97	6.47-6.62	10.15-10.29
3.05-3.37	6.83-6.92	10.48-10.61
3.96-4.19	6.97-7.30	10.69-10.80

The value of this extrapolation and the suggested time scale is that it enables one to predict the anomalies one would expect to observe over any other part of the ridge system which has been active for the past 11 m.yr., and at some other latitude, orientation, and rate of spreading. In Figure 3 it is suggested that the anomalies to the west of the

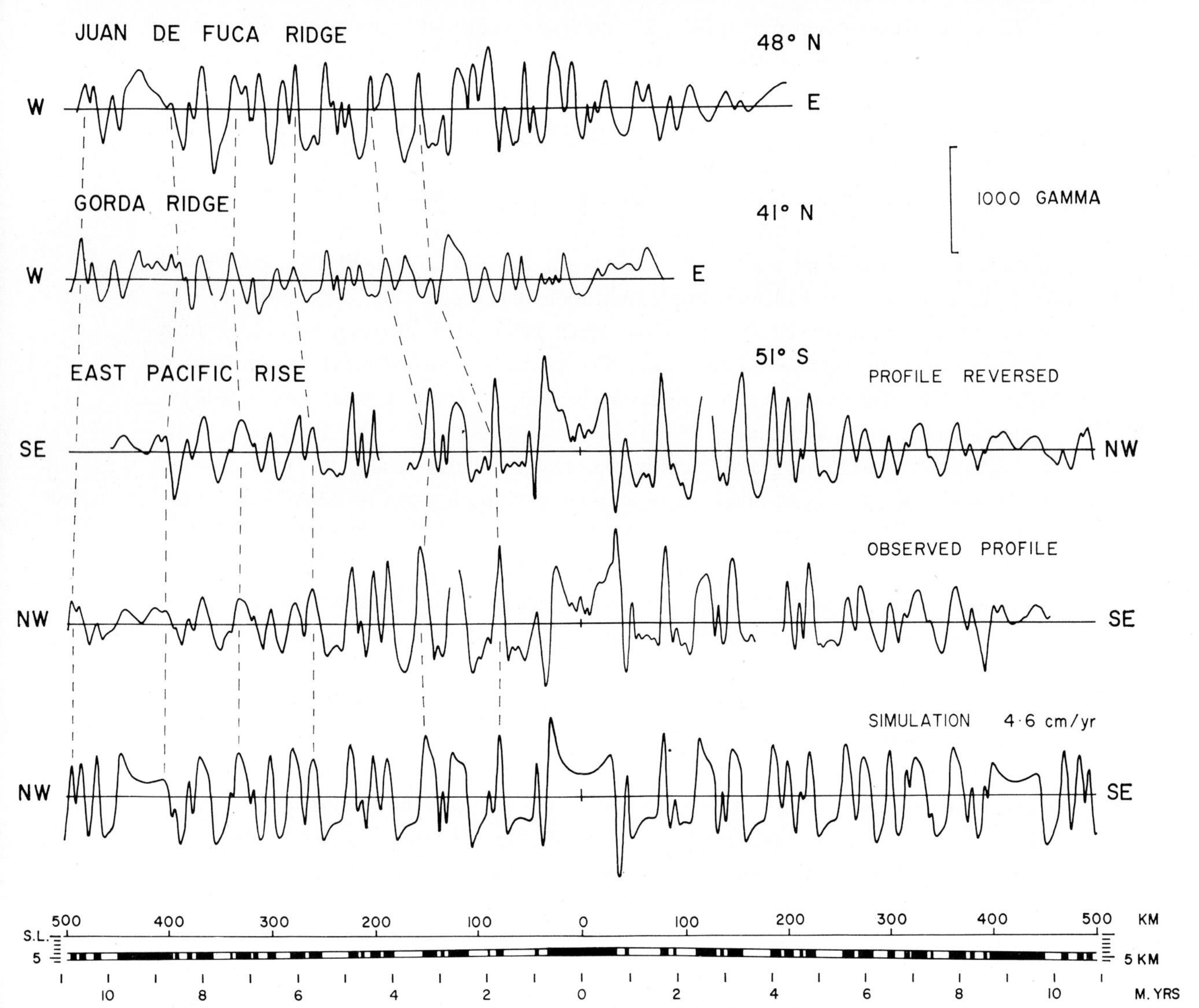

Figure 3. The *Eltanin* 19 profile across the East Pacific Rise (Pitman and Heirtzler, 1966) together with the profile reversed about its midpoint to demonstrate its symmetry, and a computed profile assuming the reversal time scale for the past 11 m.yrs. listed in Table 1, a total field intensity and dip of 48,700 gamma and —62.6° respectively, and a magnetic bearing of 102° for the profile. The profile is also compared with a composite profile across and to the northwest of the Juan de Fuca Ridge, and a profile normal to the strike of the anomalies across and to the west of the Gorda Ridge (Raff and Mason, 1961; Vacquier et al., 1961).

Juan de Fuca and Gorda ridges in the northeast Pacific correlate with those along the East Pacific Rise profile at 51°S and 11,000 km away. Such a direct comparison between these profiles is valid because their latitudes, orientations, and rates of spreading are such as to produce directly comparable anomalies from the same reversal time scale. This correlation suggests that one might assign dates to the summary of anomalies in the Juan de Fuca area given by Raff and Mason (1961), implying in turn an age for the underlying ocean crust (Fig. 4).

Thus this extrapolation presents the intriguing possibility of mapping the ocean floor if sufficiently detailed magnetic surveys are available. This particular survey was made by ship, but it is clear that an aeromagnetic survey would be perfectly adequate, and this may well prove to be the most efficient and feasible way of obtaining reasonable coverage in finite time in the future.

The strikingly simple structure of the oceanic crust implied by the "geologic map" presented in Figure 4 contrasts with the complexity of the continental crust. However, it must be borne in mind that the magnetic anomalies are quite unlike those observed over the continents, and there is every indication that the oceanic crust is in no way comparable to that of the continents (Hess, 1962, 1965). This interpretation of the Juan de Fuca area in terms of ocean floor spreading has been discussed previously by Wilson (1965b), Vine and Wilson (1965) and Vine (1966). It would appear to be consistent with the bathymetry (McManus, 1965), seismicity (Tobin and Sykes, 1967), heat flow (Von Herzen, 1964), and gravity field in the area (Dehlinger et al., 1967), in addition to the magnetics. However, other interpretations have been given by McManus (1965, 1967), Pavoni (1966), and Talwani et al. (1965).

The structural interpretation of the western margin of North America in terms of northwest-southeast ocean floor spreading and transform faults (Wilson, 1965a) would appear to be in good accord with the current seismicity. However, the magnetic anomaly pattern of Figure 4 and the ages assigned to it imply that this regime is a very recent one geologically and that prior to the Middle Pliocene the direction of spreading and presumably faulting was predominantly east-west. Thus the San Andreas fault may only have behaved as a transform fault for the past few million years, and this may be correlatable with the observation that the more recent movement on the San Andreas has been as much as an order of magnitude greater than the average rate deduced for the whole of Cenozoic time (Holmes, 1965; Hamilton and Myers, 1966). If it is assumed that the spreading from the Juan de Fuca and Gorda ridges has been accommodated by an east-west quasi-transform fault acting on the base of the continental crust, as suggested by Vine (1966), and also by movement on the San Andreas fault system, then the implication is that the total displacement on the San Andreas since Late Miocene should not exceed 180 km, and in all probability most, if not all, of this transform motion has occurred within the past 5 m.yr. at an average rate of 3 cm/yr. The quasi-transform fault concept of Vine (1966) would appear to agree well with the interpretation of Hamilton and Myers (1966) involving east-west extension and apparent right-lateral offset of features in the western cordillera of the United States.

A further extrapolation from the time scale deduced above (Table 1) is shown in Figure 5. Here an anomaly profile across the Reykjanes Ridge southwest of Iceland, on the Mid-Atlantic Ridge, has been simulated assuming the appropriate latitude, orientation of the ridge, and rate of spreading deduced previously (Fig. 2) (Vine, 1966). Thus given the rate of spreading of 1 cm/yr from the center of the observed profile, the computation is essentially dependent on the time scale derived from the South Pacific profile, nearly 15,000 km away. Even with this great an extrapolation, the computed profile shows

Figure 4. Summary diagram of total-field magnetic anomalies southwest of Vancouver Island. Areas of positive anomalies are colored and are thought to approximate to areas of normal magnetization in the oceanic crust. These areas duplicate the reversal time scale for the past 4 m.yrs. and its extension to 11 m.yrs. derived from the South Pacific, hence implying ages for the underlying crust as indicated. Straight lines delineate faults offsetting the anomaly pattern. Central red anomalies coincide with ridge crests; the Juan de Fuca to the north and the Gorda to the south. (Based on Fig. 1 of Raff and Mason, 1961.)

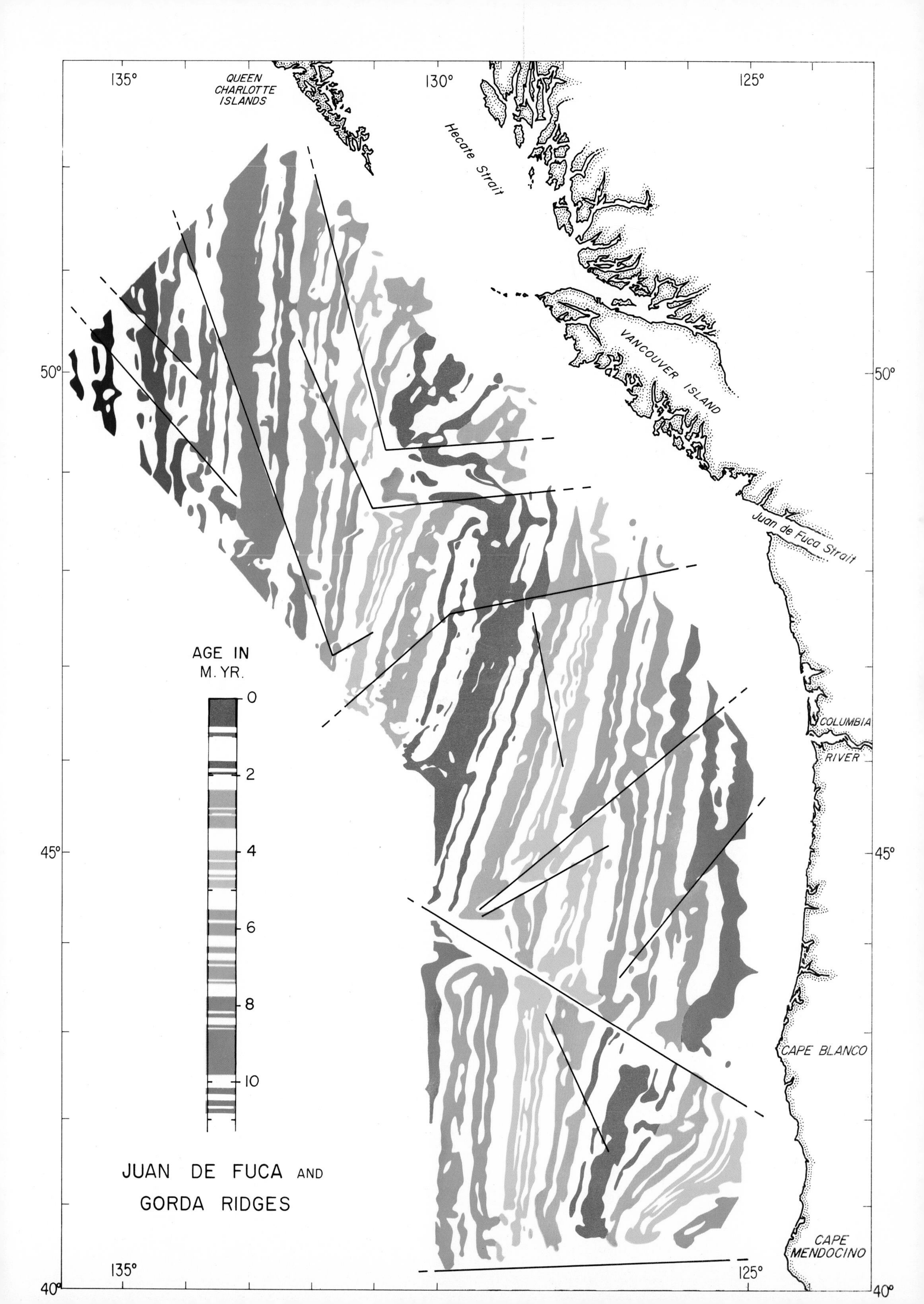
135°
130°
125°
QUEEN CHARLOTTE ISLANDS
Hecate Strait
VANCOUVER ISLAND
Juan de Fuca Strait
COLUMBIA RIVER
CAPE BLANCO
CAPE MENDOCINO
50°
45°
40°
AGE IN M. YR.
0
2
4
6
8
10
JUAN DE FUCA AND GORDA RIDGES

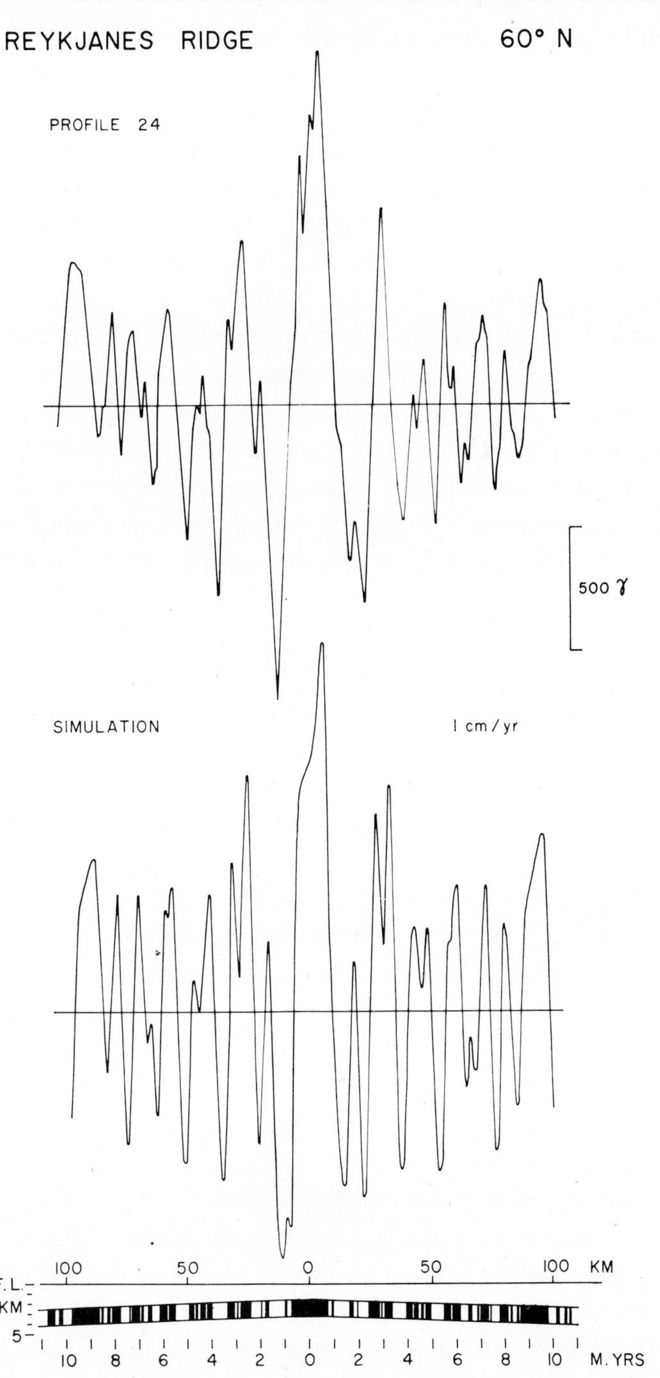

Figure 5. An observed aeromagnetic profile across the Reykjanes Ridge, southwest of Iceland, together with a simulated profile assuming the reversal time scale for the past 11 m.yr. deduced from the South Pacific (Figure 3) and listed in Table 1. Intensity and dip of the earth's field taken as 51,600 gamma and +74.3° respectively; magnetic bearing of profile 153°. (F.L. = flight level.)

striking similarities with an observed aeromagnetic profile across this ridge (Fig. 5). This observed profile is one of 58 traverses of the ridge flown by the U.S. Naval Oceanographic Office in 1963 in making a detailed aeromagnetic survey (Heirtzler, Le Pichon, and Baron, 1966). The summary map of the anomalies over this ridge presented by these authors can therefore be calibrated with a time scale in exactly the same way as was done for the Juan de Fuca area (Fig. 6). The geologic setting of the Reykjanes Ridge and the possible implications of its inferred age as regards continental drift have been discussed by Vine (1966).

Comparable simulations of observed magnetic profiles from the North Atlantic, the South Atlantic, and the northwest Indian Ocean are shown in Figure 7. The rates of spreading assumed are 1.25, 2.3, and 1.5 cm/yr respectively. The North Atlantic profile (*Vema* 17), although normal to the axis of the ridge in this area (Heirtzler and Le Pichon, 1965; Heezen and Tharp, 1965), may well not parallel the direction of spreading. The latter might be more accurately indicated by the fracture zone immediately to the south (Morgan, 1968). In this case the true rate of spreading implied would be 1.5 cm/yr for each flank of the ridge.

Extension to the Flank Areas

When the Vine-Matthews hypothesis was formulated in 1963 there was no convincing evidence available to demonstrate the symmetry and parallelism of magnetic anomalies about ridge crests, and it was certainly not realized that the central anomalies reproduce the reversal time scale, since this time-scale was not known. At that time, therefore, one of the main reasons for suggesting this model was that it might offer the most plausible explanation of the remarkably linear and continuous magnetic anomalies recorded over the western flank of the East Pacific Rise in the northeast Pacific (Raff, 1966; Peter, 1966). These well-known and enigmatic anomalies, first discovered in the late 1950's (Mason, 1958), retain their characteristic shape and spacing for thousands of kilometers along their length, interrupted only by transverse fractures which offset them (Vacquier, 1965). Now that it can be shown that these anomaly patterns can be recognized in all three ocean basins and that they too are symmetrically disposed about ridge axes (Heirtzler, this volume), it is becoming increasingly difficult to interpret these anomalies in any way other than that suggested by Vine and Matthews.

Vine (1966) suggested that flank anomalies in the northeast Pacific at 36°N could be recognized south of New Zealand on longitude 173°E. Figure 8 suggests that the same configuration of normally and reversely magnetized "blocks" within the oceanic crust, with dimensions adjusted according to an appropriate rate of spreading, might explain magnetic anomalies observed in the northeast Pacific, South Pacific, and northeast Atlantic oceans. Hence the implication that ocean floor spreading and the Vine and Matthews hypothesis are applicable to these widely separated areas for at least as far back in time as is represented by these anomalies. If the rather speculative extrapolation of Vine (1966) is followed, these anomalies correspond to underlying crust varying in age from approximately 50 to 75 m.yr. B.P. On the basis of this extrapolation and

Figure 6. Summary diagram of magnetic anomalies recorded over the Reykjanes Ridge (Heirtzler, Le Pichon, and Baron, 1966). Areas of positive anomalies are colored and are thought to approximate to areas of normal magnetization in the oceanic crust. These may be dated via the simulation shown in Figure 5 and based on the reversal time scale listed in Table 1 (cf. Figure 4). Again the central red anomaly coincides with the topographic crest of the ridge.

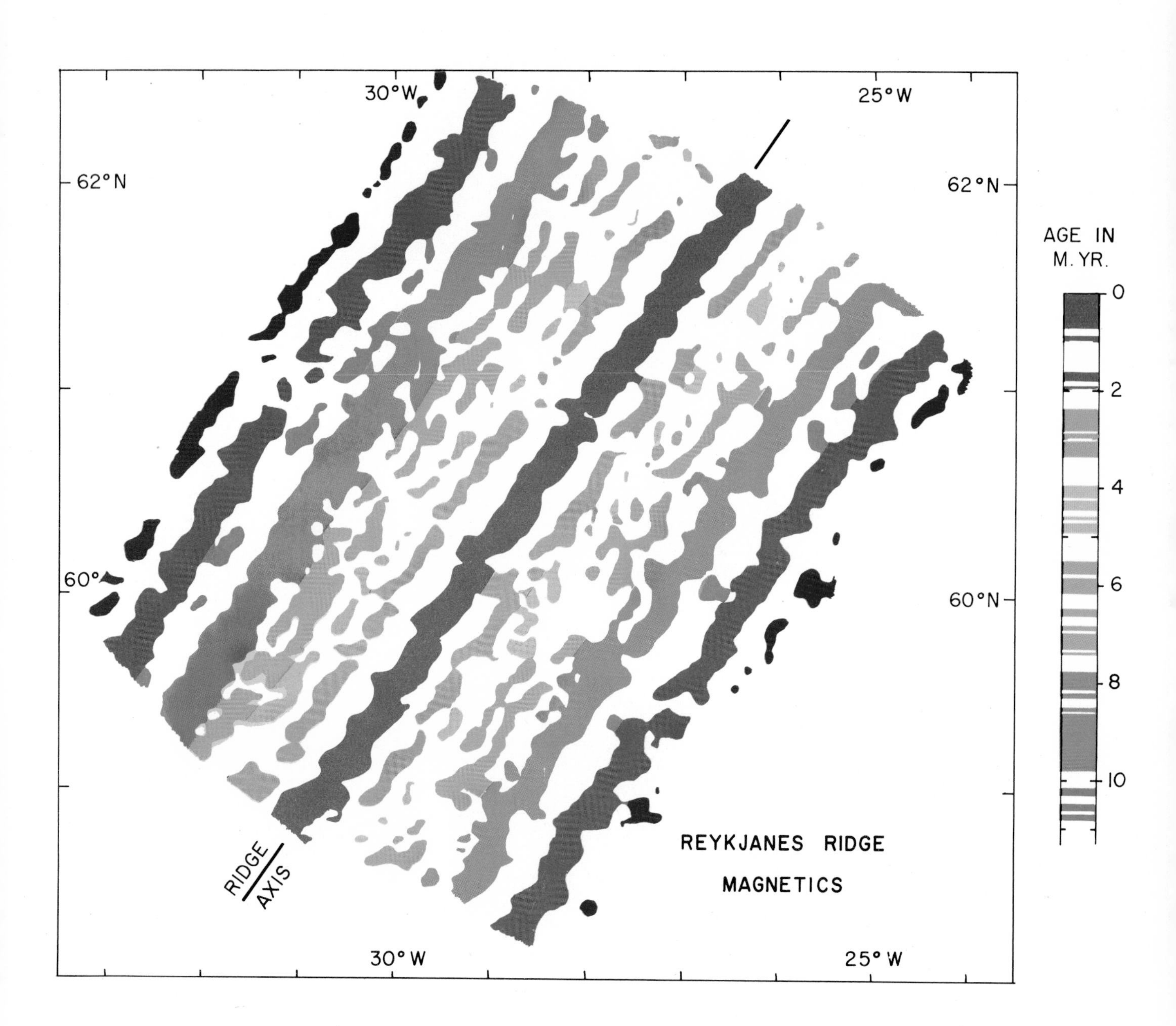
30°W
25°W
62°N
62°N
60°
60°N
AGE IN
M.YR.
0
2
4
6
8
10
RIDGE
AXIS
REYKJANES RIDGE
MAGNETICS
30°W
25°W

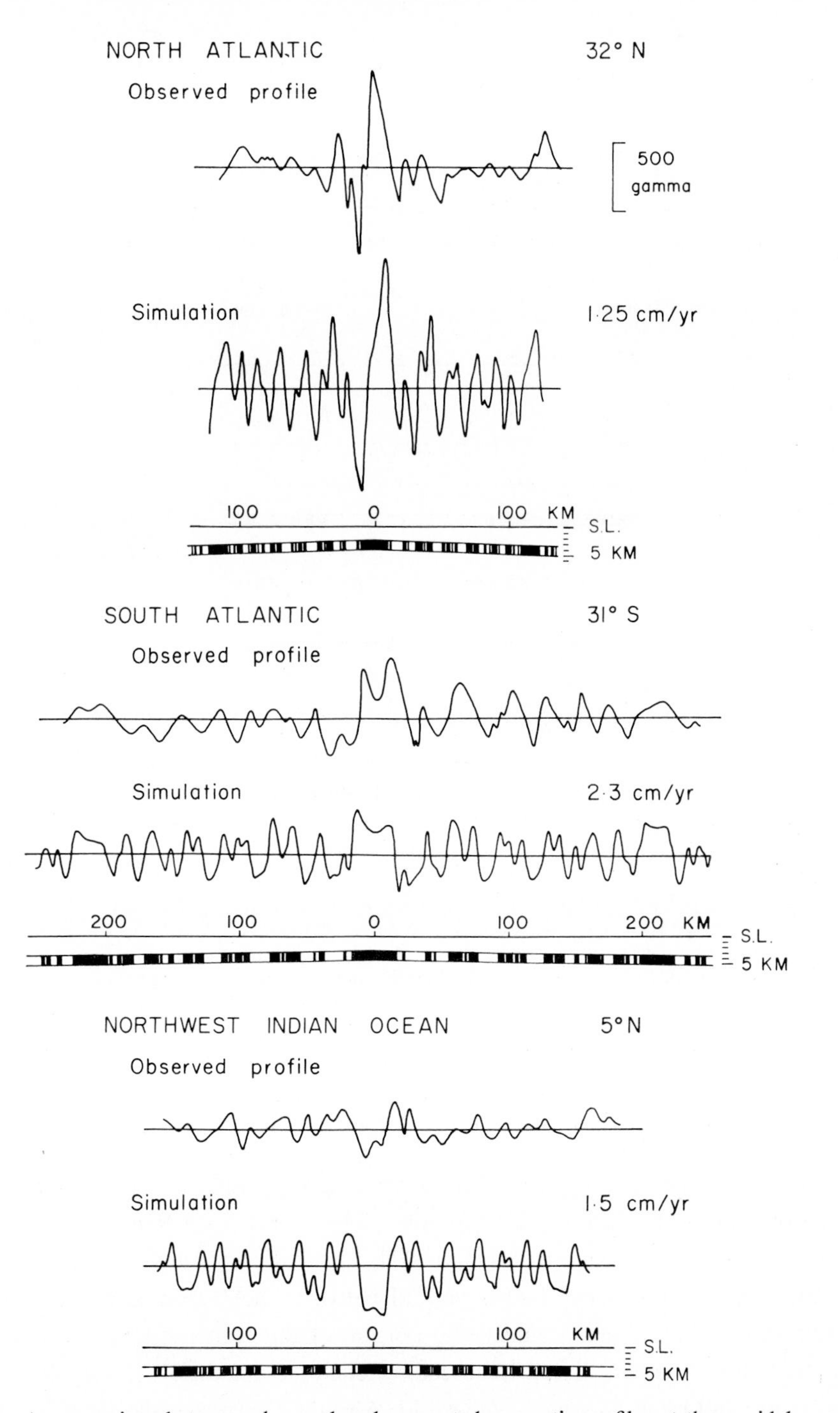

Figure 7. A comparison between observed and computed magnetic profiles at three widely separated points on the mid-ocean ridge system (cf. Figure 5). Observed profiles from Heirtzler and Le Pichon (1965) and Matthews et al. (1965). The simulations assume the reversal time scale listed in Table 1, a constant rate of spreading, and values for the intensity and dip of the earth's field and the magnetic bearing of each profile as follows: North Atlantic, 44,000 gamma, +56°, 150°; South Atlantic, 27,600 gamma, −54°, 116°; northwest Indian Ocean, 37,620 gamma, −6°, 044°.

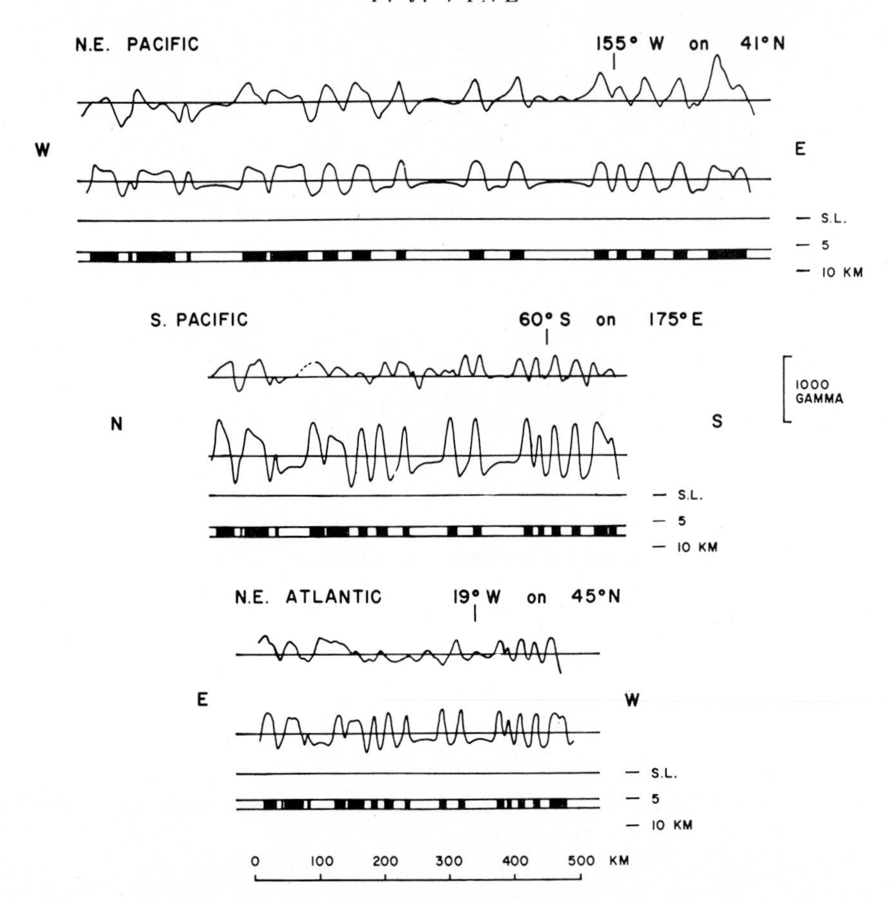

Figure 8. In the upper part of the diagram north-south boundaries between normally and reversely magnetized crust have been assigned along a profile at latitude 41° N in the northeast Pacific, in order to obtain a reasonable simulation of the observed (uppermost) anomaly profile (Raff, 1966). The same crustal configuration, contracted in length, has then been used to simulate flank anomaly profiles in the south Pacific, along approximately 175° E (Christoffel and Ross, 1965), and in the northeast Atlantic along 45° N. Hence the implication that the underlying crust in these three areas was formed by spreading during the same interval of time but at a different rate in each area. The lower, simulated profiles assume values for the intensity and dip of the earth's field and the magnetic bearing of each profile as follows: northeast Pacific, 45,000 gamma, + 58°, 074°; south Pacific, 65,000 gamma, − 75°, 146°; northeast Atlantic, 46,000 gamma, + 62°, 103°.

spreading rates deduced elsewhere (Vine, 1966) it is possible to suggest dates for the initiation of spreading in certain oceanic areas. The East Pacific Rise, for example, may have been formed within the past 80 m.yr. The Mid-Atlantic Ridge may be 60-70 m.yr. old in the vicinity of Iceland, 100 m.yr. old between the Grand Banks and Portugal, 190 m.yr. old in the remaining and wider portion of the North Atlantic, and 135 m.yr. old in the South Atlantic. Similarly spreading centered on the Carlsberg Ridge in the northwest Indian Ocean may be 75 m.yr. old. Although very speculative, it is interesting to note that these dates are in good agreement with those suggested earlier on the basis of the geology of adjacent continental margins.

Thus if such extrapolations are valid the oceans are ephemeral features of the earth's crust and the present-day ocean basins record no more, say, than 200 m.yr. of earth history, a small fraction of geologic time—a depressing possibility originally emphasized by Hess (1964). Clearly such a hypothesis is highly susceptible and vulnerable to further tests involving the dating of hard rocks and sediments from the ocean floor. As yet there appear to be few conflicts (Ewing et al., 1966; Saito et al., 1966), and some of the apparently contradictory dates sometimes quoted would appear to be unreliable. The date of 50 m.yr. given for material on Iceland by Wilson (1963) is not confirmed by more recent work (Dagley et al., 1967; Gale et al., 1966). The oldest material on Iceland is probably little more than 20 m.yrs. old (ibid.). Several authors (e.g., Saito et al., 1966; Vogt and Ostenso, 1967) have made much of the K-Ar date of 29 ± 4 m. yr. obtained from a dredged basalt sample from the median valley of the Mid-Atlantic Ridge at 46°N (Miller, 1964). An earlier date of 3 ± 2 m.yr. for this sample exists elsewhere in the literature (Muir and Tilley, 1964). It would appear that neither date is reliable in that the sample is very altered (J. A. Miller, written communication). However, some dates, for example the date of 27 ± 6 m.yr. for a sample from Cobb seamount in the vicinity of the Juan de Fuca Ridge (Budinger and Enbysk, 1967), may well be reliable, and if indicative of the age of the underlying oceanic crust (i.e., not ice-rafted material) then such a date is clearly in conflict with the concept of ocean floor spreading as interpreted above.

A Possible Mechanism

In attempting to explain the anomalously high amplitude of the central magnetic anomaly observed over ridge crests in the Atlantic and Indian oceans and at the same time to suggest a plausible mechanism for the formation of the oceanic crust by ocean floor spreading, Vine and Matthews (1963) invoked random extrusion and intrusion of material over a comparatively wide zone centrally which contaminates the adjacent "blocks" of reversed polarity with material of the current polarity, hence reducing the resultant or effective magnetization of these blocks. Such a model has been extended and made more quantitative by Loncarevic, Mason, and Matthews (1966), Matthews and Bath (1967) and most recently by Vine and Morgan (1967). The model will not be discussed in detail here but it seems important and timely to emphasize some of the conclusions drawn from it.

1. Perhaps the most fundamental and interesting result of such simulations is the indication that the "halfwidth"[1] of the zone of formation of the oceanic crust *decreases* as the inferred rate of spreading *increases*. Thus whereas a halfwidth of approximately 3 km for this zone would appear to be suitable for the simulation of magnetic profiles from the North Atlantic, a zone of halfwidth no greater than 2 km is necessary in the South Pacific if the remarkable details of the anomaly profile are to be reproduced. Thus, because of the differences in the rate of spreading and the width of the zone of intrusion, the contamination of "blocks" of a particular polarity by material of the opposite polarity will be very much greater in the Atlantic and northwest Indian oceans compared with

[1] "Halfwidth" is defined as the standard deviation of a gaussian frequency distribution of points of intrusion about the ridge axis.

the Pacific. Thus in sampling the crustal layers by means of a deep-sea drilling program, although a reliable date for the material will be very significant as regards testing the hypothesis of ocean floor spreading, the polarity of an oriented core will not necessarily be a crucial test of the Vine-Matthews hypothesis; however, the likelihood of obtaining the implied polarity from beneath a wide anomaly in the Pacific is very much greater than in the Atlantic. If, however, there are very short intervals of a particular polarity within the periods defined by the oceanic magnetic anomalies, this likelihood will be correspondingly less in both cases.

2. It can indeed be shown that this model is capable of reproducing the greater amplitude of the central anomaly (e.g., Matthews and Bath, 1967), but it is felt that an additional parameter in this connection is the viscous decay of remanent magnetization with time. It will be noted from the simulations presented in Figures 3, 5, and 7 that the amplitude of the computed anomalies is increasingly too large as one moves away from the crest of the ridge. Although this could be explained in terms of a progressive decrease in the intensity of the earth's magnetic field back in time, it is felt that the gradual decay of remanence with time is a more likely explanation.

3. An additional effect of such an injection or emplacement model would be to smooth the magnetic profiles predicted on the basis of the simple "block" model, as noted earlier. In particular, shorter polarity events will not give the effect predicted by the homogeneous "block" model.

Although such a model and mechanism is incompatible with other concepts of spreading recently formulated (Orowan, 1966; Vogt and Ostenso, 1967), it is, unlike these, capable of explaining the observed magnetic anomalies which at the present time are proving to be one of the most convincing and unambiguous lines of evidence supporting the hypothesis of the evolution of the present ocean basins by spreading during the past 200 million years.

Acknowledgments

I thank J. R. Heirtzler, D. H. Matthews, W. C. Pitman, and D. I. Ross for original magnetic data from the Reykjanes Ridge, northeast Atlantic, East Pacific Rise, and South Pacific respectively; A. D. Raff for a copy of the diagram on which Figure 4 is based; A. Cox, H. H. Hess, and W. J. Morgan for valuable discussions; and Susan Vine for preparing the manuscript and diagrams. This work was supported in part by NSF grant GP 3451, and facilities at the Bedford Institute of Oceanography, Dartmouth, Nova Scotia.

Department of Geological and Geophysical Sciences
Princeton University

Discussion

DR. ANDERSON: What is causing the spreading? Is it being pushed out from the center or is it being carried on the back of a convection cell?

DR. VINE: It is being carried on the back of a convection cell.

DR. ANDERSON: What is causing the spreading? Why is the ocean spreading? Is it being pushed out from the middle?

DR. VINE: I am not particularly worried. I think it is convection.

DR. ANDERSON: I am wondering what is driving the continents.

DR. VINE: The Atlantic cell essentially. One has to make two big distinctions, first between the oceans and the continents, and then perhaps between the Atlantic and Indian oceans, which would appear to have actively spread continents apart, and the Pacific Ocean which has not done this. Presumably the rise in the Pacific Ocean was formed within older former oceanic crust. It didn't split continents apart.

DR. WASSERBURG: Why are the fields asymmetric across the center of the ridge?

DR. VINE: It is important to emphasize that the crustal model is perfectly symmetrical about the axis of the ridge; the asymmetry of the computed, and presumably the observed, profiles is a function of the latitude and orientation. In general a symmetric body with an arbitrarily oriented inducing field will not have a symmetric anomaly.

DR. PRESS: Have you tested the symmetry statistically?

DR. VINE: I never touch statistics. I just deal with the facts.[2] (*Laughter*)

DR. MUNK: Another statistical question then: Is there a predominance of roughly north-south trending magnetic anomalies of all those you have studied, rather than being randomly distributed in all possible orientations?

DR. VINE: The anomalies are predominantly north-south. It is just possible that in the Western Pacific there is another set of linear anomalies, perhaps correlated with the Darwin Rise. At the present time we don't even know whether they are there. There are suggestions in places of east-west anomalies. These may turn out to be complications due to topography associated with multiple fractures.

DR. MUNK: If they are really rather convincingly more often north-south then you should forget about heat flow and consider that it is due to something having to do with rotation of the earth, such as slowing down of the rotation, which makes the bulge too big at any moment.

DR. VINE: This is the old question of whether spreading and drifting has been east-west. I don't support that entirely; when you consider Antarctica and the surrounding ridges, there has clearly been quite a big northerly component.

REFERENCES

Ade-Hall, J. M. (1964), The magnetic properties of some submarine oceanic lavas, *Geophys. J.* 9:85-92.

Budinger, T. F., and B. J. Enbysk (1967), Late Tertiary date from the East Pacific Rise, *J. Geophys. Res.* 72:2271-74.

[2] It is tempting, although not very profound, to quote Rutherford on this point. Rutherford once stated that if he obtained results which required statistics to interpret them he would throw them away.

Cann, J. R., and F. J. Vine (1966), An area on the crest of the Carlsberg Ridge: Petrology and magnetic survey, *Phil. Trans. Roy. Soc. Lond. A* 259:198-217.
Christoffel, D. A., and D. I. Ross (1965), Magnetic anomalies south of the New Zealand Plateau, *J. Geophys. Res.* 70:2857-61.
Cox, A., G. B. Dalrymple, and R. R. Doell (1967), Reversals of the earth's magnetic field, *Sci. Amer.* 216:44-54.
Cox, A., R. R. Doell, and G. B. Dalrymple (this volume), Time-scale for geomagnetic reversals.
Dagley, P., R. L. Wilson, J. M. Ade-Hall, G. P. L. Walker, S. E. Haggerty, T. Sigurgeirsson, N. D. Watkins, P. J. Smith, J. Edwards, and R. L. Grasty (1967), Geomagnetic polarity zones for Icelandic lavas, *Nature* 216:25-29.
Dehlinger, P., R. W. Couch, and M. Gemperle (1967), Gravity and structure of the eastern part of the Mendocino escarpment, *J. Geophys. Res.* 72:1233-47.
Dickson, G. O., and J. H. Foster (1966), The magnetic stratigraphy of a deep-sea core from the north Pacific Ocean, *Earth Planet. Sci. Letters* 1:458-62.
Ewing, J., J. L. Worzel, M. Ewing, and C. Windisch (1966), Ages of Horizon A and the oldest Atlantic sediments, *Science* 154:1125-32.
Gale, N. H., S. Moorbath, J. Simons, and G. P. L. Walker (1966), K-Ar ages of acid intrusive rocks from Iceland, *Earth Planet. Sci. Letters* 1:284-88.
Hamilton, W., and W. B. Myers (1966), Cenozoic tectonics of the western United States, *Rev. Geophys.* 4:509-49.
Heezen, B. C., and M. Tharp (1965), Tectonic fabric of the Atlantic and Indian Oceans and Continental Drift, *Phil. Trans. Roy. Soc. Lond. A* 258:90-106.
Heirtzler, J. R., and X. Le Pichon (1965), Crustal structure of the mid-ocean ridges, 3, Magnetic anomalies over the Mid-Atlantic Ridge, *J. Geophys. Res.* 70:4013-34.
Heirtzler, J. R., X. Le Pichon, and J. G. Baron (1966), Magnetic anomalies over the Reykjanes Ridge, *Deep-Sea Res.* 13:427-43.
Hess, H. H. (1962), History of ocean basins, pp. 599-620 in *Petrologic Studies*, Buddington Vol., A. E. J. Engel et al., eds., *Geological Society of America.*
Hess, H. H. (1964), Seismic anisotropy of the uppermost mantle under the oceans, *Nature* 203: 629-31.
Hess, H. H. (1965), Mid-ocean ridges and tectonics of the sea floor, pp. 317-33 in *Submarine Geology and Geophysics*, Vol. XVII, Colston Papers, W. F. Whittard and R. Bradshaw, eds., Butterworths Sci. Publ., London.
Holmes, A. (1965), *Principles of Physical Geology*, Nelson Press/Ronald Press.
Loncarevic, B. D., C. S. Mason, and D. H. Matthews (1966), Mid-Atlantic Ridge near 45° north, I, The median valley, *Can. J. Earth Sci.* 3:327-49.
Mason, R. G. (1958), A magnetic survey off the west coast of the United States. (Lat. 32°-36° N: Long. 121°-128° W), *Geophys. J.* 1:320-29.
Matthews, D. H., and J. Bath (1967), Formation of magnetic anomaly pattern of Mid-Atlantic Ridge, *Geophys. J.* 13:349-357.
Matthews, D. H., F. J. Vine, and J. R. Cann (1965), Geology of an area of the Carlsberg Ridge, Indian Ocean, *Geol. Soc. Amer. Bull.* 76:675-82.
McDougall, I., and F. H. Chamalaun (1966), Geomagnetic polarity scale of time, *Nature* 212: 1415-18.
McManus, D. A. (1965), Blanco fracture zone, northeast Pacific Ocean, *Marine Geol.* 3:429-55.
McManus, D. A. (1967), Physiography of Cobb and Gorda Rises, northeast Pacific Ocean, *Geol. Soc. Amer. Bull.* 78:527-46.
Melson, W. G., and T. H. Van Andel (1966), Metamorphism in the Mid-Atlantic Ridge, 22° N Latitude, *Marine Geol.* 4:165-86.
Miller, J. A. (1964), Age determinations made on samples of basalt from the Tristan da Cuhna group and other parts of the Mid-Atlantic Ridge, *Phil. Trans. Roy. Soc. A* 256:565-69.
Morgan, W. J. (1968), Rises, trenches, great faults and crustal blocks, *J. Geophys. Res.* 73:1959-82.
Morley, L. W., and A. Larochelle (1964), Paleomagnetism as a means of dating geological events, pp. 39-51 in *Geochronology in Canada,* F. Fitz Osborne, ed., Roy. Soc. Canada Spec. Publ. 8, Univ. of Toronto Press.
Muir, I. D., and C. E. Tilley (1964), Basalts from the northern part of the rift zone of the Mid-Atlantic Ridge, *J. Petrol.* 5:409-34.
Muir, I. D., and C. E. Tilley (1966), Basalts from the northern part of the Mid-Atlantic Ridge, *J. Petrol.* 7:193-201.
Ninkovich, D., N. D. Opdyke, B. C. Heezen, and J. H. Foster (1966), Paleomagnetic stratigraphy, rates of deposition and tephrachronology in North Pacific deep-sea sediment, *Earth Planet. Sci. Letters* 1:476-92.
Opdyke, N. D., and R. Hekinian (1967), Magnetic properties of some igneous rocks from the Mid-Atlantic Ridge, *J. Geophys. Res.* 72:2257-60.
Orowan, E. (1966), Age of the ocean floor, *Science* 154:413-16.

Pavoni, N. (1966), Tectonic interpretation of the magnetic anomalies southwest of Vancouver Island, *Pure Appl. Geophys.* 63:172-78.

Peter, G. (1966), Magnetic anomalies and fracture pattern in the northeast Pacific Ocean, *J. Geophys. Res.* 71:5365-74.

Pitman, W. C., and J. R. Heirtzler (1966), Magnetic anomalies over the Pacific-Antarctic Ridge, *Science* 154:1164-71.

Raff, A. D. (1966), Boundaries of an area of very long magnetic anomalies in the northeast Pacific, *J. Geophys. Res.* 71:2631-36.

Raff, A. D., and R. G. Mason (1961), Magnetic survey off the west coast of N. America, 40° N-52° N Latitude, *Geol. Soc. Amer. Bull.* 72:1267-70.

Saito, T., M. Ewing, and L. H. Burckle (1966), Tertiary sediment from the Mid-Atlantic Ridge, *Science* 151:1075-79.

Talwani, M., and J. R. Heirtzler (1964), Computation of magnetic anomalies caused by two-dimensional structures of arbitrary shape, pp. 464-80 in *Computers in the Mineral Industries*, Stanford Univ. Publ. Geol. Sci. 9, Pt. 1.

Talwani, M., X. Le Pichon, and J. R. Heirtzler (1965), East Pacific Rise: the magnetic pattern and fracture zones, *Science* 150:1109-15.

Tobin, D. G., and L. R. Sykes (1967), Seismicity and tectonics of the northeast Pacific Ocean, *Trans. Amer. Geophys. Un.* 48:204 (Abstr.).

Vacquier, V. (1965), Transcurrent faulting in the ocean floor, *Phil. Trans. Roy. Soc. London A* 258:77-81.

Vacquier, V., A. D. Raff, and R. E. Warren (1961), Horizontal displacements in the floor of the N.E. Pacific Ocean, *Geol. Soc. Amer. Bull.* 72:1251-58.

Vine, F. J. (1966), Spreading of the ocean floor: new evidence, *Science* 154:1405-15.

Vine, F. J., and D. H. Matthews (1963), Magnetic anomalies over oceanic ridges, *Nature* 199:947-49.

Vine, F. J., and W. J. Morgan (1967), Simulation of mid-ocean ridge magnetic anomalies using a random injection model. Program Geol. Soc. Amer. Annual Meeting, p. 228.

Vine, F. J., and J. T. Wilson (1965), Magnetic anomalies over a young oceanic ridge, off Vancouver Island, *Science* 150:485-89.

Vogt, P. R., and N. A. Ostenso (1966), Magnetic survey over the Mid-Atlantic Ridge between 42° N and 46° N, *J. Geophys. Res.* 71:4389-411.

Vogt, P. R., and N. A. Ostenso (1967), Comments on mantle convection and mid-ocean ridges, *J. Geophys. Res.* 72:2077-85.

Von Herzen, R. P. (1964), Ocean-floor heat-flow measurements west of the United States and Baja California, *Marine Geol.* 1:225-39.

Wilson, J. T. (1963), Evidence from islands on the spreading of ocean floors, *Nature* 197:536-38.

Wilson, J. T. (1965a), A new class of faults and their bearing upon continental drift, *Nature* 207:343-47.

Wilson, J. T. (1965b), Transform faults, oceanic ridges and magnetic anomalies southwest of Vancouver Island, *Science* 150:482-85.

EVIDENCE FOR OCEAN FLOOR SPREADING ACROSS THE OCEAN BASINS*

J. R. HEIRTZLER

INTRODUCTION

There have been recent studies of marine magnetic anomalies which suggest that the ocean floor may be spreading outwardly near the axis of the mid-ocean ridge system. Several studies of magnetic anomalies nearing completion at Lamont not only support axial spreading but indicate that this process extends completely from the ridge axis to the continental shelf in many areas. The preliminary results presented here are extracted from several papers in progress by the author and W. C. Pitman III, G. O. Dickson, E. Herron, and X. Le Pichon. It was felt that the implications of this work were so relevant to the present symposium that a preliminary version should be given at this time.

This contribution will briefly trace the marine magnetic program at Lamont as it bears on this problem.

REYKJANES RIDGE

In 1959 the extensive magnetic survey off the west coast of North America indicated a most unusual pattern of magnetic anomalies (Vacquier et al., 1961; Mason and Raff, 1961; Raff and Mason, 1961). A similar but less extensive and less systematic survey off the east coast of North America indicated a linear pattern of anomalies there also (Drake et al., 1963). A third and important fact known at the time was that a large magnetic anomaly seemed to be consistently associated with the axis of the Mid-Atlantic Ridge (Ewing et al., 1957; Keen, 1963). This situation suggested that a detailed magnetic survey should be made in the vicinity of the axis of the Mid-Atlantic Ridge. Should symmetric anomalies be found in such a survey it would indicate that the great elongated anomalies were of structural origin, although one did not know of any detailed symmetric structure associated with the ridge and one had no reason to suspect detailed structural differentiation off western North America.

The Reykjanes Ridge is the portion of the Mid-Atlantic Ridge southwest of Iceland and was known to have a large axial anomaly and to be within the range of electronic navigational aids. Lamont proposed a survey of that area in 1962, and the survey was carried out as a joint Lamont–U.S. Naval Oceanographic Office project in 1963. This survey had more accurate navigation than other extensive marine surveys done as of that time and did show linear and symmetric anomalies (Fig. 1; Heirtzler et al., 1966). The anomalies were determined using a regional field which was a least squares fit to a

* Lamont Geological Observatory Contribution No. 1192

simple linear function of latitudes and longitude. Apparently there is a longer-wavelength anomaly near the ridge axis because profiles show short-wavelength anomalies (Talwani et al., 1965) suppressed in the presentation of Figure 1.

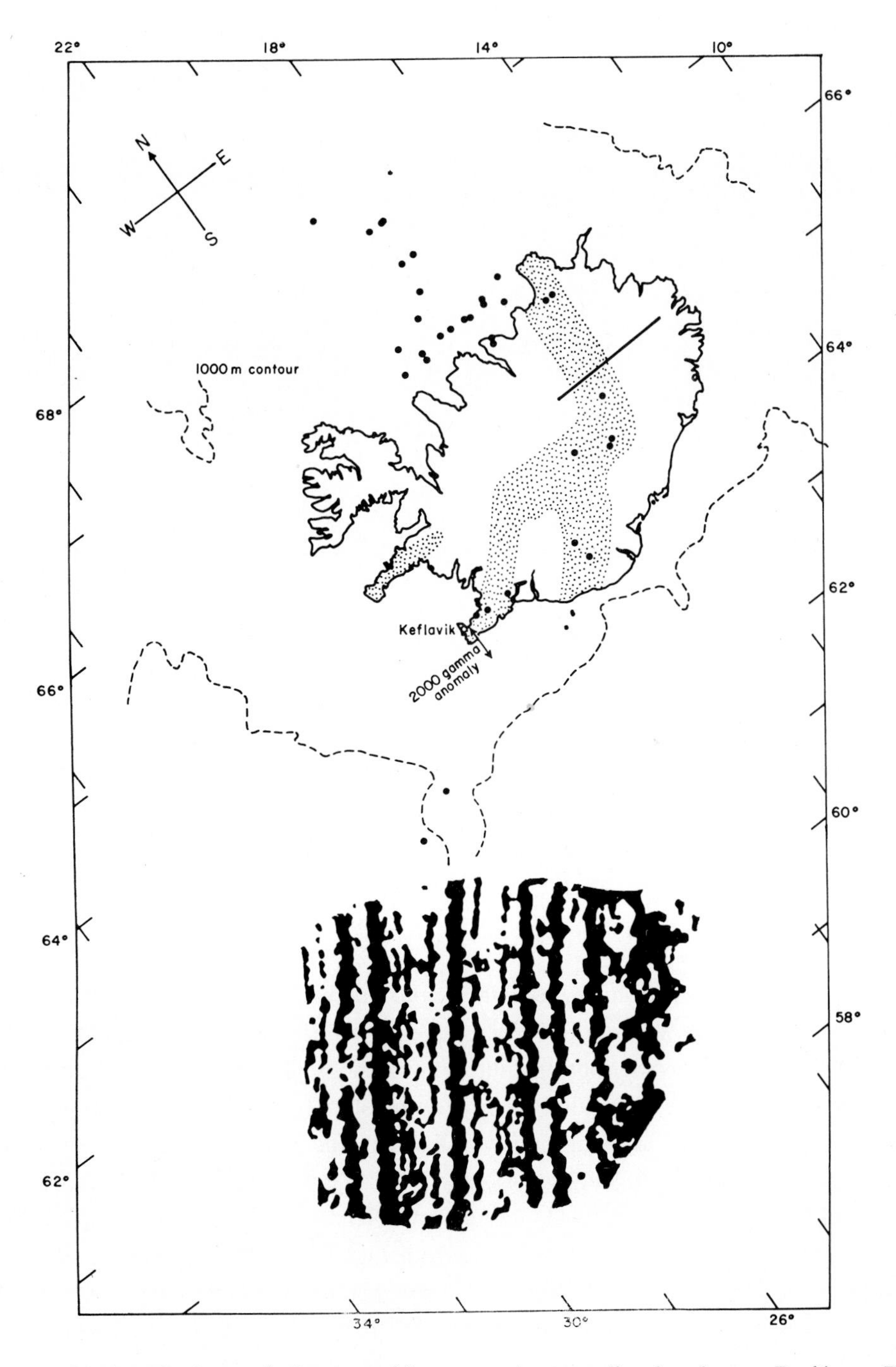

Figure 1. Black striped area indicates positive magnetic anomalies found over Reykjanes Ridge south of Iceland (after Heirtzler et al., 1966).

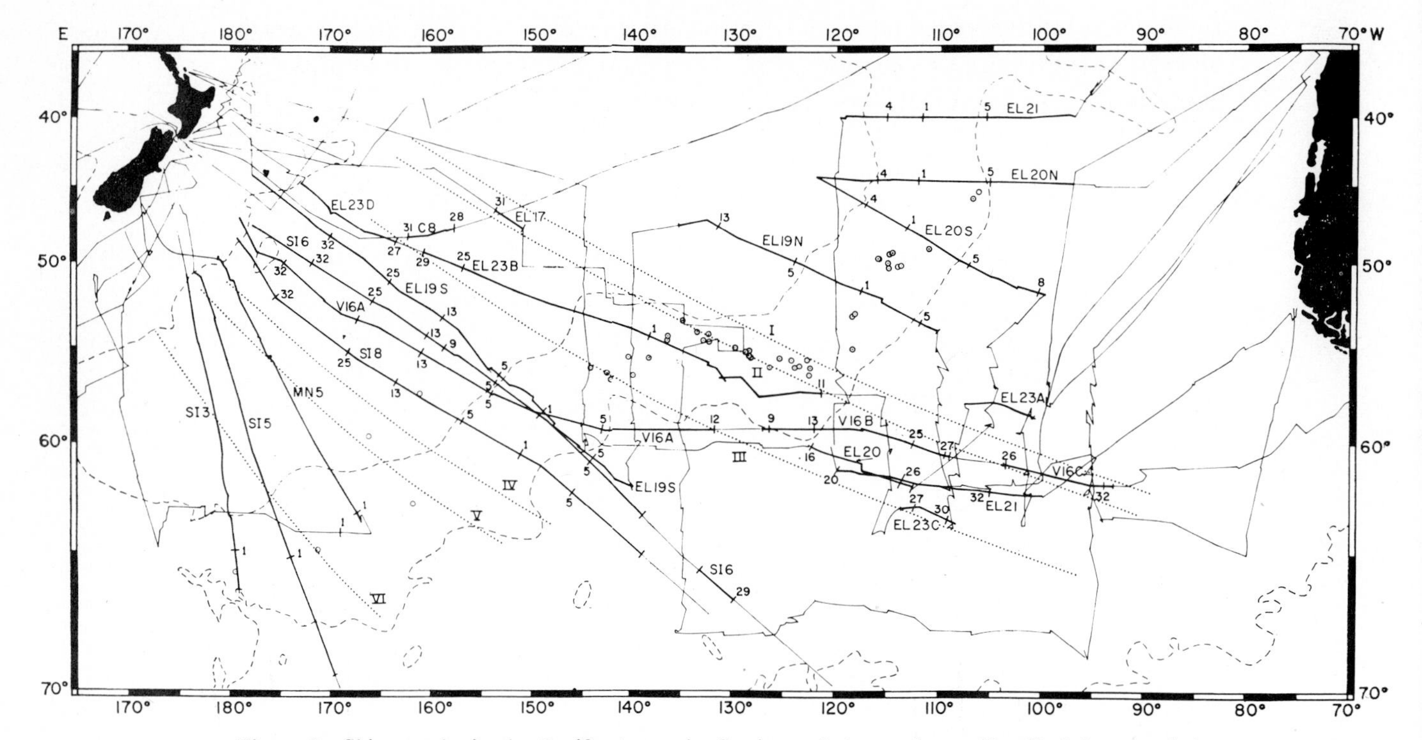

Figure 2. Ships tracks in the Pacific Antarctic. Sections of the track are identified for correlation in Figure 6. Dotted lines indicate fracture zones. Numbers along track indicate key anomalies.

At the time of the Reykjanes Ridge survey, the paper of Vine and Matthews (1963) appeared in which they proposed that the anomalies are associated with ocean floor spreading and reversals of the earth's magnetic field. The survey of the Reykjanes Ridge bore out this theory in so far as the anomalies were symmetric. At the time it was not clear that anomalies could be associated in a unique way with the known reversals of the field.

Pacific Antarctic

In early 1965 a magnetics program was begun on the U.S.N.S. *Eltanin* in the Pacific Antarctic. Figure 2 shows tracks of that ship plus an earlier track of *Vema*, the track of the Monsoon cruise of Scripps Institute of Oceanography and tracks covered by the *Staten Island* during the Operation Deep Freeze of the U.S. Navy. The ridge axis is defined in a general way as being half way between the 2,000-fathom contours. Notice the four tracks between longitudes 100° and 125°W and 40° and 55°S. Profiles over these tracks are shown in Figure 3. Pitman and Heirtzler (1966) have analyzed these profiles, and Figures 3 through 5 are taken from that paper. The lower magnetic profile in Figure 3 does, in fact, show a great deal of symmetry that is not evident from a casual inspection. Figure 4 shows the profile adjacent to the same profile plotted with the east and west end reversed. This degree of symmetry is of paramount importance, since no other geophysical observation is known to be so symmetric over such a large geographic

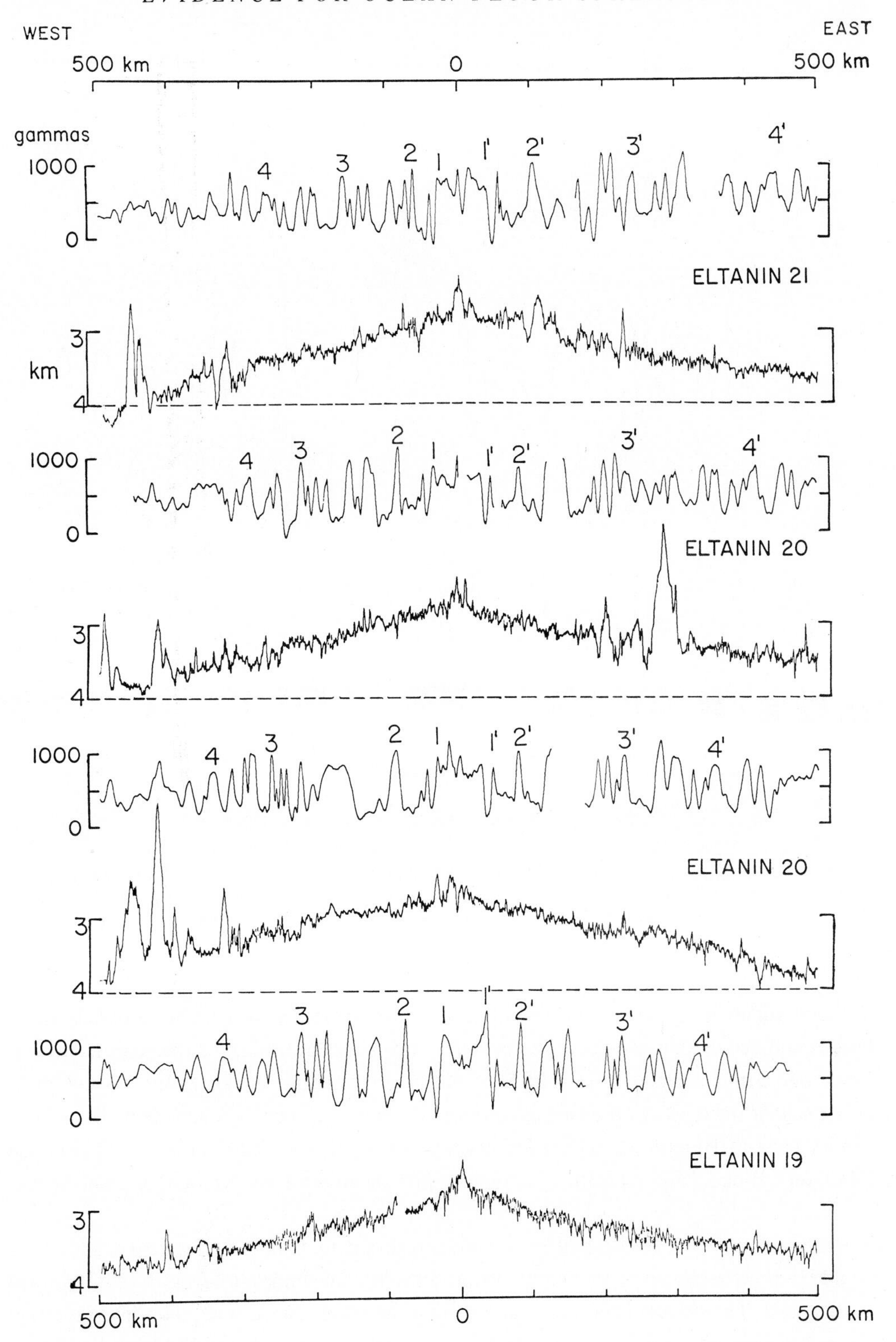

Figure 3. Total magnetic intensity and topographic profiles over the Pacific Antarctic ridge. The numbers 1 and 1′ identify the edges of the axial anomaly here, but number 1 is used to identify the axis in other figures.

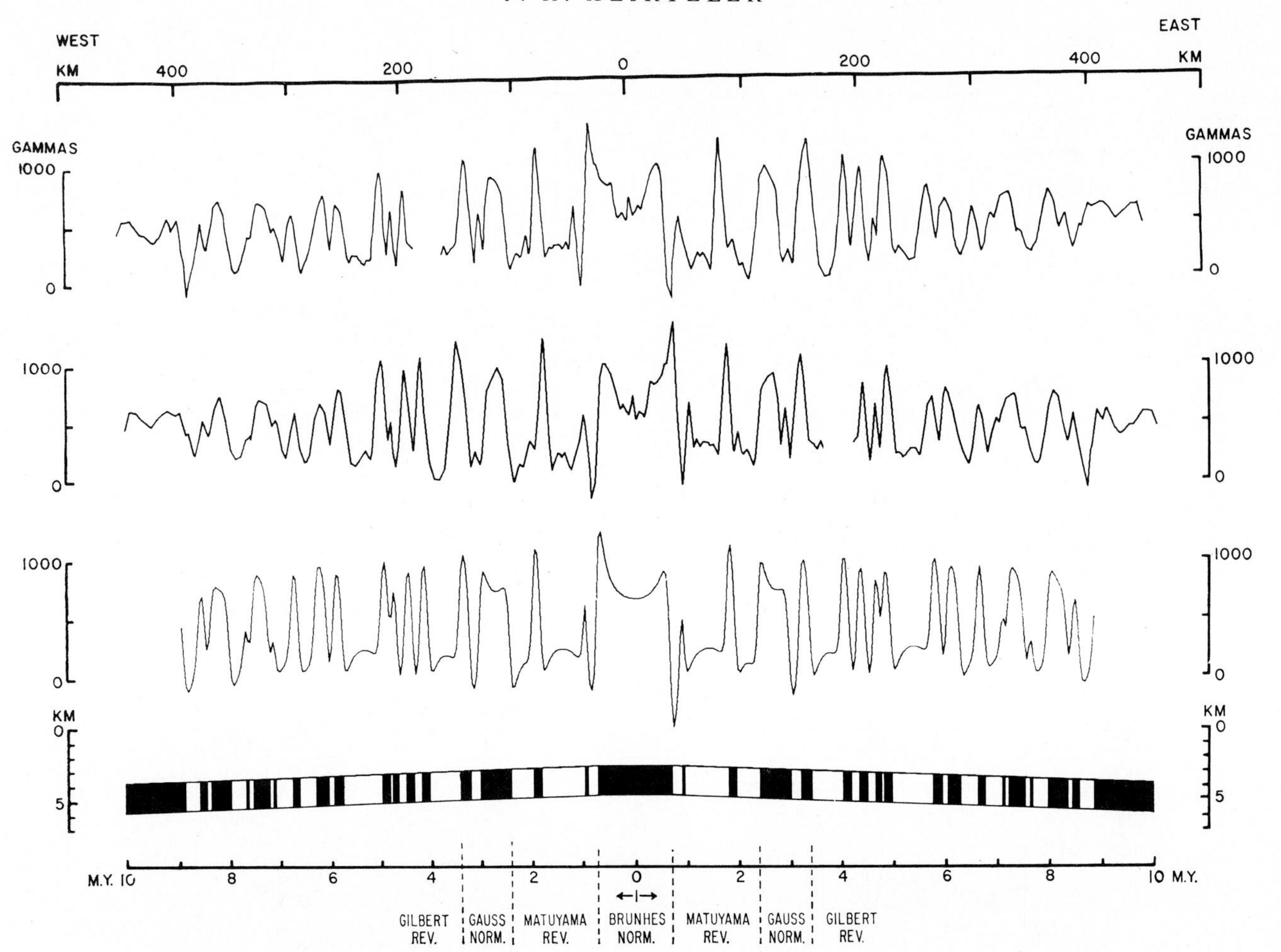

Figure 4. Center and upper magnetic profiles are observed, with east on the right, and reversed, with east on the left, respectively. The lower profile is the theoretical profile over the bodies shown. Normally magnetized bodies are black and reversely magnetized are white, and all are 2 km thick. With a spreading rate of 4.5 cm/yr the magnetized blocks correlate with the known history of reversals to the Gilbert epoch.

distance. Figure 4 shows the profile with the symmetric bodies that could be the causative bodies and shows that they can be related to reversals known from the studies of lava flows, and if this correlation is valid it would fit a spreading rate of about 4.5 cm/yr to either side of the crest. Under this assumption the model indicates a geomagnetic reversal history back to 10 m.yr. This detailed history was applied to the data collected over the Reykjanes Ridge (Fig. 5) with a spreading rate of about 1 cm/yr, and a good fit was obtained.

Vine and Wilson (1965) had previously suggested that the first few axial anomalies could be directly related to known reversals, although the Jaramillo event was not known at that time. Their work indicated that spreading is taking place now. The interpretation given in Figure 4 and substantiated in Figure 5 indicates that the Vine-Matthews theory can be applied quite literally across the entire axial zone.

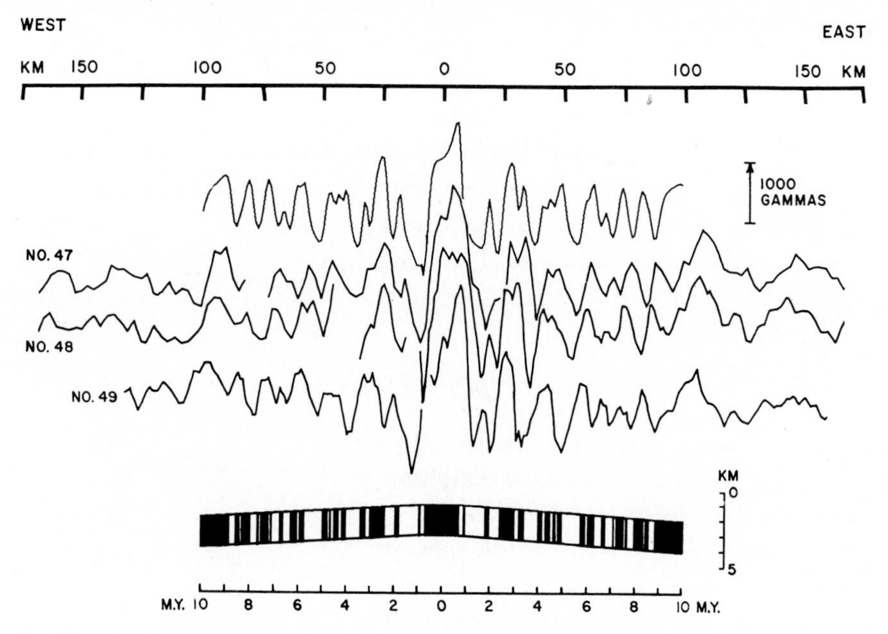

Figure 5. The geomagnetic field history as revealed in the Pacific Antarctic applied to the Reykjanes Ridge profiles 47, 48, and 49. The profile at the top is the theoretical profile for the model shown at the bottom. A spreading rate of 1 cm/yr was assumed.

Figure 6 shows magnetic profiles to the east and west of the ridge and our correlation of the anomalies. Anomalies are identified by an arbitrary numbering system. A simple linear extrapolation of the axial spreading rate illustrated in Figure 4 would yield an age of about 50 m.yr. for anomaly number 32; however, we now have reason to believe that spreading here is not linear, and the time scale has been somewhat revised. These profiles of the off-axial anomalies suggest that in many areas the spreading occurs across the entire ocean basins to the edge of continental shelves. The pattern of anomalies is not obviously coherent over the New Zealand continental shelf, just as the pattern is destroyed off the west coast of North America. The off-axial magnetic anomalies are also shown, in Figure 6, to be similar to anomalies in the northeast Pacific.

The profiles show an interruption of the magnetic anomaly pattern in places. This is manifest as a duplication of one or a few anomalies or as the absence of one or a few anomalies from the sequence. These situations are known to be characteristic of the anomaly pattern off the west coast of North America at the location of fracture zones. A similar interpretation can be made for the Pacific Antarctic. These fracture zones are illustrated in Figure 2. Some of these fracture zones are known to have a topographic expression, but whether they all do is uncertain at this time.

South Atlantic

Correlation of axial anomalies in the Pacific Antarctic with those in the North Atlantic were made by Pitman and Heirtzler (1966), and correlation of all anomalies of the

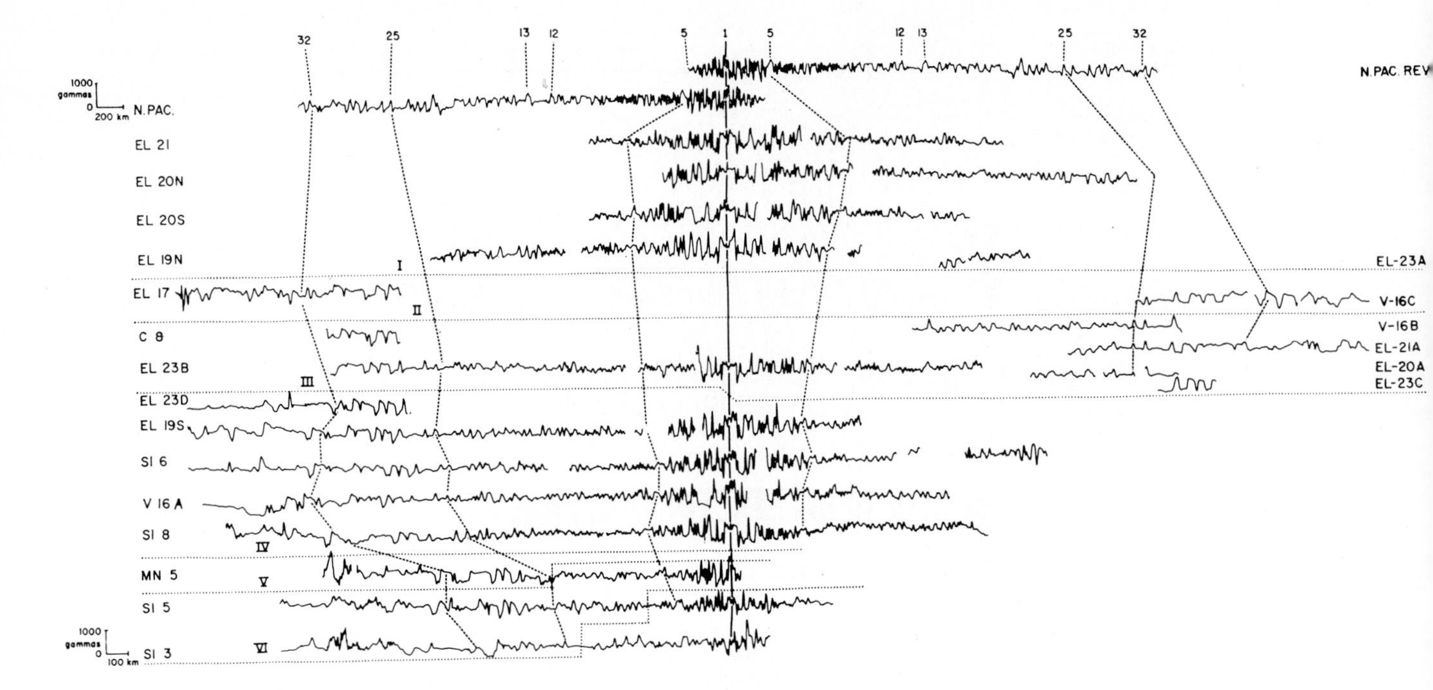

Figure 6. Observed magnetic profiles in the Pacific Antarctic (refer to Figure 2) compared to a composite profile from the northeast Pacific and that profile reversed. Dotted horizontal lines identify fracture zones. Key anomalies are numbered with an arbitrary numbering scheme.

Pacific Antarctic with those of the northeast Pacific has also been illustrated here. How do the anomalies of the Pacific Antarctic correlate with those of the other oceans? The South Atlantic is one area that has been examined.

With the number of anomalies that exist in the sequence, almost any single observed anomaly could find a match somewhere, although that would have no meaning. In general, to associate a series of magnetic anomalies with ocean floor spreading several criteria must be satisfied. These are (i) the anomalies must be linear in a direction parallel or nearly parallel to ridge axis; (ii) the anomalies must be in accord with the history of field reversals found in other oceans, modified by the presence of fracture zones.

Figure 7 shows magnetic profiles to the east and west respectively of the South Atlantic Ridge. The profiles go to the Rio Grande Rise on the west and to the Walvis Ridge on the east. The correlation of the anomalies is indicated, and their similarity to the anomalies of the northeast Pacific is illustrated. The different distance scales in the Pacific and the Atlantic would indicate a rate of about 2 cm/yr in the South Atlantic.

The locations of these correlated anomalies are shown in Figure 8. It was mentioned by Heirtzler and Le Pichon (1965) that it was difficult to correlate the axial anomalies south of about 30°S. The basin anomalies can be correlated south of 30°S even though the axial anomalies cannot.

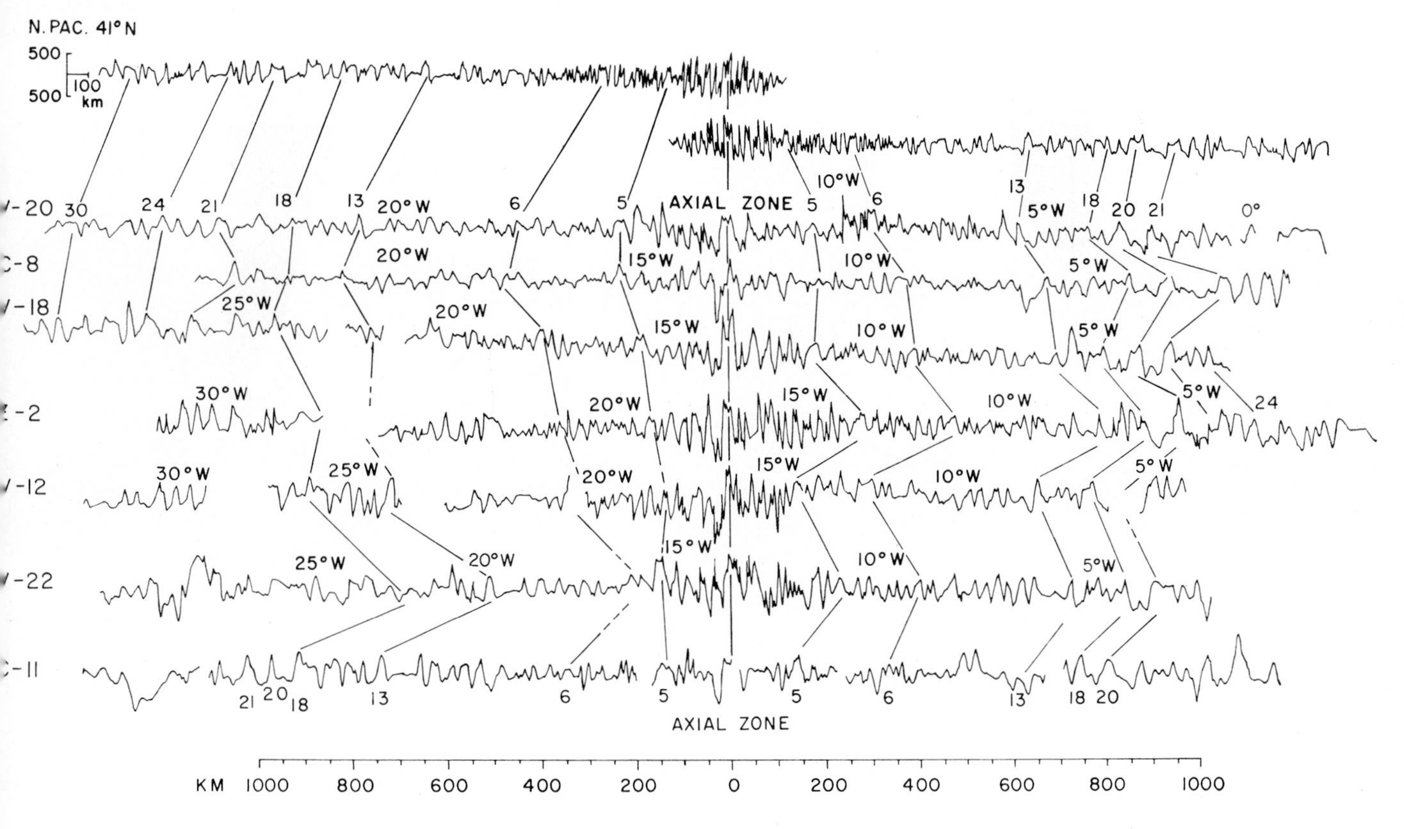

Figure 7. Observed magnetic profiles in the South Atlantic compared to the northeast Pacific.

South Indian Ocean

For that portion of the Indian Ocean south and southwest of Australia, Figure 9 shows where there is data and the location of correlated anomalies. This figure indicates several important things: The sea floor immediately south of Australia is approximately 35 m.yr. old. Also a major fracture zone, with a lateral offset of about 1,200 km, is revealed running south by southeast from Adelaide. There is a short section of ridge axis in the center of the figure that appears as though it overlaps the northerly section of axis. An overlap of ridge axis would not seem likely on the ocean floor. To put this another way, one should not cross a ridge axis twice in crossing an ocean along the spreading velocity direction.

Figure 10 shows the profiles for the tracks of the previous figure. The correlation of anomalies is poor if not impossible near the ridge axis, but starting with anomaly number 6 the recognizable linear pattern is evident to anomaly 13. The offset of 1,200 km for the fracture zone is evident from the offset in the magnetic pattern. Under the assumption of constant spreading (about 3.5 cm/yr) in this area, one can utilize anomalies 6 through 13 to identify the position of the crest of the easternmost axial section even though the ship did not cross it.

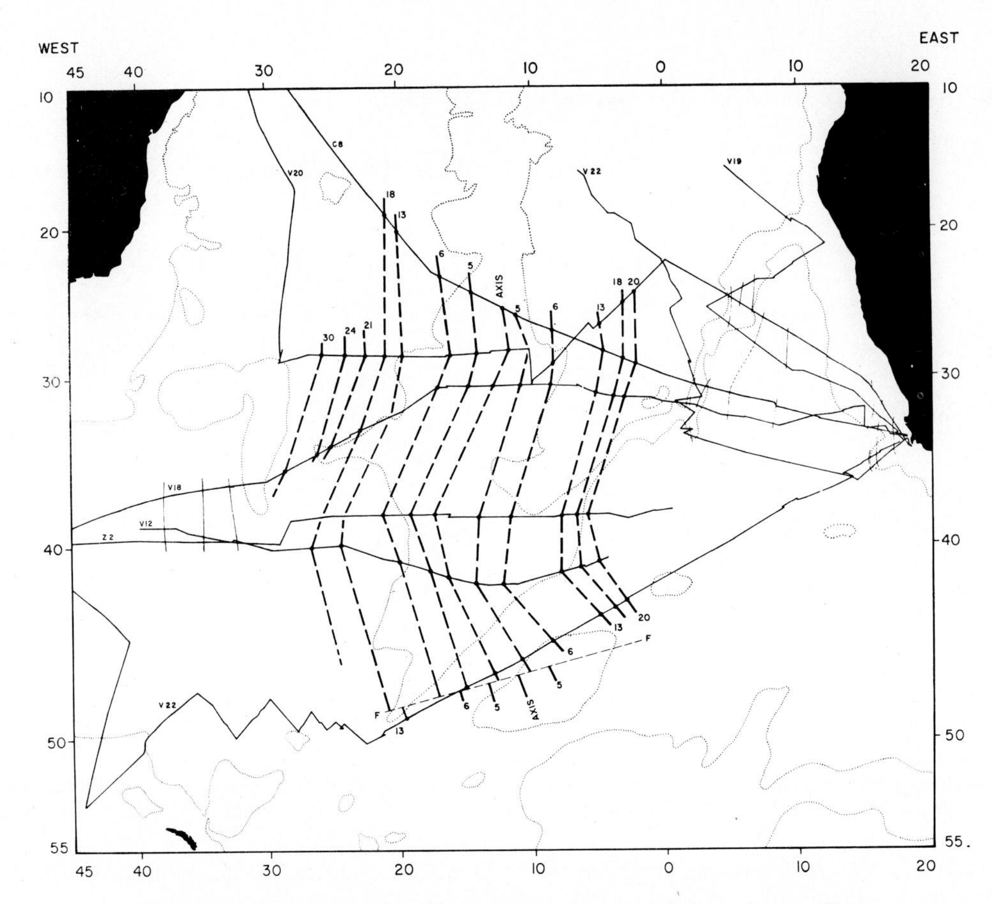

Figure 8. Tracks in the South Atlantic and the correlation of anomalies.

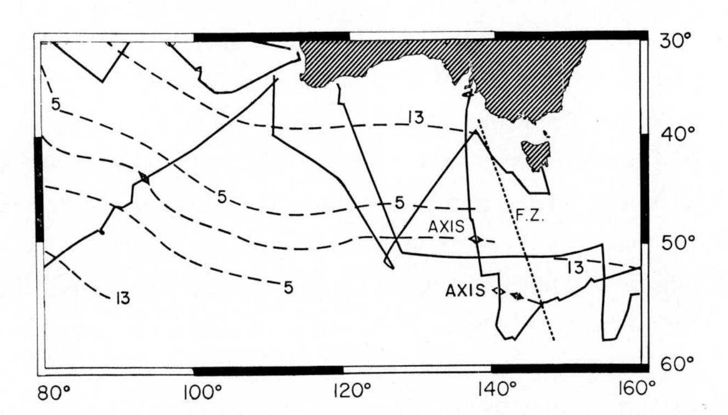

Figure 9. Tracks south of Australia and the correlation of anomalies. A major fracture zone is indicated.

The location of magnetic axis on the northernmost axial section does not coincide with the topographic axis shown on generalized bathymetric charts but is some miles to the north of it.

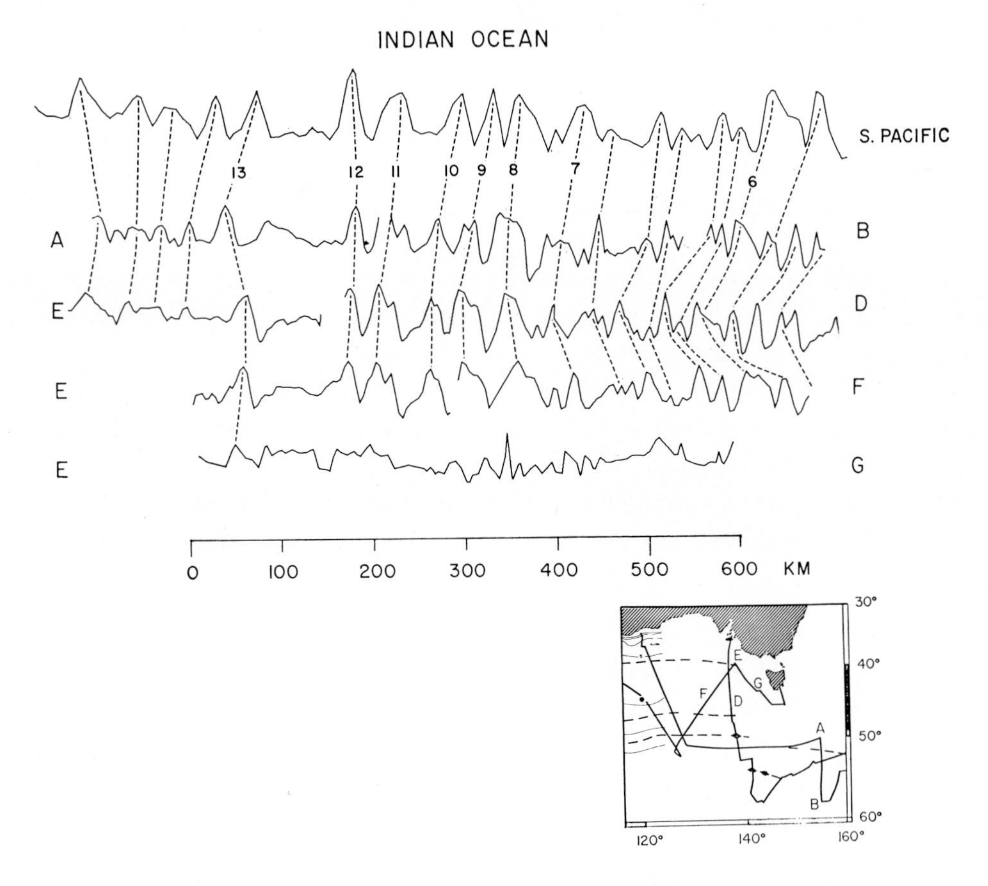

Figure 10. Magnetic profiles south of Australia with a profile from the Pacific Antarctic.

Conclusions

The interpretation of marine magnetic anomalies in terms of the Vine-Matthews hypothesis would suggest that much of the ocean floor is in relatively uniform motion to the edges of the continents. There are some marine areas that have elongated anomalies that do not fit the simple and literal interpretation of that hypothesis, as off the east coast of North America (Drake et al., 1963).

The northeast Pacific and Pacific Antarctic oceans indicate a spreading rate in the vicinity of 4.5 cm/yr; the South Indian Ocean 3.5 cm/yr; the North Atlantic Reykjanes Ridge indicates about 1 cm/yr, and the South Atlantic 2 cm/yr.

While the observations reported here indicate that large-scale and rapid motions of the oceanic crust are taking place, and other papers in this volume support the idea, there are other considerations that are not easily reconciled. Most of the difficulties are associated with the mechanics of the motion. Two of the mechanical difficulties for which detailed explanations are lacking are the following:

1. *The mechanics of the axis*: Magnetized material seems to move laterally away from a source of zero width. How can the material become magnetized at depth as it passes the Curie isotherm and not be mechanically reoriented as its motion becomes horizontal?

2. *The mechanics of the terminus*: Presumably this material turns under the edges of continents. If a horizontal velocity of about 4 cm/yr is diverted down, a very large trench would be created unless a high sedimentation rate is maintained.

It is of some importance to consider what this new evidence implies about continental drift. One area where there seems to be paleomagnetic evidence for drift since the Cretaceous and where an ocean floor velocity has been determined is Australia. The direction of drift appears to be in the direction of spreading. In the South Atlantic the direction of spreading is consistent with the separation of Africa and South America as usually depicted. If the fracture zones and spreading direction in the Pacific Antarctic can be used to indicate a separation of New Zealand from South America and Antarctica, there should be evidence of matching geology on the opposite sides of that ocean.

Acknowledgments

The author would like to emphasize that much of this material was extracted from papers in press by the author and W. C. Pitman III, G. O. Dickson, E. Herron, and X. Le Pichon, and he is indebted to them for permission to use it here.

This work has been supported by several grants from the National Science Foundation and Contracts with the Office of Naval Research.

Lamont Geological Observatory
Columbia University

References

Drake, C. L., J. Heirtzler, and J. Hirshman (1963), Magnetic anomalies off eastern North America, *J. Geophys. Res.* 68:5259-75.

Ewing, M., B. C. Heezen, and J. Hirshman (1957), Mid-Atlantic Ridge seismic belts and magnetic anomalies (Abstr.), Comm. No. 119 bis. Assoc. Seismol. Asso. Gen. U.G.G.I., Toronto.

Heirtzler, J. R., X. Le Pichon, and J. G. Baron (1966), Magnetic anomalies over the Reykjanes Ridge, *Deep-Sea Res.* 13:427-43.

Heirtzler, J. R., and X. Le Pichon (1965), Crustal structure of the mid-ocean ridges, 3, Magnetic anomalies over the Mid-Atlantic Ridge, *J. Geophys. Res.* 70:4013-33.

Keen, M. J. (1963), Magnetic anomalies over the Mid-Atlantic Ridge, *Nature* 197:888-90.

Mason, R. G., and A. D. Raff (1961), Magnetic survey off the west coast of N. America, 32°N-42°N latitude, *Geol. Soc. Amer. Bull.* 72:1259-65.

Pitman, W. C., and J. R. Heirtzler (1966), Magnetic anomalies over the Pacific Antarctic Ridge, *Science* 154:1164-71.

Raff, A. D., and R. G. Mason (1961), Magnetic survey off the west coast of N. America, 40°N-52°N latitude, *Geol. Soc. Amer. Bull.* 72:1267-70.

Talwani, M., X. Le Pichon, and J. R. Heirtzler, East Pacific Rise: The magnetic pattern and the fracture zones, *Science* 150:1109-15.

Vacquier, V., A. D. Raff, and R. E. Warren (1961), Horizontal displacements in the floor of the N.E. Pacific Ocean, *Geol. Soc. Amer. Bull.* 72:1251-58.

Vine, F. J., and D. H. Matthews (1963), Magnetic anomalies over oceanic ridges, *Nature* 199:947-49.

Vine, F. J., and J. T. Wilson (1965), Magnetic anomalies over a young oceanic ridge off Vancouver Island, *Science* 150:485-89.

TIME SCALE FOR GEOMAGNETIC REVERSALS*

BY ALLAN COX, R. R. DOELL, G. B. DALRYMPLE

The idea that the geomagnetic field changes polarity, or "reverses," was originally advanced in the early part of the twentieth century to explain the observation that many volcanic rocks are "reversely" magnetized; that is, the direction of their natural remanent magnetism is antiparallel to that of the present geomagnetic field (Brunhes, 1906; Mercanton, 1926; Matuyama, 1929). At that time there was no general theory of geomagnetism to provide a basis for determining whether reversals were physically reasonable; thus, for several decades the early work on reversals remained outside the mainstream of geophysical research.

During the past two decades, interest in the problem of reversals has been revived by advances in several fields of research. The development of the magnetohydrodynamic theory of geomagnetism has provided a basis for explaining the origin of polarity reversals. Moreover, the observational basis for reversals has been broadened through paleomagnetic investigations of igneous and sedimentary rocks on all continents. During the past few years, it has proved possible to establish a radiometric time scale for reversals by combining paleomagnetic research with age dating using the potassium-argon method. This dating technique utilizes the decay of radioactive potassium-40 to the inert gas argon-40 and, under certain circumstances, is capable of measuring the age of volcanic rocks as young as 10^4 years (McDougall, 1966; Dalrymple et al., 1967). The usefulness of the time scale in geological and geophysical research has been demonstrated by its application in determining rates of sedimentation in deep-sea cores, in inferring rates of sea floor spreading from the spacing of oceanic linear magnetic anomalies, and in correlating sedimentary deposits and volcanic formations that are difficult to date in other ways.

Reversal Time Scale

The geomagnetic reversal time scale has now been established back to about 4 m.yr. B.P. by measuring the magnetic polarities and potassium-argon ages of volcanic rocks from many different continental areas and oceanic islands (Rutten, 1959; Cox and others, 1963a, 1963b, 1964a, 1964b, 1966; McDougall and Tarling, 1963; Grommé and Hay, 1963, 1967; Evernden and others, 1964; Doell and others, 1966; Cox and Dalrymple, 1966; Chamalaun and McDougall, 1966; McDougall and Wensink, 1966; Doell and Dalrymple, 1966; McDougall and Chamalaun, 1966; McDougall and others, 1966; Dalrymple and others, 1965, 1967). The length of the polarity epochs determined by this work ranges from 0.7 to 1.75 m.yr. The epochs are both longer and more variable than previously thought on the basis of geologic estimates (Hospers, 1954; Khramov, 1958).

* Publication authorized by the Director, U.S. Geological Survey

The existence of much briefer polarity fluctuations, termed "events," with durations of the order of 10^5 years (Cox et al., 1964b) was a second unexpected discovery of this work. The internal consistency of worldwide paleomagnetic and radiometric results has established that these polarity events are produced by rapid switching of the main geomagnetic dipole. The Olduvai event, for example, is recorded in lava flows with similar radiometric ages from Africa (Grommé and Hay, 1963, 1967), Alaska (Cox et al., 1966), and the Indian Ocean (Chamalaun and McDougall, 1966).

Current research is directed toward filling gaps in the present data to determine whether additional events exist; toward obtaining better estimates of the ages of polarity transitions (Cox and Dalrymple, 1967); and toward extending the time scale back beyond 4 m.yr. (McDougall et al., 1966; Dalrymple et al., 1967). In the more recent work, there has been an attempt to obtain better resolving power by using only data which satisfy minimum standards of reliability and precision (Doell et al., 1966; McDougall and Chamalaun, 1966). The precision of the dating, which is especially important, depends in part on the analytical precision of the dating experiments and in part on the amount of atmospheric argon present in the rock. Atmospheric argon may exceed the amount of radiogenic argon present and lead to dating errors of 0.5 m.yr. or more, errors much too large for the ages to be useful in studies of reversals. Some arbitrary cut-off level based on dating precision is needed, and one that has proved useful is to reject data with ages having standard deviations greater than 0.1 m.yr. in the age range 0 to 2 m.yr., and greater than 5% for ages greater than 2 m.yr. It is equally important that the paleomagnetic and dating samples are from the same volcanic cooling unit, and that the measurements are made on material known to yield valid paleomagnetic and radiometric results. About 25% of the published data have been rejected by these criteria (Doell et al., 1966; McDougall and Chamalaun, 1966; Cox and Dalrymple, 1967). The remaining data, which provide the basis for the present reversal time scale, are shown in Figure 1.

Dating errors for the select data set increase from an average value of about 0.05 m.yr. for ages less than 2 m.yr. to about 0.25 m.yr. for ages of 10 m.yr. (Table 1). As the dating errors increase, the exact positions of polarity-epoch boundaries become increasingly uncertain and polarity events become more difficult to distinguish from dating errors (Cox and Dalrymple, 1967). Dalrymple et al. (1967) found that it was possible to obtain the ages of older isolated polarity transitions such as the "A 12" boundary (Fig. 1) where lava flows with opposite polarities occur in superposition. However, at present it is not possible to determine whether such transitions are the boundaries of epochs or events. The main results to come from the studies of rocks with ages between 4 and 10 m.yr. are (1) that the complexity of the polarity structure seems to persist back to about 10 m. yr. B.P., and (2) that the radiometric polarity time scale probably cannot be extended in detail much beyond 5 or 6 m.yr. This does not preclude the possibility of dating certain distinctive polarity transitions or defining longer periods of uniform polarity (Dalrymple et al., 1967).

Table 1. Standard deviations of potassium-argon dates used for determining reversal time scale

Radiometric ages (million years)	Average standard deviation (million years)
0 to 1	0.042
1 to 2	0.051
2 to 3	0.059
3 to 4	0.099
4 to 6	0.136
6 to 8	0.221
8 to 10	0.247

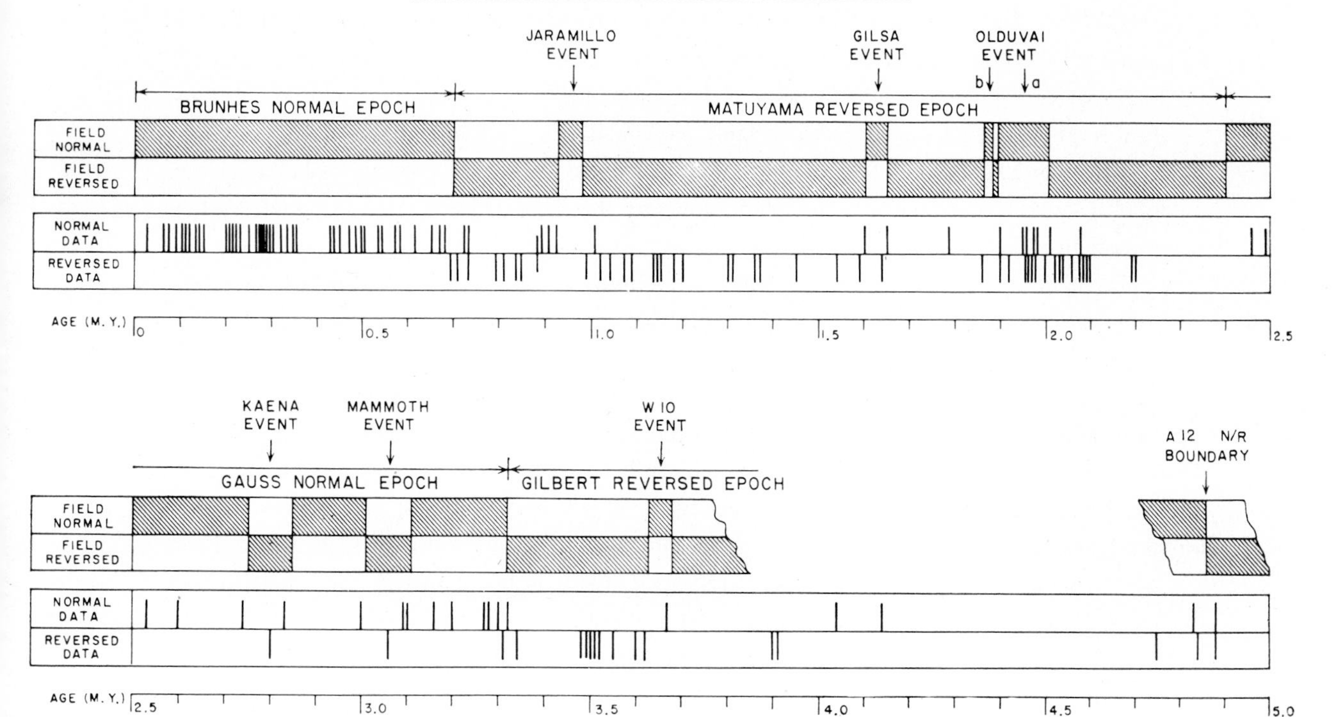

Figure 1. Time scale for geomagnetic reversals. Each short vertical bar represents the polarity and K-Ar age measurement on one volcanic unit. An intermediate direction of magnetization is represented by a bar that crosses the line between normal and reversed data. Only those data which satisfy minimum reliable criteria are shown. The inferred time scale (solid bars) corresponds to the best estimates given in Tables 2 and 3. The duration of the Jaramillo event and Olduvai "b" event are based on results from deep-sea sediments.

Comparison of the Radiometric Reversal Time Scale with Data from Other Sources

Additional information about geomagnetic polarity changes is provided by the remanent magnetism of deep-sea sediment cores (Harrison and Funnell, 1964; Dickson and Foster, 1966; Harrison, 1966; Harrison and Somayajulu, 1966; Ninkovich et al., 1966; Opdyke et al., 1966; Watkins and Goodell, 1967) and by magnetic profiles over mid-oceanic ridges as interpreted by the theory of sea floor spreading (Pitman and Heirtzler, 1966; Vine, 1966). Because these reversal sequences lack an absolute time base, they must be calibrated against known points of the potassium-argon reversal time scale. In Tables 2 and 3, the results from the sediment cores were calibrated using 0.70 m.yr. for the age of Brunhes-Matuyama boundary. The results from the mid-oceanic magnetic anomalies were calibrated against the age of 3.32 m.yr. for the Gauss-Gilbert boundary. This boundary was used for calibration rather than the boundary at 0.70 m.yr. because it provides a longer time base and hence minimizes errors due to variations in spreading rate. At most of the boundaries, the agreement between the three sets of results is within the limits of experimental error.

All three sets of results were combined to obtain the best estimates given in Tables 2 and 3 for the ages of polarity transitions. The best estimates of the duration of the events are provided by the results from deep-sea sediments because errors due to variations in the rate of sedimentation are probably smaller than errors in radiometric dating. For epoch boundaries and the midpoints of polarity events, where large cumulative errors may result from variations in sedimentation rate, the best estimates are provided by the radiometric dates.

The split of the Olduvai normal event into a main "a" pulse (Fig. 1) lasting 0.11 m.yr. followed, after a reversed interval lasting 0.01 m.yr., by a short "b" pulse lasting 0.025 m.yr. is based on results by Ninkovich et al. (1966) on core V20-119 from the North Pacific. Although events this closely spaced cannot be resolved from the present potassium-argon data, the rather wide spread in the ages of normally magnetized rocks associated with the Olduvai event are consistent with a subdivision of the event into at least two pulses. A split of the Olduvai into two more widely spaced events, the younger termed the Gilsa event, has been suggested by McDougall and Wensink (1966) on the basis of a potassium-argon date from Iceland. However, there is no record of this event in the mid-oceanic magnetic profiles, and the evidence from deep-sea sediment cores is ambiguous (Dickson and Foster, 1966; Ninkovich et al., 1966; Opdyke et al., 1966; versus Watkins and Goodell, 1967).

Two events, Mammoth and Kaena, may also exist within the Gauss normal epoch. There are two datum points with reversed polarities and radiometric ages separated by 0.26 m.yr., and Vine (1966) infers a double event from the mid-oceanic magnetic profiles, as do Dickson and Foster (1966) from deep-sea sediment cores. Finally, several events within the Gilbert reversed epoch have been identified on the basis of mid-oceanic magnetic profiles (Pitman and Heirtzler, 1966) and deep-sea sediments (Dickson and Foster, 1966). The boundary of the youngest of these may correspond to the normal-to-reversed transition found from potassium-argon dating and designated by the site number W10 (Dalrymple et al., 1967).

Time-Dependent Statistical Properties

The main time-dependent statistical properties of geomagnetic reversals are summarized in Table 4. The values for the interval from the present back to 3.32 m.yr.B.P.

Table 2. Comparison of the ages of polarity epoch boundaries as determined by K-Ar dating, deep-sea sediment cores, and midoceanic magnetic anomalies

Epoch boundary	K-Ar	Deep-sea cores[1]	Magnetic anomalies[2]	Best estimate
Brunhes/Matuyama (N/R)	0.70 m.yr.	0.70 m.yr.	0.68 m.yr.	0.70 m.yr.
Matuyama/Gauss (R/N)	2.4	2.27	2.34	2.4
Gauss/Gilbert (N/R)	3.32		3.32	3.32

[1] Data from Ninkovich and others (1966). Rate of sedimentation assumed constant and calibrated using an age of 0.70 million years for the Brunhes/Matuyama boundary.
[2] Data from Pitman and Heirtzler (1966). Rate of sea-floor spreading assumed constant and calibrated using an age of 3.32 million years for the Gauss/Gilbert boundary.

Table 3. Ages of events as determined from K-Ar dating, deep-sea sediment cores, and midoceanic magnetic anomalies

Polarity	Event	K-Ar	Sediment cores[1]	Magnetic anomalies[2]	Best estimate
N	Jaramillo:				
	Age of midpoint	0.94 m.yr.	0.92 m.yr.	0.90 m.yr.	0.94 m.yr.
	Duration	0.10	0.05	0.05	0.05
N	Gilsa:				
	Age of midpoint	(1.62)			(1.62)
N	Olduvai "b":				
	Age of midpoint		1.67		1.87
	Duration		0.024		0.024
N	Olduvai "a":				
	Age of midpoint				1.95
	Duration				0.11
N	Olduvai "a" + "b":				
	Age of midpoint	1.95	1.72		1.95
	Duration	0.16	0.14		0.14
R	Kaena:				
	Age of midpoint	2.80			2.80
	Duration				(0.10)
R	Mammoth:				
	Age of midpoint	3.06			3.06
	Duration	0.07			(0.10)
R	Kaena + Mammoth				
	Age of midpoint			3.03	
	Duration			0.20	
N	W10[4]				
	Age of beginning	3.7	4.1[3]	3.91	3.7
	Duration			0.16	0.16

[1] Data from Ninkovich and others (1966). Rate of sedimentation assumed constant and calibrated using an age of 0.70 million years for the Brunhes/Matuyama boundary.
[2] Data from Pitman and Heirtzler (1966). Rate of sea-floor spreading assumed constant and calibrated using an age of 3.32 million years for the Gauss/Gilbert boundary.
[3] Data from Dickson and Foster (1966).
[4] Cochiti normal event of Cox and Dalrymple (1967), *Earth Planet. Sci. Letters* 3:173-177.

are based on the reversal time scale defined by potassium-argon dating, assuming a split of the Olduvai event into "a" and "b" parts (Table 3) rather than into more widely spaced Olduvai and Gilsa events. For the interval from 3.32 to 9.8 m.yr. B.P., the values in Table 4 were calculated from the reversal time scale of Pitman and Heirtzler (1966), which is based on sea floor spreading adjusted to agree with the potassium-argon age of 3.32 m.yr. for the Gauss-Gilbert boundary.

Table 4. Time-dependent statistical properties of intervals between successive polarity changes. Data from 0 to 3.32 million years are by K-Ar dating. Data from 3.32 to 9.76 million years are interpreted from magnetic profiles across mid-oceanic ridges (Pitman, and Heirtzler, 1966).

	Range of ages, m.yr.		
	0 to 3.32	3.32 to 9.76	0 to 9.76
Normal intervals			
Time field was normal, m.yr.	1.63	3.13	4.76
Percent of total time	49	49	49
Number of normal intervals	7	13	20
Mean length of normal intervals, m.yr.	0.23	0.24	0.24
Reversed intervals			
Time field was reversed, m.yr.	1.69	3.31	5.00
Percent of total time	51	51	51
Number of reversed intervals	6	13	19
Mean length of reversed intervals, m.yr.	0.28	0.25	0.26
All intervals			
Number of intervals	13	26	39
Mean interval length, m.yr.	0.26	0.25	0.25

The average statistical properties for the two sets of results are remarkably similar (Table 4). Moreover, the frequency distributions of the time intervals between successive reversals (Fig. 2) show a similar proportion of short and long polarity intervals for the two sets of results. This statistical homogeneity suggests that both sets of observations are due to the same process of geomagnetic reversals. It further suggests that the rate of sea floor spreading was constant during the interval from 3.32 to 9.8 m.yr. and that in both studies the time spanned is sufficiently large to yield meaningful statistical results.

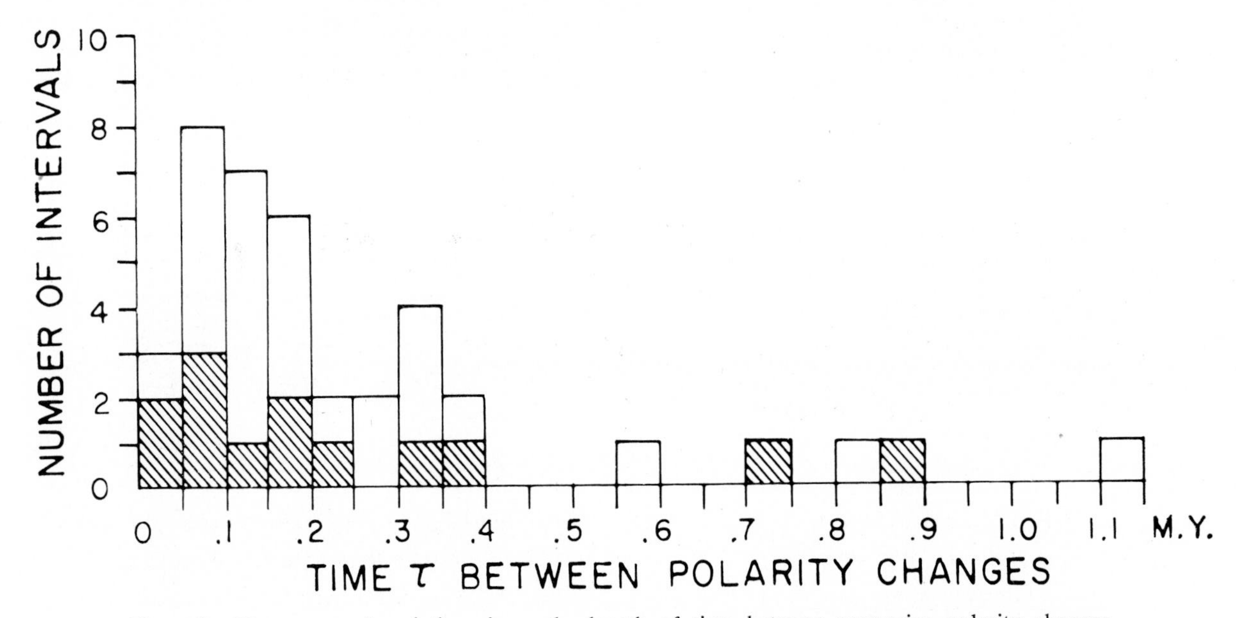

Figure 2. Histograms of variations in τ, the length of time between successive polarity changes. *Solid squares*: data for interval from present back to 3.32 m.yr. B.P. based on potassium-argon dating ($\tau > 0.15$ m.yr.) and deep-sea sediments ($\tau \leq 0.15$ m.yr.). *Open squares*: data for interval from 3.32 m.yr. to 9.8 m.yr. B.P. based on mid-oceanic magnetic anomalies (Pitman and Heirtzler, 1966).

The present study shows that the total length of time during which the field was reversed is equal to the total length of time when it was normal. This important result does not agree with an earlier finding by Valiev (1960), who concluded from a paleomagnetic study of sediments that reversed polarity epochs during the Pliocene were 2.8 times longer than normal epochs. On the other hand, support for the present conclusion is provided in a recent study by Dalrymple et al. (1967) of 36 lava flows with ages distributed rather irregularly in the interval 3.32 to 9.8 m.yr. Of these flows, 19 are normal and 17 are reversed, suggesting nearly equal total time intervals of normal and reversed polarity.

The results summarized in Figure 2 raise the question of the reality of epochs and events as distinct physical entities. The idea that events may exist arose in the course of working out the reversal time scale from potassium-argon dating, where the events appeared as anomalous data points within sequences of otherwise consistent data which defined the polarity epochs. In using reversals for purposes of stratigraphic correlation or to establish rates of sedimentation in deep-sea cores, it has been found that the polarity epochs can almost always be recognized, whereas often there is no paleomagnetic record of some of the events. Thus the two-level system of names shown in Figure 1 has proved useful in stratigraphy and will probably continue to be used. However, as more data are acquired and more events discovered, it appears increasingly probable that the length of time between successive reversals may have a continuous, rather than bimodal, frequency distribution.

A probability model for reversals which yields a continuous frequency distribution has been derived by Cox (in preparation). This distribution fits the data summarized in Figure 2 reasonably well, except for polarity intervals in the range 0 to 0.05 m.yr., where the predicted frequency is 8 and the observed is only 3. This suggests that several as yet unrecognized polarity events with durations of 0.05 m.yr. or less may yet be found in the interval from 0 to 10 m.yr.

U.S. Geological Survey
Menlo Park, California

REFERENCES

Brunhes, Bernard (1906), Recherches sur la direction d'aimantation des roches volcaniques(¹), *J. Physique* (4th ser.) 5:705-24.

Chamalaun, F. H., and Ian McDougall (1966), Dating geomagnetic polarity epochs in Réunion, *Nature* 210:1212-14.

Cox, Allan, and G. B. Dalrymple (1966), Palaeomagnetism and potassium-argon ages of some volcanic rocks from the Galapagos Islands, *Nature* 209:776-77.

Cox, Allan, and G. B. Dalrymple (1967), Statistical analysis of geomagnetic reversal data and the precision of potassium-argon dating, *J. Geophys. Res.* 72:2603-14.

Cox, Allan, R. R. Doell, and G. B. Dalrymple (1963a), Geomagnetic polarity epochs and Pleistocene geochronometry, *Nature* 198:1049-51.

Cox, Allan, R. R. Doell, and G. B. Dalrymple (1963b), Geomagnetic polarity epochs—Sierra Nevada II, *Science* 142:382-85.

Cox, Allan, R. R. Doell, and G. B. Dalrymple (1964a), Geomagnetic polarity epochs, *Science* 143:351-52.

Cox, Allan, R. R. Doell, and G. B. Dalrymple (1964b), Reversals of the earth's magnetic field, *Science* 144:1537-43.

Cox, Allan, D. M. Hopkins, and G. B. Dalrymple (1966), Geomagnetic polarity epochs—Pribilof Islands, *Geol. Soc. Amer. Bull.* 77:883-910.

Dalrymple, G. B. (1967), Potassium-argon dating in Pleistocene correlation, *INQUA* (Intern. Assoc. Quaternary Res., Salt Lake City, 1965), Univ. Utah Press.

Dalrymple, G. B. (1968), Potassium-argon ages of Recent rhyolites of the Mono Craters, California, *Earth Planet. Sci. Letters* 3:289-298.

Dalrymple, G. B., Allan Cox, and R. R. Doell (1965), Potassium-argon age and paleomagnetism of the Bishop Tuff, California, *Geol. Soc. Amer. Bull.* 76:665-73.

Dalrymple, G. B., Allan Cox, R. R. Doell, and C. S. Grommé (1967), Pliocene geomagnetic polarity, *Earth Planet. Sci. Letters* 2:163-173.

Dickson, G. O., and J. H. Foster (1966), The magnetic stratigraphy of a deep-sea core from the North Pacific Ocean, *Earth Planet. Sci. Letters* 1:458.

Doell, R. R., and G. B. Dalrymple (1966), Geomagnetic polarity epochs—A new polarity event and the age of the Brunhes-Matuyama boundary, *Science* 152:1060-61.

Doell, R. R., G. B. Dalrymple, and Allan Cox (1966), Geomagnetic polarity epochs—Sierra Nevada data, 3, *J. Geophys. Res.* 71:531-41.

Evernden, J. F., D. E. Savage, G. H. Curtis, and G. T. James (1964), Potassium-argon dates and the Cenozoic mammalian chronology of North America, *Amer. J. Sci.* 262:145-98.

Grommé, C. S., and R. L. Hay (1963), Magnetization of basalt of Bed I, Olduvai Gorge, Tanganyika, *Nature* 200:560-61.

Grommé, C. S., and R. L. Hay (1967), Geomagnetic polarity epochs—New data from Olduvai Gorge, Tanganyika, *Earth Planet. Sci. Letters* 2:111-15.

Harrison, C. G. A. (1966), The paleomagnetism of deep sea sediments, *J. Geophys. Res.* 71:3033-43.

Harrison, C. G. A., and B. M. Funnell (1964), Relationship of palaeomagnetic reversals and micropalaeontology in two late Caenozoic cores from the Pacific Ocean, *Nature* 204:566.

Harrison, C. G. A., and B. L. K. Somayajulu (1966), Behaviour of the earth's magnetic field during a reversal, *Nature* 212:1193-95.

Hospers, Jan (1954), Reversals of the main geomagnetic field III: *Koninkl. Nederlandse Akad. Wetensch. Proc.*, ser. B, 57:112-21.

Khramov, A. N. (1958), Paleomagnitnaya korrelyatsiya osadochnykh tolsch, *Vses. Neft. Nauchno-Issled. Geologoraz Inst. Tr.*, vyp. 116, 219 pp.; English transl. 1960, Palaeomagnetism and stratigraphic correlation (A. J. Lojkine, Canberra, Australian Natl. Univ., 178 pp.).

McDougall, Ian (1966), Precision methods of potassium-argon isotopic age determination on young rocks, pp. 279-304 in *Methods and Techniques in Geophysics*, Vol. 2, S. K. Runcorn, ed., London, Interscience Publishers.

McDougall, Ian, and D. H. Tarling (1963), Dating of polarity zones in the Hawaiian Islands, *Nature* 200:54-56.

McDougall, Ian, and F. H. Chamalaun (1966), Geomagnetic polarity scale of time, *Nature* 212: 1415-18.

McDougall, Ian, and H. Wensink (1966), Paleomagnetism and geochronology of the Pliocene-Pleistocene lavas in Iceland, *Earth Planet. Sci. Letters* 1:232-36.

McDougall, Ian, H. L. Allsopp, and F. H. Chamalaun (1966), Isotopic dating of the newer volcanics of Victoria, Australia, and geomagnetic polarity epochs, *J. Geophys. Res.* 71:6107-18.

Matuyama, Motonori (1929), On the direction of magnetization of basalt in Japan, Tyôsen, and Manchuria, *Japan Imp. Acad. Proc.* 5:203-05.

Mercanton, P. L. (1926), Magnétisme terrestre—Aimantation de basaltes groenlandais, *Compt. Rend. Acad. Sci.* [Paris] 182:859-60.

Ninkovich, D., N. Opdyke, B. C. Heezen, and J. H. Foster (1966), Paleomagnetic stratigraphy, rates of deposition and tephrachronology in North Pacific deep-sea sediments, *Earth Planet. Sci. Letters* 1:476-92.

Opdyke, N. D., B. Glass, J. D. Hays, and J. H. Foster (1966), Paleomagnetic study of Antarctic deep-sea cores, *Science* 154:349-57.

Pitman, W. C., III, and J. R. Heirtzler (1966), Magnetic anomalies over the Pacific-Antarctic ridge, *Science* 154:1164-71.

Rutten, M. G. (1959), Paleomagnetic reconnaissance of mid-Italian volcanoes, *Geologie en Mijnbouw* 21:373-74.

Valiev, A. A. (1960), Paleomagnetic subdivision of the Marguzar section of the Cenozoic continental molasse beds, *Akad. Nauk SSSR Izv. Ser. Geol.* 7:974-76 (in Russian); English translation Amer. Geophys. Union, 1961.

Vine, F. J. (1966), Spreading of the ocean floor—New evidence, *Science* 154:1405-15.

Watkins, N. D., and H. G. Goodell (1967), Confirmation of the reality of the Gilsa geomagnetic polarity event, *Earth Planet. Sci. Letters* 2:123-29.

SOME REMAINING PROBLEMS IN SEA FLOOR SPREADING

H. W. MENARD

The newly discovered symmetry of magnetic anomalies and seismicity of transform faults presents convincing evidence of sea floor spreading (Hess, 1962; Wilson, 1963a; Vine, 1966). However, in this forward thrust in our understanding a few puzzles have been left behind, and these are the subject of this paper. I shall deal with problems of the terminations of fracture zones and apparent complications in sea floor spreading.

Terminations of Fracture Zones

Magnetic anomalies in the northeastern Pacific appear to be offset for very long distances along fracture zones (Vacquier, et al., 1961). These offsets may be explained in different ways. Crustal blocks may move in opposite directions in an absolute sense; this would be the explanation of classical tectonics but does not appear to be applicable. Differential movements may occur with the same absolute sense of motion. This was originally proposed on the basis of differential crustal stretching away from the rise crest (Menard, 1960) but now seems attributable to differential rates of crustal creation and spreading, following Hess (1962). Finally, Wilson (1965) suggests that the apparent offset is not the result of differential movement at all but rather of uniform spreading away from two rises which were initially separated by the amount of the apparent offset. A combination of the last two origins is also discussed by Wilson.

Some insight regarding the nature of the offsets on these particular fracture zones, and perhaps others, can be gained by considering their terminations. Presumably, if no differential movement has occurred, the simple fracture zones on the crest of the rise merely terminate without any change in form.

However, numerous fracture zones on the eastern edge of the East Pacific Rise appear to be branches of the Mendocino, Murray, Molokai, Clarion, and Clipperton fracture zones (Menard, 1967). These branches closely resemble those at the ends of great continental wrench faults. Thus some differential movement probably has occurred. This is further confirmed by the fact that the branches are concentrated in those crustal blocks which appear to have moved farthest from the crest. On the other hand, the fracture zones cross, but do not offset in any significant way, straight volcanic ridges such as the Hawaiian and Line Islands (Malahoff, et al., 1966; Menard, 1967). This suggests either that differential movement has been minor or that the volcanoes formed after such movement ceased.

It appears that some differential movement has been dissipated on fracture zone branches in the northeastern Pacific. It is not established whether the differential movement was large or whether the apparent offsets of magnetic anomalies are largely attributable to the initial positions of parts of the rise crest.

Was the East Pacific Rise originally relatively straighter and more continuous? An alternative to Wilson's hypothesis that the rise was initially offset is that it was once straight and has become offset. This would not invalidate his concept of transform faults in any way but would merely attempt to describe a pre-history before his concept begins to act. In some ways it might even strengthen his concept if the *ad hoc* initial offsets along pre-existing fracture zones, which he assumes for simplicity, could be shown to be the consequence of an early stage of rise development and convection.

The chief reason for believing that the East Pacific Rise was once straight and continuous is that the great fracture zones throughout the world are essentially perpendicular to the oceanic rise-ridge system with which they are associated (Menard, 1966). This gives very strong evidence that they were produced by the same convection that formed the rises. Moreover, they are remarkably close to perpendicular to the magnetic anomalies characteristic of the same rises and ridges. Most of the great fracture zones of the northeastern Pacific are also remarkably straight and close to parallel, as are the magnetic anomalies on the flanks of the rise which they cut. The simplest explanation for this very regular orthogonal pattern is that it was produced by an initially continuous and relatively straight convection cell which has since broken into smaller cells.

Smaller convection cells, bounded by fracture zones, must have had very elongate shapes if they produced the spreading demonstrated by symmetrical magnetic anomalies. The separation between great fracture zones is only a small fraction of the distance that the sea floor has spread from the crest of the East Pacific Rise. In some other regions, such as the southwestern Pacific discussed by Heirtzler (this volume), the required elongation of the convection cells is even more extreme. This elongation is eliminated if in the initial stage the convection cell was continuous in large regions.

Has the crest of the East Pacific Rise migrated? The crest is now broken into fragments, and sea floor spreading continues, so small convection cells now exist. Both crest and cell have migrated if the crest was once relatively straight and the convection cell once continuous. Several lines of evidence indicate that at least minor migration has occurred. The Juan de Fuca and Gorda ridges, which are active centers of spreading, have rotated relative to the magnetic anomalies to the west (Wilson, 1965). The two ridges are now connected by an active transform fault, the Blanco fracture zone. The peculiar geometry of the region appears to require that the crest of the Gorda Ridge has migrated east relative to other anomalies (Fig. 1). The symmetrically disposed magnetic anomalies were once generated at the crest. Inasmuch as some are much longer than the crest, and the anomalies cannot stretch longitudinally, it follows that the crest was formerly at the position of the long anomalies and has migrated and become shorter. The absolute movement is not defined.

Evidence of migration of the crest of the East Pacific Rise is also available from the Gulf of California to Easter Island. In this vast region the greater fracture zones are not seismically active and some do not offset the crest of the rise (Menard, 1967).

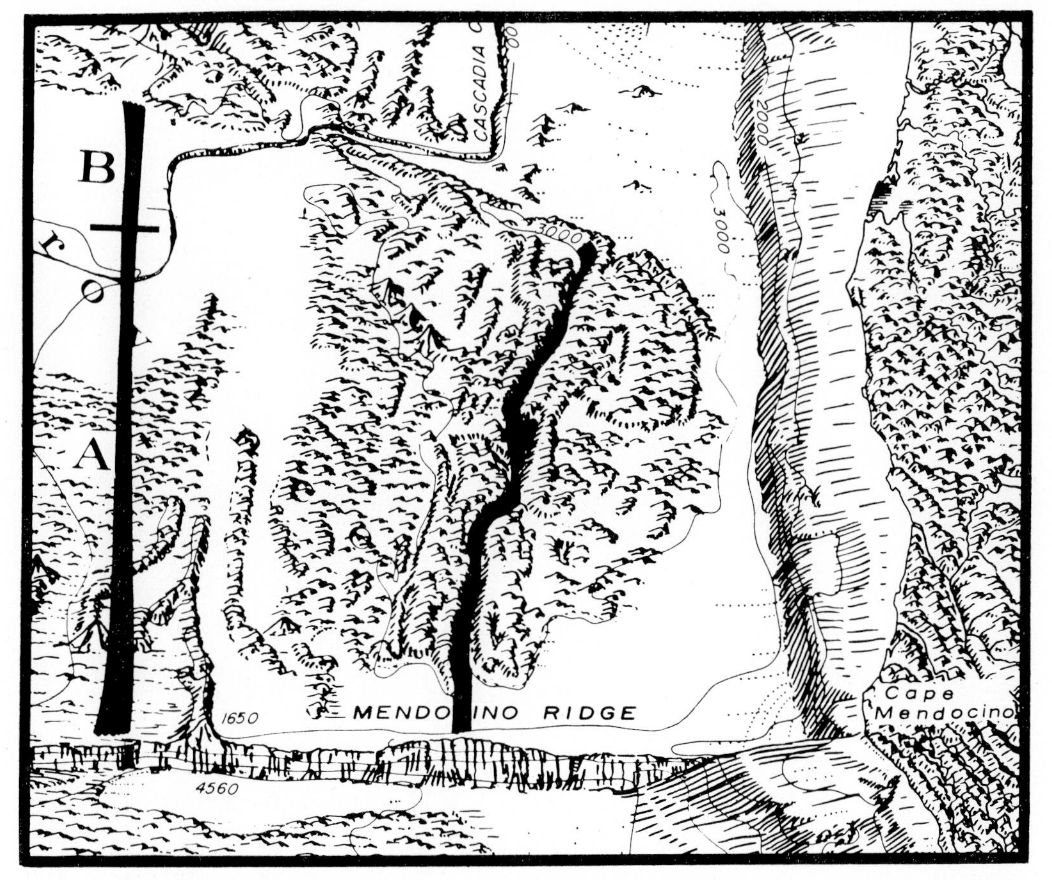

Figure 1. Black line (right), the graben at the crest of the Gorda Ridge; left broad line is a magnetic anomaly produced by spreading from that crest. The length of the crest is "A," the excess length of the magnetic anomaly is "B." Considering that this ridge is bounded by the Mendocino fracture zone on the south and the Blanco fracture zone on the north, the difference in length suggests that the crest has migrated relatively east and has grown shorter. After Menard, 1964.

However an extensive group of shorter fracture zones do offset the crest and are seismically active. As Sykes (this volume) has shown, some of these fracture zones have the first motion of transform faults. Although the magnetic anomalies are not yet mapped, the location, orientation, and activity of these small fracture zones make it probable that they are equivalent to the transform faults connecting fragments of the Gorda and Juan de Fuca ridges (Fig. 2). These small fracture zones are not parallel, and thus the geometry is much the same as between the Mendocino and Blanco fracture zones, where migration of the crest has occurred. If all the apparent offsets are removed, the crest in this region becomes a relatively straight line. It cannot be demonstrated that such a line once actually existed. However, the hypothesis that a relatively straight and continuous convection cell once existed and has since decayed into smaller cells in this region is capable of explaining a substantial number of otherwise unrelated facts. Vine (1966) has suggested that just such a decay has been engendered in the region of the Gorda and Juan de Fuca ridges because the East Pacific Rise has been overridden by the western

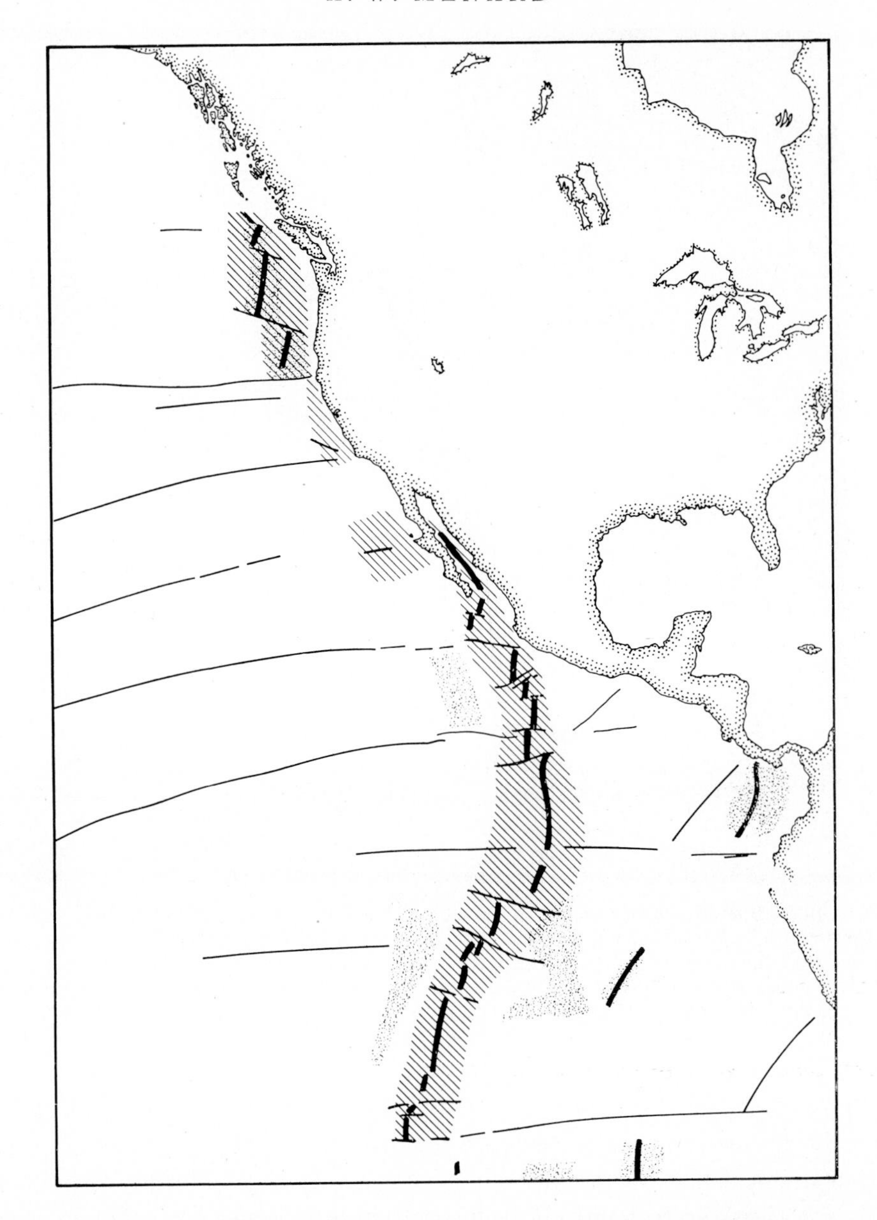

Figure 2. Distribution of a zone of disturbance on the East Pacific Rise. The crest of various rises is shown by a heavy line. West of the shaded and stippled zone the great fracture zones and the orthogonal magnetic anomalies are remarkably straight, although the anomalies are offset. The shaded zone contains seismically active transform faults and short segments of magnetic anomalies which are not parallel to the long straight anomalies farther west. Stippled regions contain linear ridges and troughs, generally parallel to the rises.

drift of North America. It appears that this decay has occurred in the central part of the rise from 45°N to at least 20°S latitude. If so, it indicates that the decay is not related to the motion or location of North America; rather it suggests that the convective overturn is approaching completion. This is also indicated by the declining spreading rates

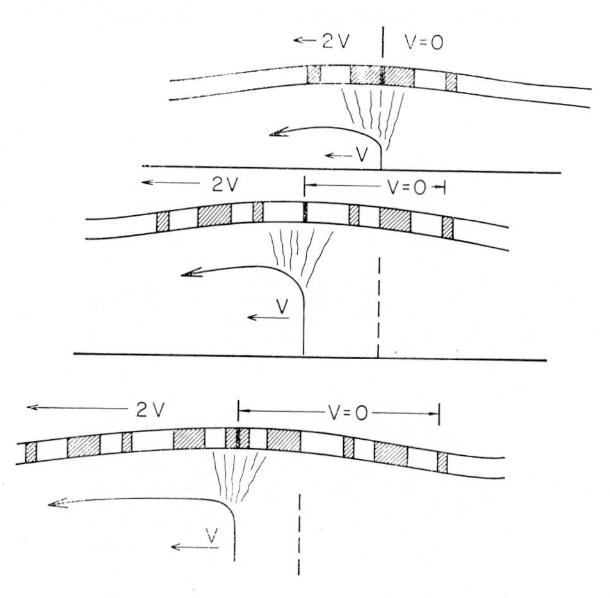

Figure 3. A means of producing symmetrical magnetic anomalies without moving one flank of a rise. The underlying convection cell migrates at the apparent spreading rate. One side spreads at twice the apparent rate and the other is fixed. This is one extreme possibility, the other extreme is that the convection is fixed and the two flanks spread at the same rate.

compared to the South Pacific (Vine, 1966; Heirtzler, this volume). It confirms the physiographic evidence that the northern part of the rise is in a later stage of development than the southern (Menard, 1964).

Can symmetrical magnetic anomalies be produced by one-sided convection? If the crests, and thus the centers of convection, migrate, how is it possible to obtain symmetrical magnetic anomalies which appear to have moved away from the crest? One possibility is indicated by Figure 3. If the crest migrates at the same rate that the sea floor appears to spread, one magnetic anomaly remains fixed and the other spreads at twice the apparent rate of spreading. On the basis of the anomalies alone, the end result cannot be distinguished from uniform spreading from a fixed crest. Such one-sided convection is not unknown in nature (Wilson, 1963a). As a means of producing symmetrical anomalies it appears less probable than symmetrical convection. With regard to some other related phenomena, however, it appears more probable. The Juan de Fuca Ridge provides a test. Magnetic anomalies symmetrical to the ridge extend under the Cascadia abyssal plain and the adjacent continental slope (Raff and Mason, 1961). The plain has not been disturbed in any way by sea floor spreading. This can be explained if the crest has migrated away from the continent and the crust does not continue to move under the plain. Another characteristic of the Juan de Fuca Ridge is that volcanoes are very common on the west flank but virtually absent on the east flank (Fig. 4, after Menard, 1964). Sub-bottom profiling shows that the sediment of the plain is relatively thin and thus has not buried and concealed volcanoes on the east flank (Shor, personal communication, 1966). The asymmetrical distribution of the volcanoes can be explained

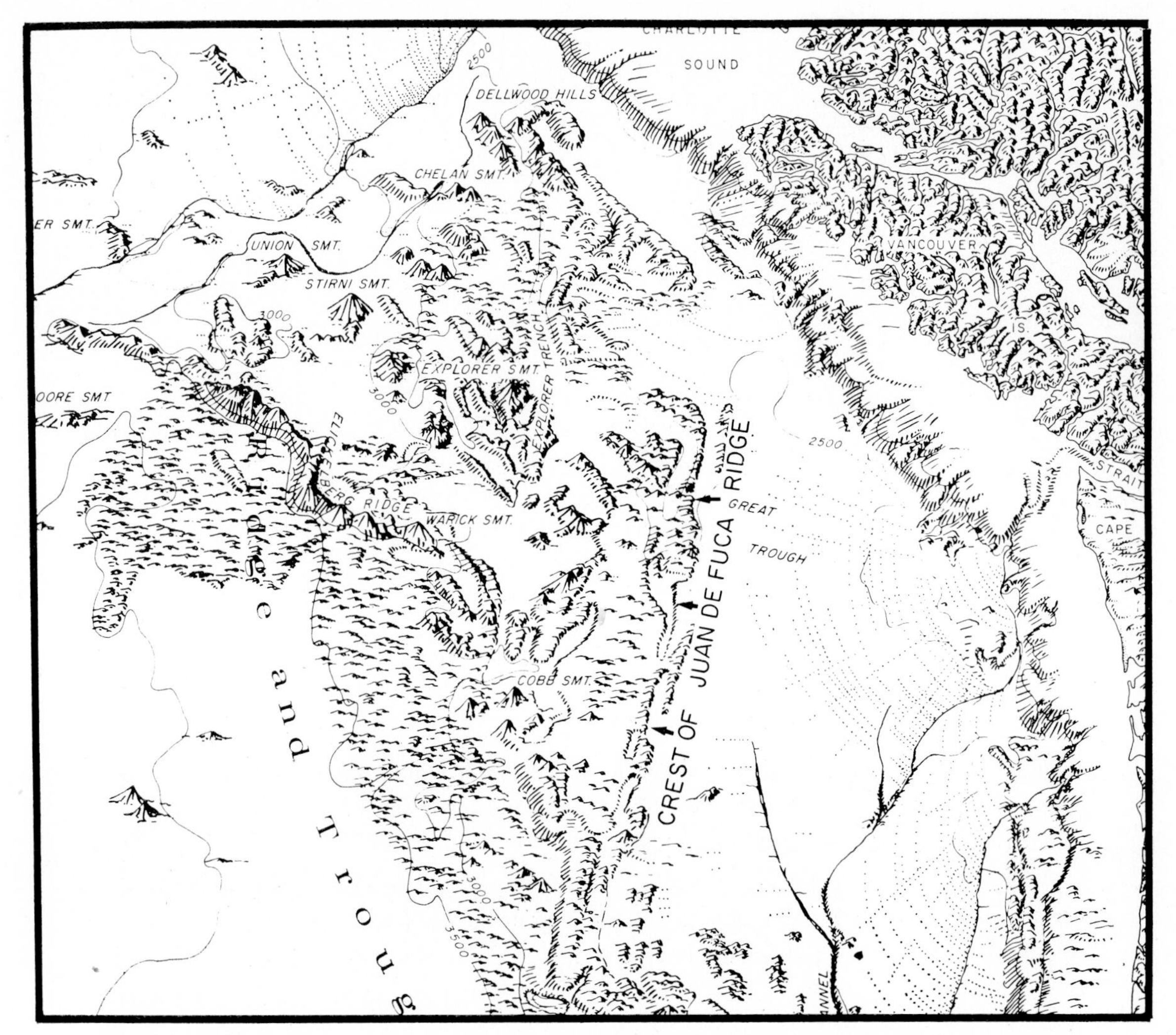

Figure 4. The distribution of large volcanoes on the Juan de Fuca Ridge. The striking asymmetry suggests more heating on the west flank and thus asymmetrical and migrating convection. After Menard, 1964.

by asymmetrical convection. If the convection is all toward the west and migrates west, heating under the western flank would far exceed normal and vulcanism might be expected. The eastern flank, in contrast, would not be heated as much, and volcanoes might be correspondingly rare. Relatively high heat flow might occur at the surface east of the crest until the new crust and the second layer have cooled, but the duration of exceptional heating would be comparatively brief.

This discussion indicates that asymmetrical, and therefore migrating, convection may explain the transition from a proposed early stage of large convection and simple crustal deformation to the smaller convection cells and complex deformation now observed. It has been assumed that the large convection was fixed in position and sea floor spreading occurred uniformly on each side. The possibility of migration of large cells also deserves consideration. Some migration of the mid-ocean ridges which almost encircle Africa is required to fit the continents together—unless the earth has expanded (Heezen, 1960). However, the convection cell under the Mid-Atlantic Ridge need not have moved, and even if it has the movement probably was west away from Asia and Africa. Thus, North and South America apparently have drifted toward the East Pacific Rise or even over it (Wilson, 1963b). Why did not spreading from the East Pacific Rise prevent this drifting? One-sided convection may explain this difficulty. Oceanic convection, and particularly convection rising under continents, appears to be symmetrical. Perhaps convection at the margin of an ocean basin is sometimes asymmetrical and toward the basin. If the northern half of the East Pacific Rise was formed by such convection, nothing would resist the westward movement of North America (Fig. 5).

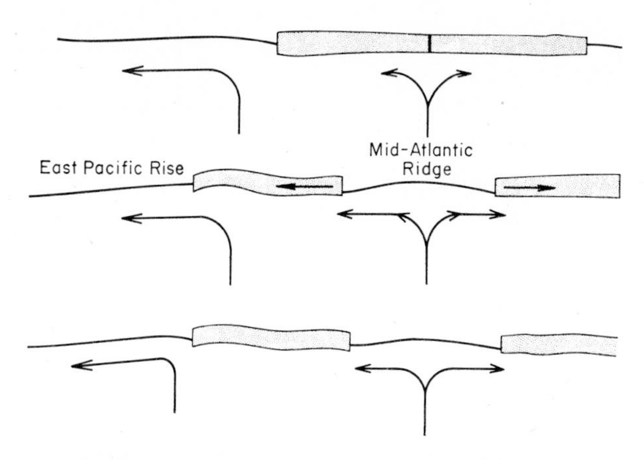

Figure 5. A possible mechanism whereby an asymmetrical and migrating East Pacific Rise would not impede the westward drift of North America from the Mid-Atlantic Ridge.

Does convective upwelling occur under the flanks of rises? The apparent offset of magnetic anomalies south of the Murray fracture zone increases abruptly about half-way between California and Hawaii (Raff, 1962). I have attempted to show that this offset can only be explained by differential movement away from the crest of the East Pacific Rise (Menard, 1964, 1966). The new evidence of the symmetry of magnetic anomalies and of sea floor spreading provides a strong confirmation of this hypothesis. I proposed that mantle convection flowing west under this region had broken and stretched the crust and had thereby accommodated the differential movement. However, this "disturbed zone" also contains linear magnetic anomalies (Raff, 1966) and more probably was formed by sea floor spreading with generation of new crust. This spreading occurred under the flank of the rise. If it had occurred at the crest and represented normal spreading, the same "disturbed zone" would be recorded in magnetic profiles elsewhere in the world and it is uniformly missing (Heirtzler, this volume and personal communication, 1966). Thus it appears that convective upwelling and spreading occur, although rarely, on the flanks of a large convection cell.

CONCLUSIONS

The sea floor spreading theory with its confirming corollaries of transform faults and symmetrical magnetic anomalies makes possible a new and deeper understanding of marine geology and global tectonics. This paper is not an attempt to retard acceptance of the theory. Rather it tries to point out that a number of observations which appear to be incompatible with the theory are not. All that is required is that the theory be slightly broadened to include complexities of the type which are commonly observed in convective processes, whether in the atmosphere, oceans, or, apparently, in the mantle.

Institute of Marine Resources and
Scripps Institute of Oceanography
University of California, La Jolla

DISCUSSION

DR. HEIRTZLER: I would like to say one thing regarding the possible displacement of established ridges. Yesterday I showed curves which showed the differential displacement rate in different areas. That was a gross picture; if you look at the details you see all kinds of interesting things in the magnetic pattern, which I am sure many of you will recognize. The symmetric pattern in the South Pacific is not by any means typical. The fact that it can exist is extremely important, but this is not typical.

You can have, apparently, spreading rates that vary linearly with distance along the ridge. As you go further from the ridge, the magnetic pattern changes so that it is not parallel to the ridge and is longer than the section of the ridge from which it came.

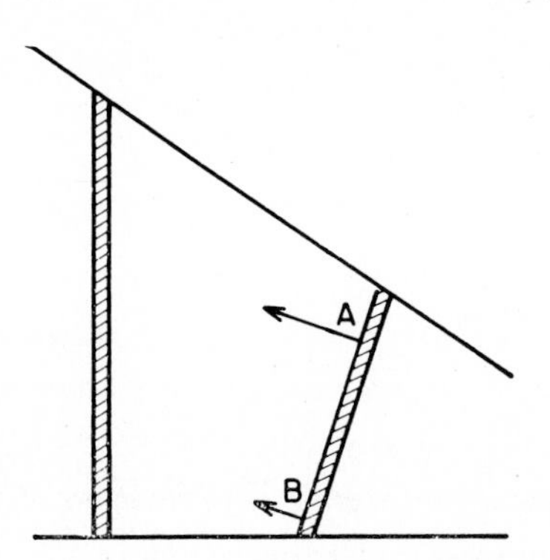

Say you have a section of ridge that is roughly like this sketch. If you have a spreading rate at *A* that is 4 cm/yr, and one at *B* that is 2 cm/yr, the pattern is like the one you see. This particular anomaly is further from the ridge at *A* than it is at *B*, and is obviously longer than the section from which it came.[1]

DR. MENARD: No, it is the "obviously longer" part that bothers me. What you are saying is that a section of ridge, of length x is capable of producing a magnetic anomaly of length $x + y$.

[1] Dr. Heirtzler goes into this in more detail in *Science* 154 (1966): 1164-71.

DR. HEIRTZLER: Yes.

DR. GAST: It has to be rotated.

DR. BULLARD: How can you elongate it?

DR. VINE: With regard to the Juan de Fuca area I think it is clear from the color slide [Fig. 4, Vine, this volume] that there is no single process going on. There are several processes at work accommodating a change in spreading direction from east-west to spreading approximately parallel to the San Andreas fault system at present. In the region of the Juan de Fuca Ridge the accommodation appears to be in the form of faulting and apparent rotation of crustal blocks. About the Gorda Ridge the form of magnetic anomalies presumably results from a change in spreading rate along the ridge crest. In both cases it seems probable that the length of a particular anomaly at the present day reflects the length of the ridge crest at the time the anomaly was formed.

The framework of ocean floor spreading, i.e., the formation of new crust, is a comparatively new geologic concept and opens up new possibilities. Transform faults and the symmetry of oceanic magnetic anomalies are corollaries we have heard much of. Another possibility is that a fracture or fault does not have to be continuous. A rate of spreading differential on either side of the fault, varying with time, at times zero, will give rise to discontinuous faults and varying offsets along them, as seems to be the case in the Juan de Fuca area.

DR. MENARD: It certainly helps to think about these things at leisure. I agree completely that a feature is produced by a ridge of the same length.

DR. HEEZEN: Consider the Atlantic Ocean, in particular the well-discussed area between South America and Africa. The Atlantic is a long, symmetric ocean. The topographic trend and, presumably, the magnetic trend parallel the ridge. If we assume any old continental drift model, a series of positions of the two continents may be postulated as a function of time. And if you plot the trajectories of arbitrary points at the edges of these moving continents, you find that they are parallel to the fracture zones.

This ocean had, as far as I am concerned, offsets built in from the very beginning. The ridge was never straight. So the hypothesis that at some time the anomaly had to be in a straight line, and was later displaced, can't be entertained as a universal possibility. It could happen somewhere, but in the Atlantic it wouldn't work.

DR. MENARD: Bruce and I are always running into difficulties in that our oceans aren't exactly the same.

DR. HEEZEN: I think the Atlantic is more regular than the Pacific.

DR. MENARD: Not the spacing of the fracture zones.

DR. HEEZEN: No. You must explain, in the Pacific, that there has been some shifting of the ridge. Since it is not median in the ocean anyway there is no reason to believe that it stayed in one place. You can create a little ridge here for a while and there for a while and so forth. You can have a situation where the spreading rates differ greatly on the two sides of the ridge.

But in the Atlantic it goes up the middle. It is so symmetrical that we have a more systematic pattern. We just don't happen to have any detailed surveys. If we had a survey, my prediction would be that the magnetic pattern in the Atlantic would turn out to be much more regular than in the Pacific.

DR. HALES: I think one point that Bruce is bringing out is that the two sides of the South Atlantic are essentially pre-Cambrian blocks and they are fixed. There are no strike-slip faults running through them from east to west, giving you a rigid framework. Whether they are close together or not, you can't play with the borders.

REFERENCES

Heezen, B. C. (1960), The rift in the ocean floor, *Sci. Amer.* 203:98-110.

Hess, H. H. (1962), History of ocean basins, in *Petrologic Studies: a Volume to Honor A.F. Buddington*, Geological Society of America, 660 pp.

Malahoff, A., W. E. Strange, and G. P. Woollard (1966), Molokai fracture zone: continuation west of the Hawaiian Ridge, *Science* 153:521-22.

Menard, H. W. (1960), The East Pacific Rise,. *Science* 132:1737-46.

Menard, H. W. (1964), *Marine Geology of the Pacific*, McGraw-Hill Book Company, New York.

Menard, H. W. (1966), Sea floor relief and mantle convection, pp. 315-64 in *Physics and Chemistry of the Earth*, Vol. 6, Pergamon Press, New York.

Menard, H. W. (1967), Extension of the northeastern-Pacific fracture zones, *Science* 155:72-74.

Pitman, W. C. III, and J. R. Heirtzler (1966), Magnetic Anomalies over the Pacific Antarctic Ridge. *Science* 154:1164-71.

Raff, A. D., and R. G. Mason (1961), Magnetic survey off the west coast of North America, 40° N. latitude to 52° N. latitude, *Geol. Soc. Amer. Bull.* 72:1267-70.

Raff, A. D. (1962), Further magnetic measurements along the Murray fault, *J. Geophys. Res.* 67: 417-18.

Raff, A. D. (1966), Boundaries of an area of very long magnetic anomalies in the northeast Pacific, *J. Geophys. Res.* 71:2631-36.

Vacquier, V., A. D. Raff, and R. E. Warren (1961), Horizontal displacements in the floor of the northeastern Pacific Ocean, *Geol. Soc. Amer. Bull.* 72:1251-58.

Vine, F. J. (1966), Spreading of the ocean floor: new evidence, *Science* 154:1405-15.

Wilson, J. T. (1963a), A possible origin of the Hawaiian Islands, *Can. J. Phys.* 41:863-70

Wilson, J. T. (1963b), Continental drift, *Sci. Amer.* 208 (4):86-100.

Wilson, J. T. (1965), Transform faults, oceanic ridges, and magnetic anomalies southwest of Vancouver Island, *Science* 150:482-85.

HEAT FLOW THROUGH THE OCEAN FLOOR AND CONVECTION CURRENTS*

X. LE PICHON

ABSTRACT

The contributions of Heirtzler and Vine have shown that the pattern of magnetic anomalies over the mid-ocean ridges can be interpreted according to the spreading floor hypothesis, thus providing strong support for the horizontal mobility of the oceanic crust. This hypothesis involves the transfer of large volumes of material from deep in the mantle to the vicinity of the sea floor, which should result in the release of a much larger amount of heat over the ridges than over the adjacent basins. The number of reliable heat flow measurements over the Mid-Atlantic Ridge and the East Pacific Rise and the adjacent basins is now large enough to estimate the excesses of heat released over both ridges and compare them with those which should result from the spreading rates deduced from magnetics. The excess of heat over the Mid-Atlantic Ridge is of the order of 40 cal/sec/cm of ridge length, most of this heat being released in a narrow axial zone. In contrast, the excess of heat released over the East Pacific Rise is nearly 10 times as large. The results are in qualitative agreement with the much faster spreading rate attributed to the East Pacific Rise. However, quantitatively, the excesses of heat released are much smaller than those computed for models of convection which fit the magnetic data, unless only a small portion of the heat lost by the underlying mantle moving away from the axis escapes by conduction through the sea floor.

Lamont Geological Observatory
Columbia University

* This paper appears in full in the *Journal of Geophysical Research* 71:5321-55, 1966.

SEISMOLOGICAL EVIDENCE FOR TRANSFORM FAULTS, SEA FLOOR SPREADING, AND CONTINENTAL DRIFT*

LYNN R. SYKES

Introduction

During the last few years there has been considerable interest in the hypothesis that displacements as large as hundreds or thousands of kilometers have occurred on the ocean floor. The discovery of a large number of fracture zones on the mid-oceanic ridges (Menard, 1955, 1965, 1966; Heezen and Tharp, 1961, 1965a) has led to a renewed interest in the existence of large horizontal displacements and to reconsiderations of various hypotheses of sea floor growth, continental drift, and mantle convection currents. Evidence that magnetic anomalies on the ocean floor can be identified with past reversals in the earth's magnetic field (Vine and Matthews, 1963; Vine and Wilson, 1965; Cann and Vine, 1966; Pitman and Heirtzler, 1966; Vine, 1966; Heirtzler, this volume) has added strong support to the hypothesis of ocean floor spreading as postulated by Hess (1962) and Dietz (1961).

The offsets of magnetic anomalies and bathymetric contours at prominent fracture zones have been used as arguments for transcurrent fault displacements as great as 1,000 km (Vacquier, 1962; Menard, 1965). Nevertheless, several problems are raised about interpretations of this type. The inferred magnitude and sense of displacement are not always constant along a given fracture zone (Menard, 1965; Talwani et al., 1965), and some fault zones appear to end quite abruptly (Wilson, 1965a). Even if displacements of this magnitude do not end abruptly, it is still difficult to imagine how the resulting strains can be accommodated on an earth of finite size. Seismic activity along fracture zones is concentrated almost exclusively between the two crests of the mid-oceanic ridge (Sykes, 1963, 1965); very few earthquakes are found on the other portions of fracture zones. These observations indicate that the entire ridge is not being offset by transcurrent faulting (simple offset) at the present time.

Wilson (1965a, b) has presented arguments for a new class of faults, the transform fault. As an alternative proposal to transcurrent faulting, however, Talwani et al. (1965) have assumed that major faulting on fracture zones is normal faulting. Wilson and Talwani et al. assume that the various segments of the mid-oceanic ridge were never displaced at all but developed at their present locations. Hence, these theories and the transcurrent fault hypothesis predict different types of relative motion on the seismically active portions of fracture zones.

Since considerable interest now centers on the problem of large-scale deformation of the sea floor, it seemed timely to inquire if the nature of relative displacements could be

* Lamont Geological Observatory Contribution 1199

ascertained from an analysis of the seismic waves that are generated by earthquakes on the mid-oceanic ridges. Fortunately this interest in large horizontal displacements coincided with the advent of a significantly new source of seismological data. Long-period seismograph records from more than 125 stations of the World-Wide Standardized Seismograph Network (WWSSN) of the U.S. Coast and Geodetic Survey, about 25 stations of the Canadian network, and about 20 stations that cooperate with the Lamont Geological Observatory now furnish data of greater sensitivity, greater reliability, and broader geographical coverage than were available in most previous investigations of the mechanisms of earthquakes.

This paper presents data from 30 earthquakes on the mid-oceanic ridges and on the extensions of this ridge system into East Africa, western North America, and the Arctic. Eleven of these events were located on the Mid-Atlantic Ridge, eight occurred on the East Pacific Rise and in the Gulf of California, and seven were situated in East Africa, the Gulf of Aden, and adjacent oceanic areas.

Sykes (1967a) studied the mechanisms for 17 of these 30 events. This paper summarizes the results of the former study and presents solutions for an additional 13 events. Many of these additional events were located in the Gulf of California and in the Gulf of Aden. This paper also examines the mechanism and distribution of earthquakes in the two gulfs in the light of the hypotheses of sea floor spreading, transform faults, and continental drift.

Previous investigations of the mechanisms of earthquakes for events on the mid-oceanic ridges are described by Sykes (1967a). That paper also discusses the analysis of the seismic data in considerably more detail than will be done in this presentation.

These investigations revealed two principal types of mechanisms for earthquakes on the mid-oceanic ridges. Earthquakes on fracture zones are characterized by a predominance of strike-slip motion; earthquakes located on the crest of the ridge but apparently not situated on fracture zones are characterized by a predominance of normal faulting. The inferred axes of maximum tension for these latter events are approximately perpendicular to the local strike of the ridge.

This study of the mechanisms of earthquakes is in agreement in every case with the sense of motion predicted by Wilson (1965a, b) for transform faults. The results support the hypothesis of ocean floor growth at the crest of the mid-oceanic ridge. The sense of displacement indicated by these studies of earthquakes is opposite to that expected for a simple offset of the ridge crest. The uniformity of the mechanisms of earthquakes throughout the mid-oceanic ridge system indicates that the processes of transform faulting and sea floor spreading probably represent the long-term behavior of this world-wide tectonic system. The similarity of the shapes of the continental margins on either side of the Gulf of Aden and the Gulf of California and the similarity between the shapes of the margins and the configuration of the ridge crest in both gulfs strongly suggest that these features opened by growth of new sea floor and by displacements on transform faults.

Table 1. Summary of Earthquake Locations and other Pertinent Data

Event Number	Figure Numbers	Region	Latitude	Longitude	Date	Origin Time h	m	s	Depth km	Magnitude*
1	4, 10	Romanche fracture zone, equatorial Atlantic	00.17S	18.70W	15 Nov 1965	11	18	46.8	0	5.6
2	10	equatorial Atlantic, fracture zone 'Z' of Heezen et al. [1964b]	07.45N	35.82W	3 Aug 1963	10	21	30.4	0	6.1
3	10	equatorial Atlantic, fracture zone 'Z' of Heezen et al. [1964b]	07.80N	37.35W	17 Nov 1963	00	47	58.8	0	5.9
4	10	Vema fracture zone, equatorial Atlantic	10.77N	43.30W	17 Mar 1962	20	47	33.4	31	
5	5, 11	North Atlantic unnamed fracture zone	23.87N	45.96W	19 May 1963	21	35	45.4	0	6.0
6	7, 11	North Atlantic	31.03N	41.49W	16 Nov 1965	15	24	40.8	0	6.0
7	11	North Atlantic	32.26N	41.03W	6 Aug 1962	01	35	27.7	0	
8	6, 11	North Atlantic unnamed fracture zone	35.29N	36.07W	17 May 1964	19	26	16.4	0	5.6
9	3, 12	off north coast of Iceland unnamed fracture zone	66.29N	19.78W	28 Mar 1963	00	15	46.2	0	
10	8, 13	Laptev Sea, near Arctic shelf of Siberia	78.12N	126.64E	25 Aug 1964	13	47	13.8	0	6.1
11		Lake Kariba, Southern Rhodesia	16.64S	28.55E	23 Sep 1963	09	01	51.1	0	5.8
12		Lake Kariba, Southern Rhodesia	16.70S	28.57E	25 Sep 1963	07	03	48.8	0	5.8
13	9	Western Rift, East Africa	00.81N	29.93E	20 Mar 1966	01	42	46.8	0	6.1
14	16, 17	Rivera fracture zone, East Pacific rise	18.87N	107.18W	6 Dec 1965	11	34	48.9	0	5.9
15		Easter fracture zone East Pacific rise	26.87S	113.58W	7 Mar 1963	05	21	56.6	0	

* Body-wave magnitude of U.S. Coast and Geodetic Survey.

Analysis of Data

DATA USED

In this investigation first motions of the phases *P* and *PKP* were examined for 29 earthquakes that occurred between March 1962 and August 1966. Seismograms used in this study were supplied by the WWSSN, the Canadian network, and about 20 stations that operate long-period seismographs in cooperation with the Lamont Geological Observatory. The geographical distribution of stations is generally more complete for earthquakes that occurred after 1962. Data from the Lamont network were also used to study the first motions of an earthquake that occurred in the Gulf of Aden during 1959. Epicentral and other pertinent data for these 30 earthquakes are listed in Table 1.

RELIABILITY OF DATA

An important factor in the use of these new data is the availability of seismograph records. Anyone who has previously attempted to obtain records from more than 100

Table 1. (Continued) Summary of Earthquake Locations and other Pertinent Data

Event Number	Figure Numbers	Region	Latitude	Longitude	Date	Origin Time h	m	s	Depth km	Magni-tude*
16		Eltanin Fracture zone East Pacific rise	55.35S	128.24W	3 Apr 1963	14	47	50.4	0	5.8
17	18, 19	Macquarie ridge, south of New Zealand	49.05S	164.26E	12 Sep 1964	22	06	58.5	0	6.9
18	10	Romanche fracture zone Equatorial Atlantic	00.49S	19.95S	16 Aug 1965	12	36	24.3	40	6.1
19	10	Mid-Atlantic Ridge	15.96N	46.79W	2 June 1965	23	40	22.5	23	5.6
20**		Unnamed Fracture zone Mid-Atlantic Ridge	52.9N	34.2W	5 July 1965	08	31	58.9	33	5.7
21**		Baffin Island, Canada	71.4N	73.3W	4 Sep 1963	13	32	12.3	33	5.9
22	14	Owen Fracture Zone Arabian Sea	12.86N	58.04E	13 Feb 1963	01	34	34.1	0	
23	14	Alula-Fartak trench Gulf of Aden	13.98N	51.71E	21 Dec 1959	11	19	15.1	0	
24	14	Owen Fracture zone Arabian Sea	21.86N	62.23E	30 Mar 1966	04	18	33.9	0	5.6
25	14	Red Sea	17.05N	40.58E	11 Nov 1962	15	15	28.0	0	
26	17	Gulf California	21.26N	108.75W	22 May 1966	07	42	42.9	0	5.5
27	17	Gulf California	21.36N	108.65W	23 May 1966	11	51	27.3	33	5.6
28	17	Gulf California	26.26N	110.22W	5 July 1964	19	07	58.2	28	6.0
29	17	Gulf California	29.68N	113.74W	18 Nov 1963	14	38	26.2	0	5.7
30	17	Gulf California	31.72N	114.42W	7 Aug 1966	17	36	22.8	0	6.3

* Body-wave magnitude of U. S. Coast and Geodetic Survey.

** Epicenters from cards (Preliminary Determination of Epicenters) of U. S. Coast and Geodetic Survey; other locations recomputed by author.

different stations will appreciate that the availability of these data is a great asset. All of the readings used in this study were made by the author. A better distribution of stations both in distance and in azimuth can now be obtained with the long-period seismographs in the three networks. First motions of the *P* phase are usually more reliable on the long-period records. Because the networks have a greater sensitivity to small earthquakes, reliable solutions can be determined for a large number of earthquakes with magnitudes as small as about 5.5 to 6. This is an important factor in investigations of the mechanisms of earthquakes on the mid-oceanic ridges, since events of magnitude larger than 7 (approximately the smallest magnitude for which determinations could be made previously) occur in these regions only about once during a period of a few years.

In addition to the factors already mentioned the calibration and polarity of the WWSSN stations are better known. In this study the number of inconsistent readings of first motion is less than 1%. In many previous investigations 15% to 20% of the data were often inconsistent with the inferred quadrant distribution of first motion (Hodgson and Adams, 1958). Thus, the previous inconsistencies seem to reflect a weakness in the older data rather than a weakness in the theoretical models that predict a quadrant distribution of first motions.

TRANSFORM AND TRANSCURRENT FAULTS

Wilson (1965a, b) has recently proposed a separate class of horizontal shear faults, the transform fault. Dextral and sinistral transform faults of the ridge-ridge type and their transcurrent counterparts are illustrated in Figure 1. In the transcurrent models it

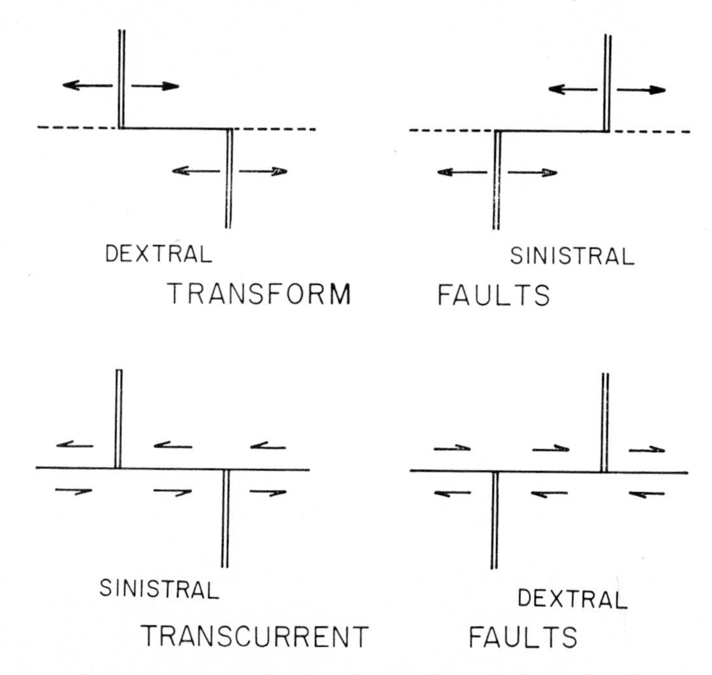

Figure 1. Sense of displacements associated with transform faults and with transcurrent faults. Double line represents crest of mid-oceanic ridge; single line, fracture zone. Terms "dextral" and "sinistral" denote sense of motion on active portions of faults.

is tacitly assumed that the faulted medium is continuous and conserved, whereas in the transform hypothesis the ridges expand to produce new crust (Wilson, 1965a). Thus, a transform fault may terminate abruptly at both ends even though great displacements may have occurred either on the central portions of the fracture zone or at some time in the past on portions of the fault that are no longer located between the two ridge crests (Fig. 1, dashed segments).

Wilson (1965a) noted that another significant difference was that the strike-slip motion on transform faults (upper half Fig. 1) is the reverse of that required to offset the ridge (lower half Fig. 1). Another difference is that present seismicity should be largely confined to the region between the ridge crests for transform faulting but should not exhibit such an abrupt decrease at the ridge crests for transcurrent faulting.

In this paper the term "fault plane" is used to describe a zone of shear displacement or shear dislocation. No attempt is made to define the physical mechanism of deformation more specifically.

EXAMPLES OF STRIKE-SLIP FAULTING AND NORMAL FAULTING ON THE MID-OCEANIC RIDGE SYSTEM

Figure 2 illustrates the locations of the 30 earthquakes for which mechanism solutions were obtained in this study. The numbers beside the epicenters refer to the data in Table 1. Events 1 through 9 and 18 through 20 are located on the Mid-Atlantic Ridge, earthquake 10 is on the extension of the mid-oceanic ridge system in the Arctic, events

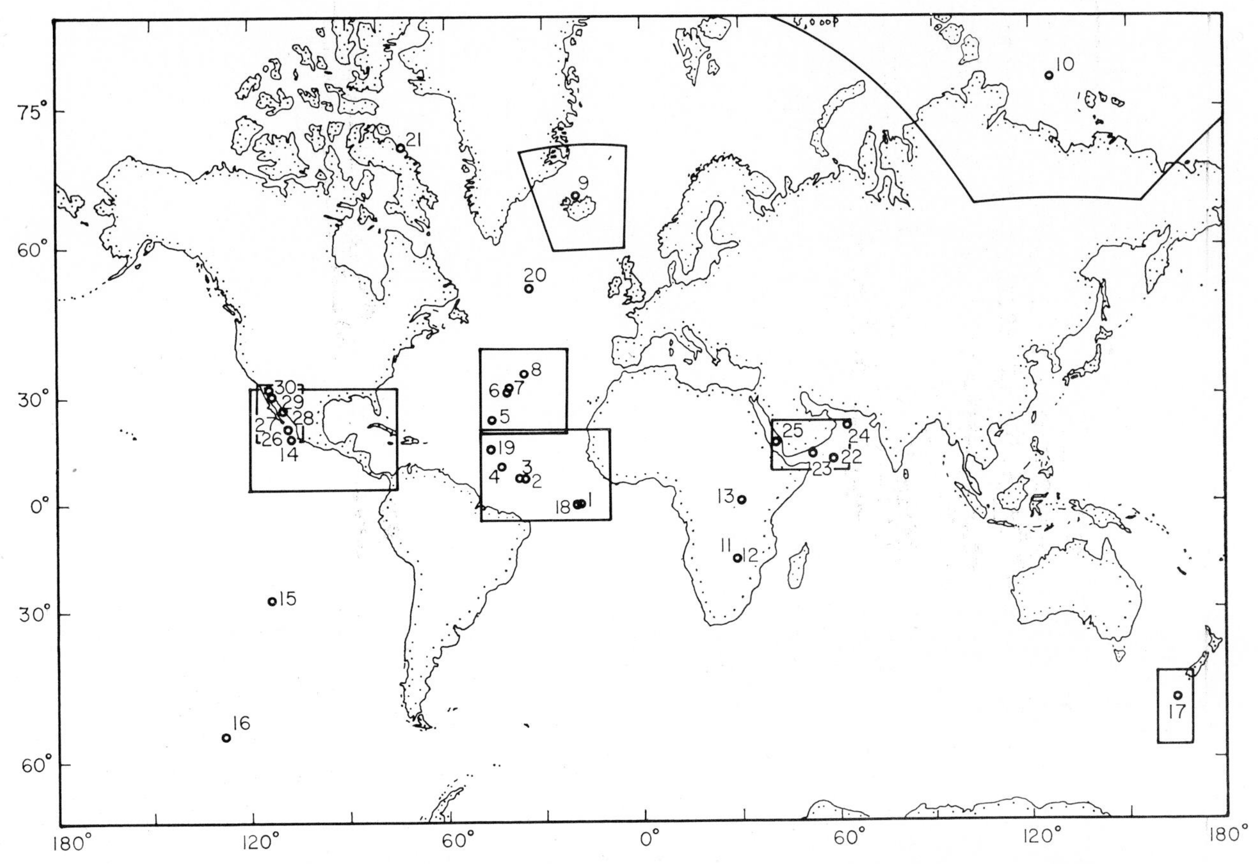

Figure 2. Earthquakes on mid-oceanic ridge system for which mechanism solutions were obtained. Numbers beside epicenters refer to designations and data in Table 1. Inserts denote areas that are shown in greater detail in Figures 10 through 18.

11 through 13 are in East Africa, 14 through 16 and 26 through 30 are on the East Pacific Rise and in the Gulf of California, 22 through 25 are in the Gulf of Aden and adjacent areas, event 17 is on the Macquarie ridge, and event 21 is located in Canada near the east coast of Baffin Island.

Examples of earthquakes that are characterized by a predominance of strike-slip faulting are shown in Figures 3 through 6. Solutions for earthquakes with a large component of normal faulting are illustrated in Figures 7 through 9. An equal-area projection of the lower hemisphere of the radiation field was used throughout this investigation. In these examples, solid circles denote compressions; open circles indicate dilatations. When

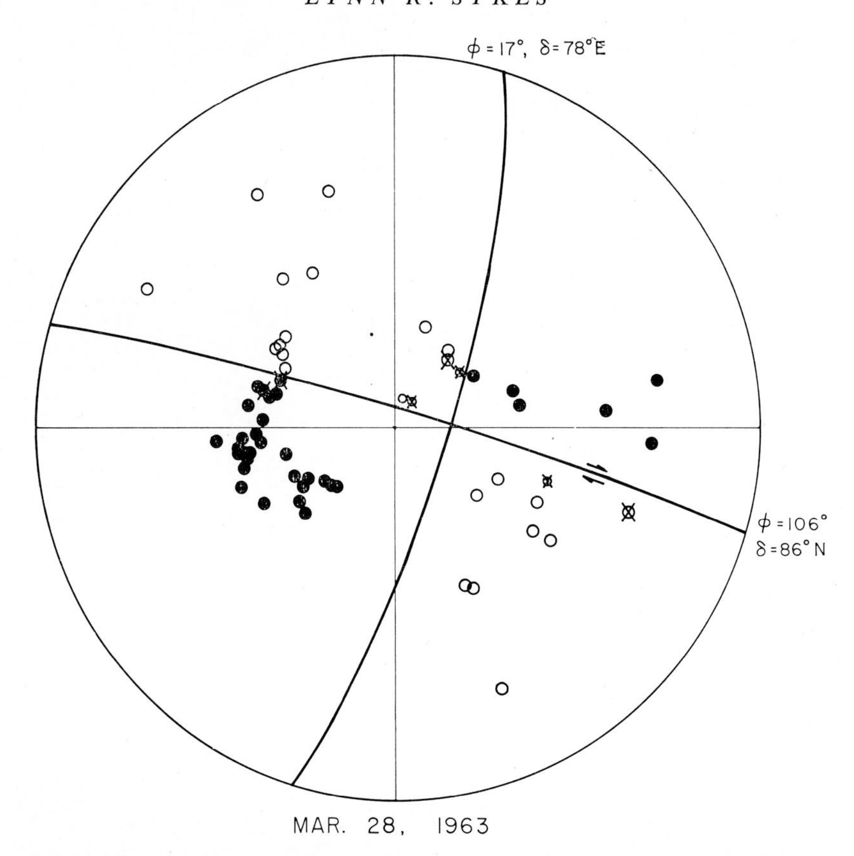

Figure 3. Mechanism determination for the shock of March 28, 1963, on the Mid-Atlantic Ridge near Iceland (event 9 in Table 1 and in Figs. 2 and 12). Diagram is an equal-area projection of the lower hemisphere of the radiation field. Solid circles, compressions; open circles, dilatations; crosses, wave character on seismograms indicates station is near nodal plane. Smaller symbols represent poorer data. φ and δ are strike and dip of the nodal planes. Arrows indicate sense of shear displacement on the plane that was chosen as the fault plane.

long-period records were available, it was often possible to tell from the size of the *P* wave relative to that of other phases whether the station was in the vicinity of one of the two nodal planes. Arrivals of this type are denoted by an × on the figures. As an examination of the figures reveals, these arrivals are often a good qualitative indication that the station was, in fact, near a nodal plane. Even when first motions cannot be ascertained, this information is an additional constraint on the orientation of the nodal planes.

A quadrant distribution of first motions seems to be an excellent approximation to the data in Figures 3 through 9. Only five readings are inconsistent with this interpretation. For the 30 events examined in this investigation less than 1% of the data are inconsistent with the inferred quadrant distribution of first motion. Seismograms from these inconsistent stations are of marginal quality and should not be taken as evidence against a quadrant distribution.

In Figures 3 through 6 two steeply dipping nodal planes can be drawn to separate the quadrants of compressional and dilatational first motion. These solutions, which are

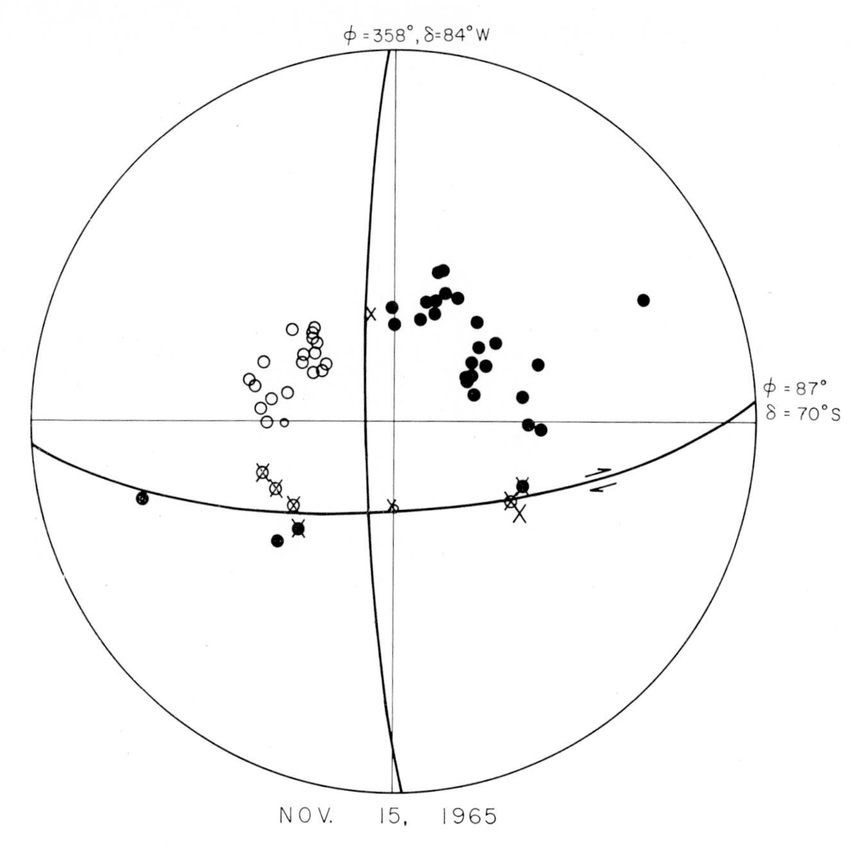

Figure 4. Mechanism solution for event 1. Table 1 and Figures 2 and 10 give epicentral and other pertinent data. Symbols same as Figure 3.

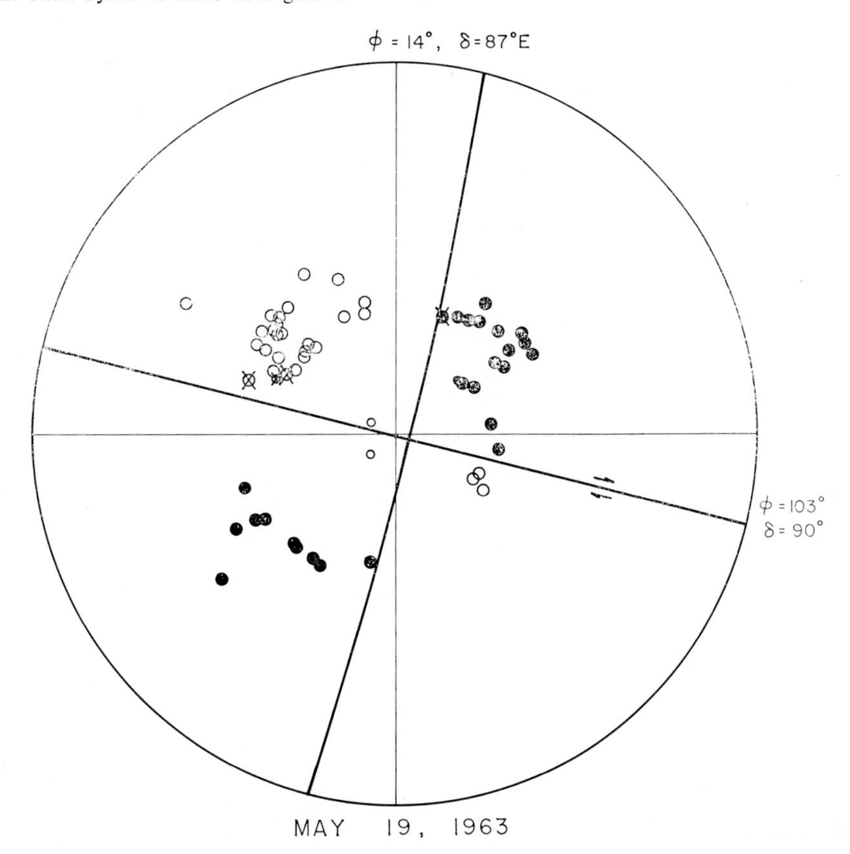

Figure 5. Mechanism solution for event 5. Table 1 and Figures 2 and 11 give position and other pertinent data. Symbols same as Figure 3.

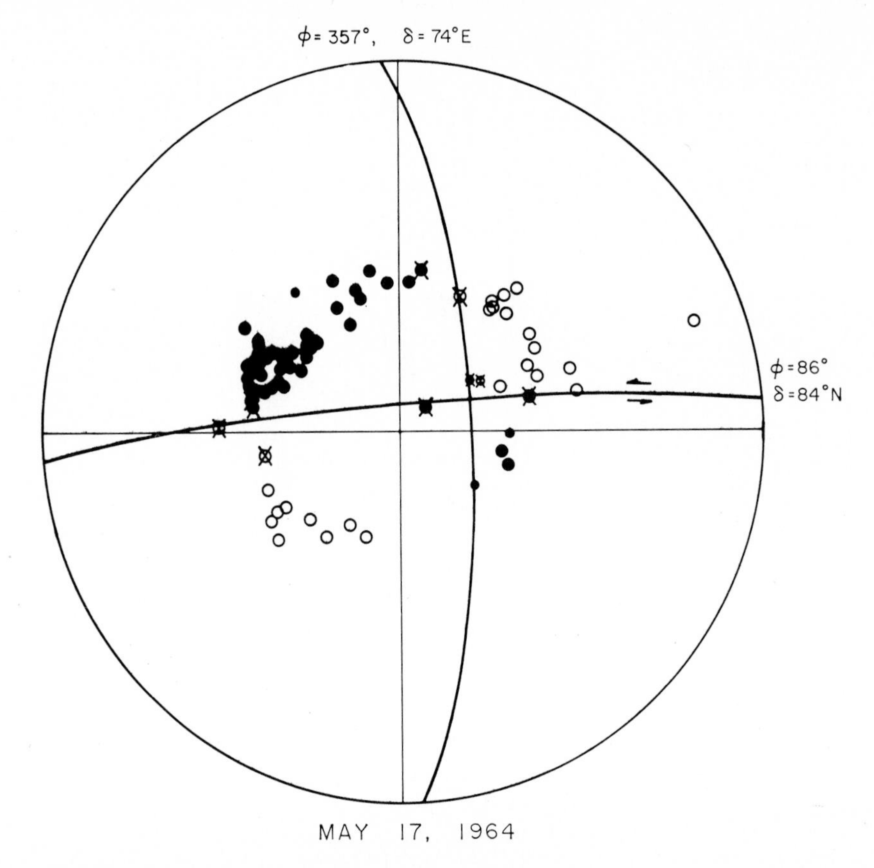

Figure 6. Mechanism solution for event 8. Table 1 and Figures 2 and 11 contain position and other pertinent data. Symbols same as Figure 3.

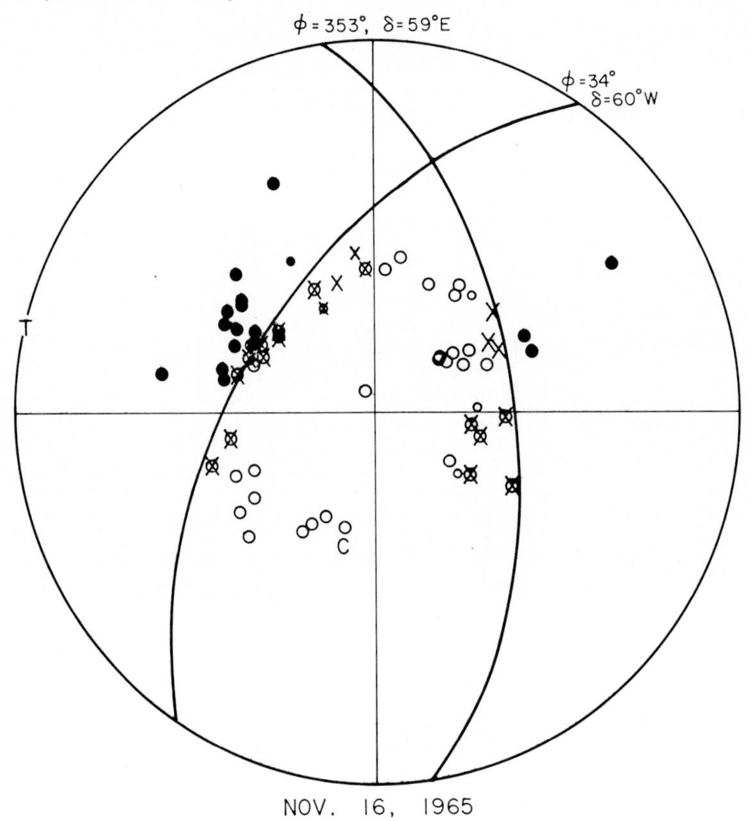

Figure 7. Mechanism solution for event 6. Table 1 and Figures 2 and 11 give epicentral and other data. T and C are inferred axes of maximum tension and compression, respectively. Other symbols same as Figure 3.

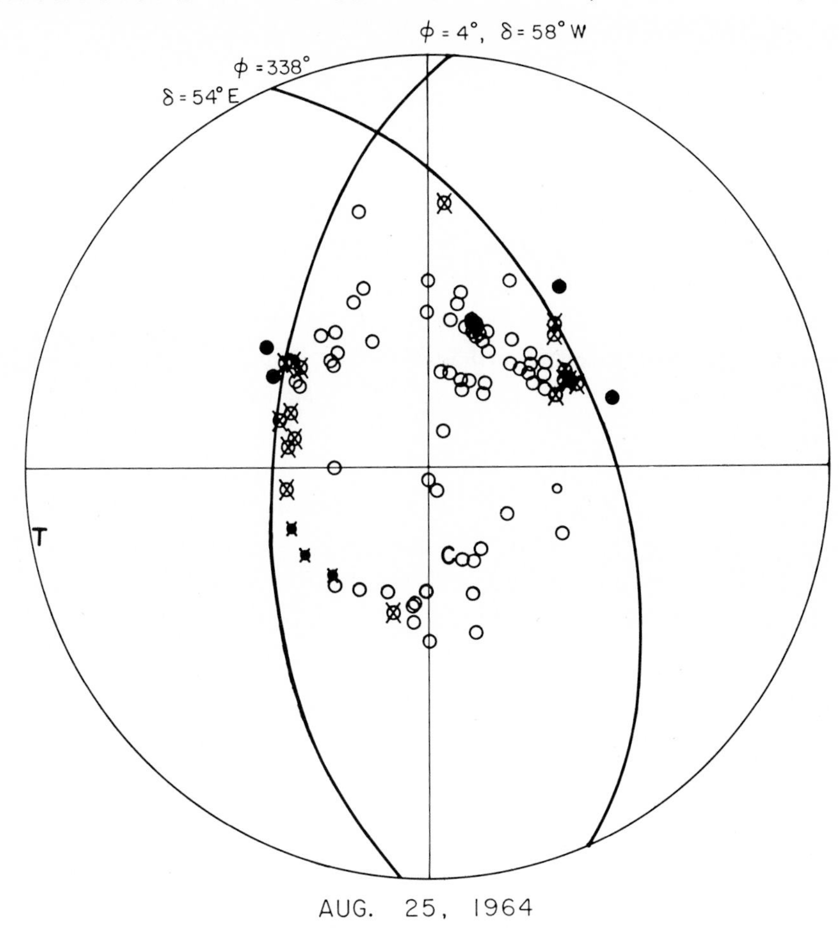

Figure 8. Mechanism solution for earthquake (event 10) near the continental shelf of northern Siberia. Symbols same as Figures 3 and 7. Location of event shown in Figures 2 and 13.

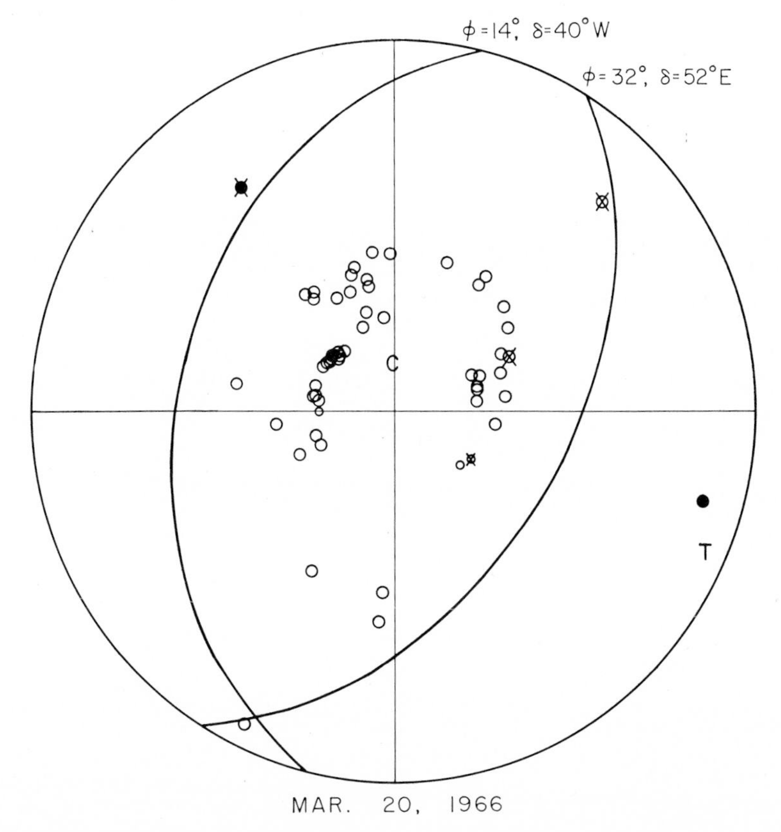

Figure 9. Mechanism solution for East African earthquake of March 20, 1966 (event 13). Symbols same as Figures 3 and 7; Table 1 and Figure 2 denote location and other data.

characterized by a predominance of strike-slip motion (regardless of which of the two nodal planes is chosen as the fault plane), can be readily distinguished from solutions such as those in Figures 7 through 9, which display an entirely different pattern of first motions. In the latter examples all of the more distant stations recorded dilatations. Thus, these three earthquakes are characterized by a large component of normal-fault motion on planes that strike approximately parallel to the axis of the ridge. Although the seismic data by themselves do not indicate which of the two nodal planes is the fault plane, the axis of maximum tension is determined uniquely; in each of the three examples this axis is nearly horizontal and is approximately perpendicular to the strike of the mid-oceanic ridge system.

Field observations for the Uganda earthquake of March 20, 1966, are in agreement with the mechanism that was deduced in Figure 9. Wohlenberg (1966) described fault displacements associated with this earthquake that start at about 0.7°N, 29.8°E and continue for about 40 km in a northerly direction. He reports a vertical displacement of 30 to 40 cm, with the western side moving up relative to the eastern side. No horizontal displacement was observed. Loupekine (1966) reported that the fault breakage extended for about 15 to 20 km in a direction NNE. He also describes a throw of 6 ft (2 m), with the down throw on the eastern side. These observations are in close agreement with the inferred mechanism if the nodal plane that strikes about N 14°E is chosen as the fault plane.

Transform Faults on the Mid-Atlantic Ridge

FRACTURE ZONES

Between 15°N and 5°S the crest of the Mid-Atlantic Ridge is displaced to the east a total of nearly 4000 km (Fig. 10). The apparent displacement is such that the ridge crest maintains its median character throughout the North and South Atlantic oceans. Hess (1955) identified the Romanche trench (near events 1 and 18 in Fig. 10) as a part of a major fracture zone and noted that St. Paul's Rocks appear to be located at the western end of a great fault scarp. Heezen and Tharp (1961) and Heezen et al. (1964a, b) recently mapped a whole series of fracture zones in the equatorial Atlantic (Fig. 10). In this area Heezen and Tharp (1965b) concluded that the ridge crest had been offset by *sinistral* transcurrent faults. They noted that the fracture zones appeared to be parallel to inferred flow lines for the continental drift of Africa relative to South America. It is difficult to imagine, however, how this drift could be accomplished by a simple offset (transcurrent faulting) without distorting and fragmenting the two continents.

SEISMICITY

Earthquake epicenters between 5°S and 41°N in the Atlantic were relocated for the period 1955 to 1965; the results are presented in Figures 10 and 11. The methods and computer programs used in these relocations were similar to those described in previous seismicity studies (Sykes, 1963, 1965, 1966; Sykes and Ewing, 1965; Sykes and Landisman, 1964; Tobin and Sykes, 1966).

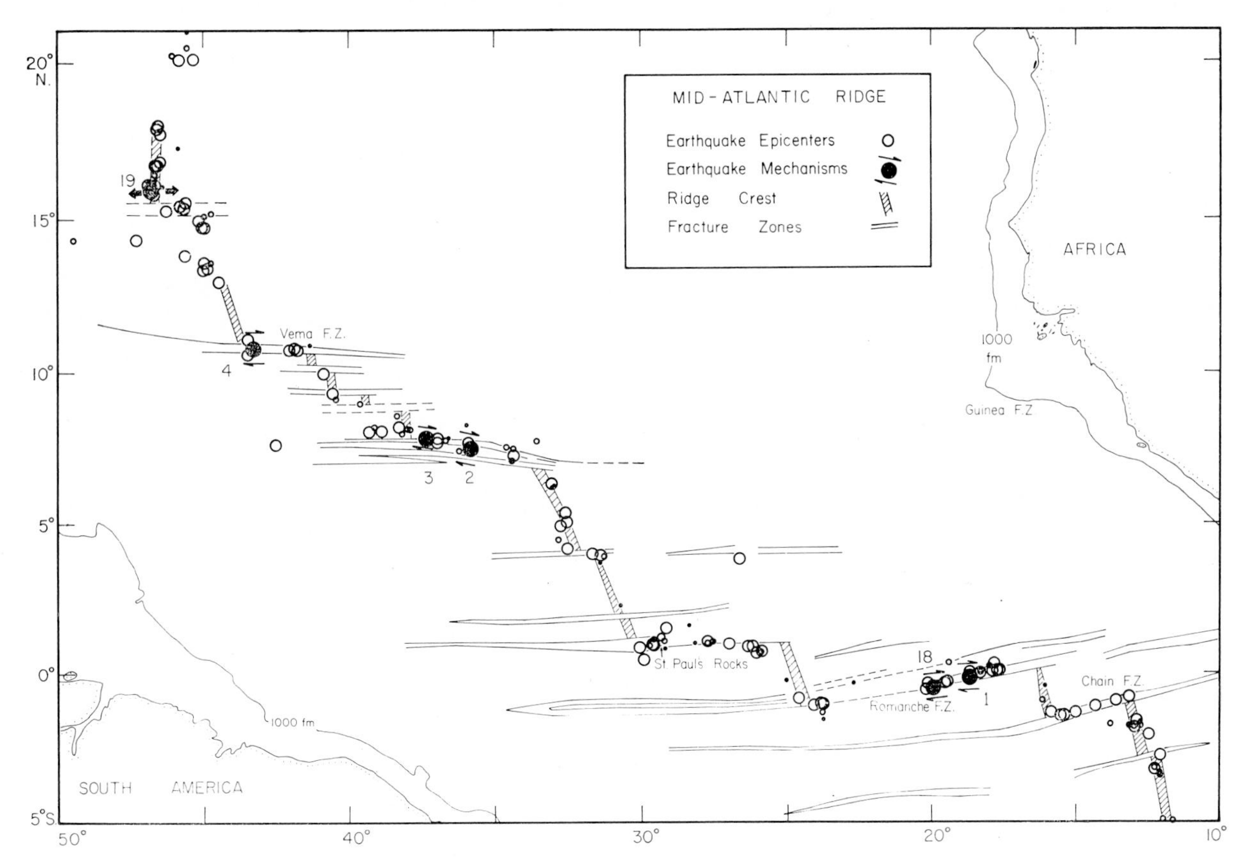

Figure 10. Relocated epicenters of earthquakes (1955-1965) and mechanism solutions for six earthquakes along the equatorial portion of the Mid-Atlantic Ridge. Ridge crests and fracture zones from Heezen et al. (1964a, b). Events 1, 2, 3, 4, and 18 were characterized by a predominance of strike-slip faulting; sense of shear displacement and strike of inferred fault plane are indicated by the orientation of the set of arrows beside each of these mechanisms. Numbers beside mechanism solutions refer to data in Table 1. Large circles denote more precise epicentral determinations; smaller circles, poorer determinations. Event 19 was characterized by a large component of normal faulting; the thick arrows denoted the inferred axis of maximum tension.

The distribution of earthquakes is similar to that found in other portions of the mid-oceanic ridge (Sykes, 1963, 1965)—nearly all of the activity on each fracture zone is confined to the region between the ridge crests. This is particularly well illustrated for the Chain fracture zone near 1°S, 15°W (Fig. 10). The distribution of epicenters also argues that Heezen and Tharp's (1961, 1965b) interpretations of the pattern of ridges and fracture zones are largely correct. Although a few earthquakes are found on other portions of fracture zones, the number is relatively small. The abrupt cessation of activity at the ridge crest appears to be a strong argument for the transform-fault hypothesis.

CHOICE OF THE FAULT PLANE

Events 1 through 4 and 18 (Fig. 10) are each located on a prominent fracture zone on the Mid-Atlantic Ridge. From analyses of the polarization of the *S* waves for several

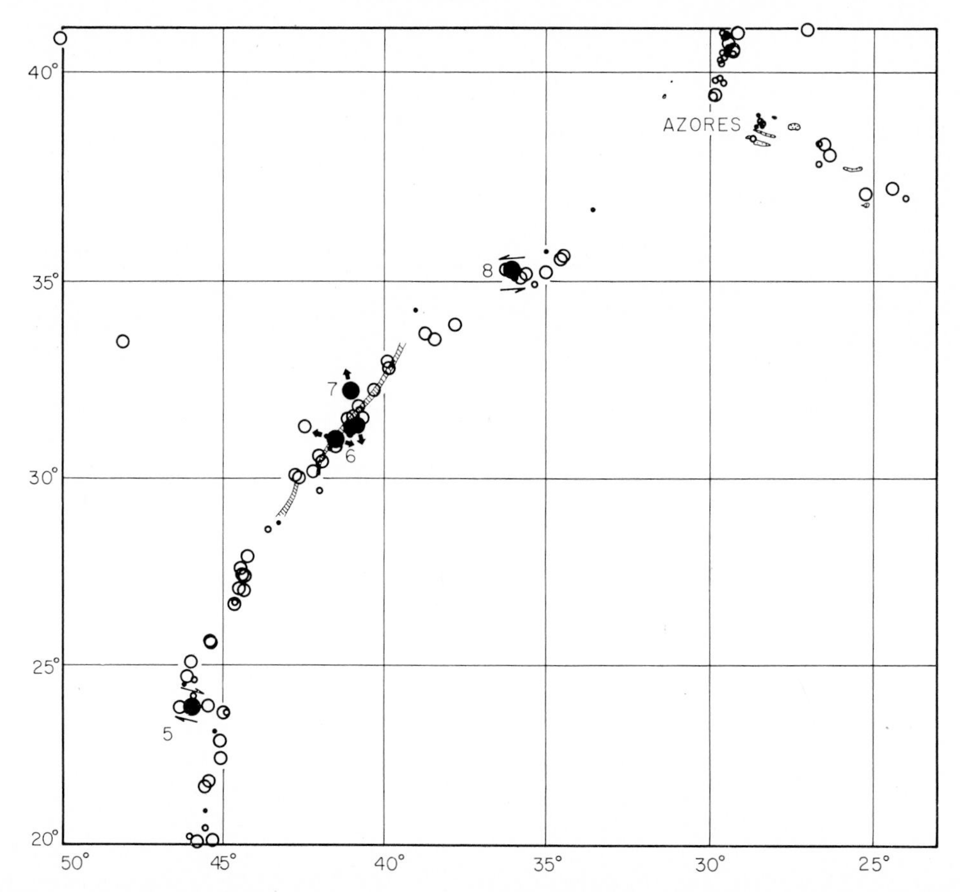

Figure 11. Relocated epicenters of earthquakes (1955-1965) and mechanism solutions for four earthquakes along a portion of the Mid-Atlantic Ridge. Rift valley in Mid-Atlantic Ridge is denoted by diagonal hatching and is from Heezen and Tharp (1965b). Other symbols same as Figure 10. For events 6 and 7 thick arrows denote the inferred axes of maximum tension. Axis of tension poorly determined for event 7.

shocks on the mid-oceanic ridge Stauder and Bollinger (1966) showed that the mechanisms are of the double-couple type. Thus, *S* waves do not furnish criteria for choosing which of the two nodal planes is the fault plane. In each of these five cases the choice is between a predominance of dextral strike-slip motion on a steeply dipping plane that strikes approximately east or a predominance of sinistral strike-slip motion on a steeply dipping plane that strikes nearly north.

Although a unique choice of the fault plane cannot be made from the first-motion data or from an analysis of *S* waves, the easterly striking plane seems to be overwhelmingly favored in each of these cases for the following reasons:

1. For each earthquake the epicenter is located on a prominent fracture zone; the easterly striking nodal plane nearly coincides with the strike of the fracture zone (Fig. 10). Fortunately, in these and in the other solutions for the Mid-Atlantic Ridge the easterly striking nodal plane can be estimated with greater precision than the northerly striking plane.

2. Earthquake epicenters are aligned along the strike of the fracture zones.

3. The linearity of fracture zones suggests a strike-slip origin (Menard, 1965). The bathymetry and other morphological aspects are similar to the morphology of the great strike-slip fault zones on continents (Allen, 1965).

4. Many of the rocks from St. Paul's Rocks (Fig. 10) are described as dunite-mylonites (Washington, 1930a, b; Tilley, 1947; Wiseman, 1966). Rocks dredged from other fracture zones (Shand, 1949; Tolstoy, 1951; Quon and Ehlers, 1963) as well as samples from cores taken in fracture zones (Heezen et al., 1964a) exhibit a similar petrology and provide evidence for intense shearing stresses in the vicinity of fracture zones.

5. The choice of the north-striking plane would indicate strike-slip motion nearly parallel to the ridge axis. On the contrary, earthquakes on the ridge axis but not on fracture zones are characterized by a predominance of normal faulting.

6. If the east-striking plane is chosen, the sense of relative motion on the fracture zone reverses when the apparent offset of the crests of the ridge is interchanged (Fig. 1).

CORRECT SENSE OF MOTION

The determination of the sense of strike-slip motion on fracture zones has been one of the most important topics of this paper. Thus, it is important to inquire if the sense of motion inferred from studies of first motions could be in error by large amounts or even possibly reversed. Although geologic and geodetic evidence for displacements are not available for many earthquakes, evidence of this kind is in close agreement with the solutions inferred from first motions in the case of several prominent earthquakes. Very good agreement with geologic evidence was found in investigations of the mechanisms of the Fairview Peak earthquake of December 1954 (Romney, 1957), the southeast Alaska earthquake of 1958 (Stauder, 1960) and the Parkfield earthquake of 1966 (McEvilly, 1966; Harding and Rinehart, 1966). Hodgson (1957) has cited other examples of good or fair agreement between first-motion studies and geologic and geodetic evidence. Hence, it seems highly unlikely that the inferred sense of motion on fracture zones is incorrect.

NORTH ATLANTIC

Figure 11 indicates the locations of four mechanism determinations for the Mid-Atlantic Ridge in the North Atlantic. The mechanisms for events 5, 8, and 6 are shown in Figures 5, 6, and 7. Two of the solutions are characterized by a predominance of strike-slip motion on steeply dipping planes; the others are characterized by a large component of normal faulting.

The observed radiation field for event 5 is similar to that for events 1 through 4 and 18 (Fig. 10). The pattern of epicenters near 24°N and the observed mechanism for event 5 are indicative of a *dextral* transform fault. Nearly all of the first motions observed for event 8 (Fig. 6), however, are *opposite* in sense to those observed for events 1 through 5. The overall strike of the ridge does not appear to change appreciably between events 5 and 8. The east-west alignment of epicenters near 35°N, the bathymetry in this region

(Tolstoy and Ewing, 1949; Tolstoy, 1951; Heezen et al., 1959), and the mechanism of event 8 are indicative of a *sinistral* transform fault that strikes approximately east. P. J. Fox (personal communication) has identified a prominent fracture zone near 35°N from topographic evidence.

Solutions for events 6, 7, and 19 (Figs. 10 and 11), however, are characterized by a large component of normal faulting. Although a fracture zone has been mapped near 30.0°N (Tolstoy and Ewing, 1949; Tolstoy, 1951; Heezen and Tharp, 1965b), no major fracture zones seem to intersect the ridge near these three events. In the region near events 6 and 7 the position of the rift valley was mapped with enough precision so that offsets greater than a few tens of kilometers should have been resolvable.

Talwani (1964) suggested that all earthquakes on the mid-oceanic ridge may be related to fracture zones that are not prominent enough to have been detected yet on the basis of bathymetry. The existence of two types of earthquake mechanisms, however, suggests that this is not the case. Heirtzler et al. (1966) concluded that although earthquakes have been detected south of Iceland (Fig. 12) on the Reykjanes Ridge (a portion of the Mid-Atlantic Ridge), there is no bathymetric or magnetic evidence for offsets of the ridge between 60°N and 63°N.

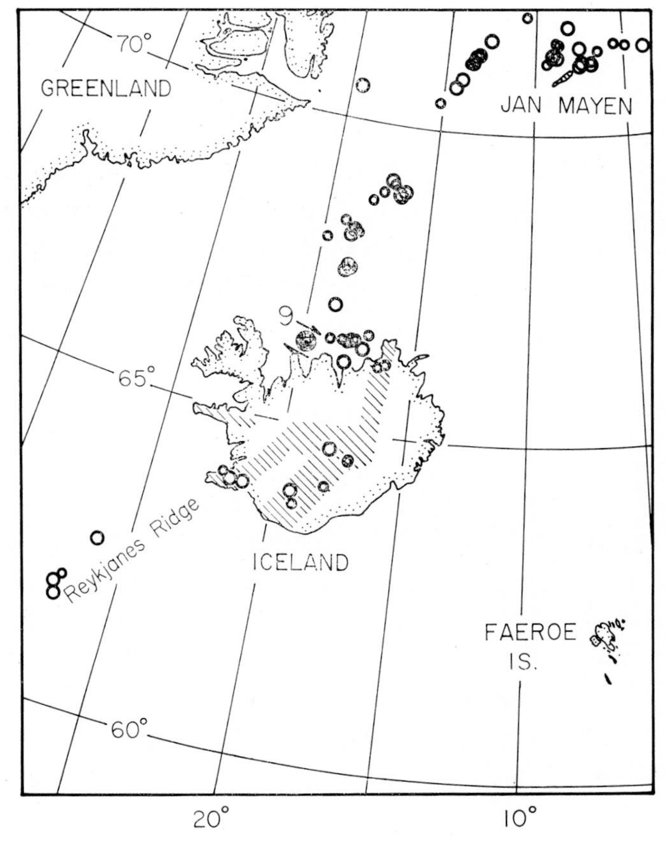

Figure 12. Relocated epicenters of earthquakes along a portion of the Mid-Atlantic Ridge near Iceland (after Sykes, 1965). Diagonal hatching denotes regions of postglacial volcanic activity in Iceland (after Bodvarsson and Walker, 1964). Other symbols same as Figure 10.

Event 20 occurred on one of a series of large fracture zones that intersect the crest of the Mid-Atlantic Ridge near the southern end of the Reykjanes Ridge. In this region the axis of the south end of the ridge is displaced to the east about 800 km. The inferred mechanism for this event was characterized by a predominance of dextral strike-slip motion. The nodal plane that was chosen as the fault plane strikes about N 74°W.

ARCTIC BASIN AND BAFFIN BAY

Event 10, which is located near the edge of the continental shelf of northern Eurasia, is characterized by a large component of normal faulting (Fig. 8). The epicenter of this event is located on the extension of the mid-oceanic ridge system into the Arctic basin and into northern Siberia (Fig. 13). The seismic zone (Sykes, 1965), which is believed to define the crest of the ridge in this region, is nearly perpendicular to the inferred axis of maximum tension.

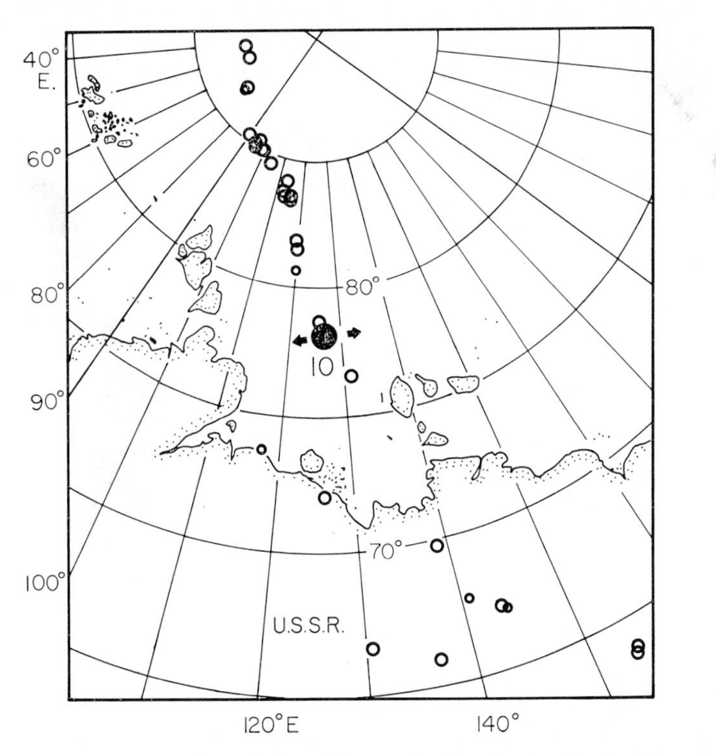

Figure 13. Epicenters along a portion of the mid-oceanic ridge in the Arctic (after Sykes, 1965). Thick arrows denote inferred axis of maximum tension for event 10. Other symbols same as Figure 10.

Event 21, an earthquake near the east coast of Baffin Island (Fig. 2), was also characterized by a predominance of normal faulting. The two nodal planes of the inferred mechanism strike N 87°W and N 74°W and dip 70°S and 20°N respectively. The axis of tension plunges 24° in an azimuth of 186°. The separation of Baffin Island and the rest of North America from Greenland has been a recurrent feature in many hypotheses of continental drift (Wilson, 1965c). Zones of minor seismicity are found in Baffin Bay, Hudson Strait, and along the Mid-Labrador Sea Ridge (Gutenberg and

Richter, 1954; Sykes, 1965). These areas, however, are not as active as the crest of the Mid-Atlantic Ridge. The reduced seismicity in the Labrador Sea, the lack of a prominent central magnetic anomaly (Vine, 1966), and the burial of the Mid-Labrador Sea Ridge by sediments (Drake et al., 1963) indicate that if Greenland is still moving with respect to North America, the rate of displacement is probably considerably less than that for much of the mid-oceanic ridge system.

East Africa, the Gulf of Aden, and the Arabian and Red Seas

The Gulf of Aden, the Red Sea, and the Gulf of California are often cited as examples of features that have been developed by horizontal displacements of a few hundred kilometers. Some authors, however, have explained these features as grabens that were developed by vertical movements alone. Others have been impressed by the similarity of the shape of the coastline on either side of these narrow bodies of water; hence, they have stressed the view that the crustal displacements were mainly horizontal. The hypotheses of dominantly vertical and dominantly horizontal motions will now be examined with evidence from the mechanism and distribution of earthquakes. This evidence strongly favors the hypothesis of large horizontal movements in these areas.

Although observational data from seismograph instruments are necessarily limited to the last few tens of years, the uniformity of the inferred mechanisms throughout the mid-oceanic ridge system suggests that transform faulting and sea floor spreading have been occurring on the ridge system for long periods of time. If the Gulf of Aden, the Red Sea, and the Gulf of California opened by a simple process of sea floor spreading from segments of ridge crest that have been arranged en echelon since the initiation of drift, then the shape of the initial break should be reflected both in the shape of the continental margins on either side of these features and in the present configuration of the ridge crest. Transform faults should connect points that were together before these features were opened. If these predictions can be verified for features a few hundred kilometers apart, then similar observations for the ocean basins can be used with greater force as arguments for continental displacements comparable in size to the widths of these basins.

CONTINUATION OF THE MID-OCEANIC RIDGE SYSTEM INTO EAST AFRICA

Wiseman and Sewell (1937), Ewing and Heezen (1956), Matthews (1963), and Laughton (1966a, b) have shown that the East African rift valleys are connected to the mid-oceanic ridge by way of a median ridge in the Gulf of Aden. The identification of the Gulf of Aden and the East African rift valleys as portions of the mid-oceanic ridge system is supported by the similarity of morphological features in the three areas and by the continuity (Fig. 14) of a belt of earthquake epicenters (Rothé, 1954; Ewing and Heezen, 1956; Matthews, 1963; Sykes and Landisman, 1964). The East African rift valleys and the Red Sea and Gulf of Aden depressions are situated near the crests of broad zones of uplift that are similar in shape and extent to the bathymetric anomaly that defines the mid-oceanic ridge. Compressional velocities below the *M* discontinuity in the Gulf of Aden and in the central rift zone in the Red Sea are anomalously low and are similar to the velocities reported along many profiles near the crest of the mid-oceanic ridge (Drake and Girdler, 1964; Girdler, 1966; Laughton, 1966a, b). Heat flow in these regions (Von Herzen, 1963; Sclater, 1966; Von Herzen and Langseth, 1966) is anomalously high and is similar to that detected on many other portions of the ridge system. Linear magnetic anomalies, which are also prominent in the two areas (Girdler, 1966; Laughton, 1966a, b; Vine, 1966), suggest that the two features opened by growth of new sea floor. Hence, considerable evidence indicates that the Gulf of Aden and the central graben in the Red Sea represent extensions of the mid-oceanic ridge system. Crustal thicknesses in these regions are more typical of those found in oceanic areas than of those reported for continental regions. These observations support the idea that these features were formed by the horizontal separation of continental blocks.

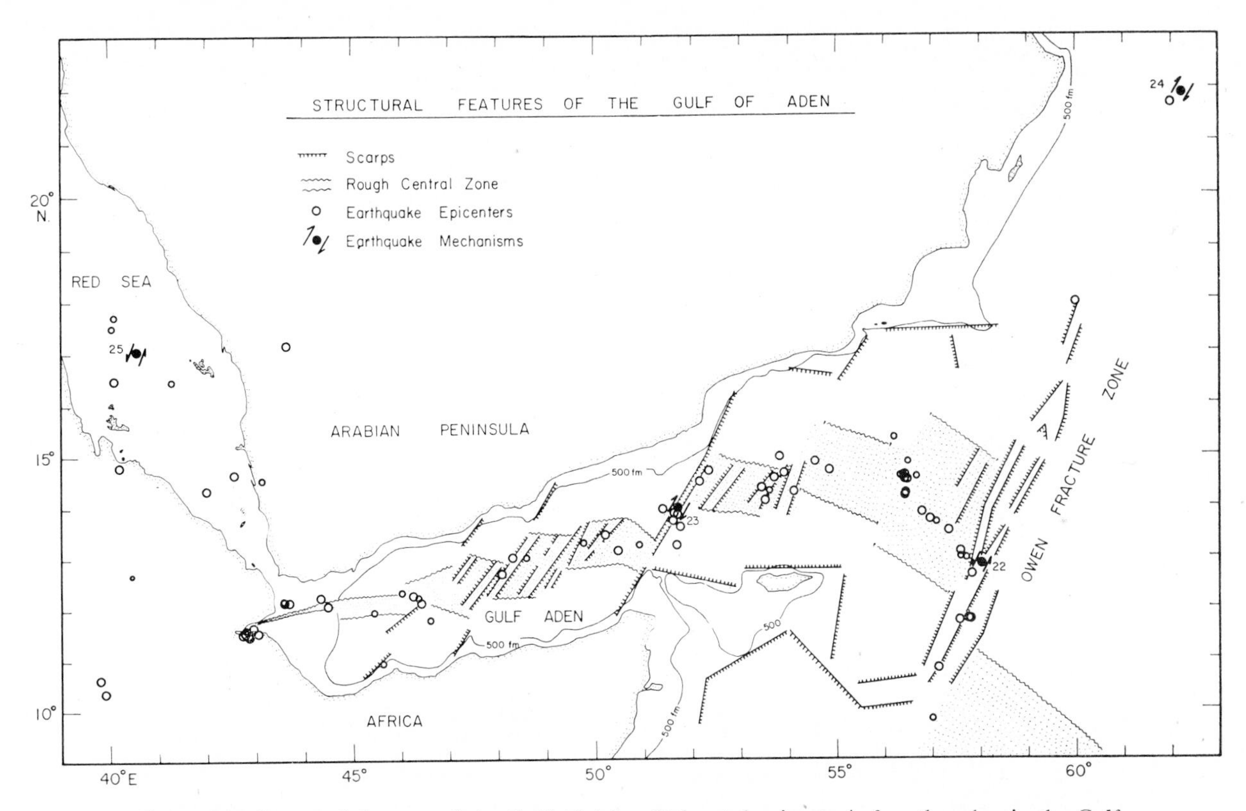

Figure 14. Structural features of the Gulf of Aden. Relocated epicenters of earthquakes in the Gulf of Aden for the period 1955-1966. Epicenters for period 1955 to 1963 from Sykes and Landisman (1964); more recent locations recomputed by author. Scarps and rough zone after Laughton (1966a). The position of the rough central zone is not precisely known between 54° E and 57° E because of insufficient bathymetric data. In this area the position of the rough central zone (indicated by question marks) was inferred from the locations of earthquakes. The Owen fracture zone continues to the NNE and passes near event 24.

Figure 14 illustrates several of the large-scale structural features in the Gulf of Aden as compiled by Laughton (1966a). The fault scarps striking NNE are some of the most prominent physiographic features in the Gulf; some of the deepest soundings in the Gulf occur in these fracture zones. Laughton (1966a) observed that these major fault zones do *not* continue inland on either side of the Gulf of Aden. The two margins of the Gulf (as denoted by the 500-fm contour) as well as the rough zone in the center of the Gulf exhibit a similar offset at these fracture zones. These observations and the data cited previously suggest that the fracture zones are transform faults and that the rough central zone is an active, growing ridge (Wilson, 1965a; Laughton, 1966b). Laughton (1966a, b) has shown that an extremely good fit is obtained when the continental margins on the two sides of the Gulf are matched at the 500-fm contour. Pre-Miocene geological features exhibit a remarkable continuity in his reconstruction (Fig. 15).

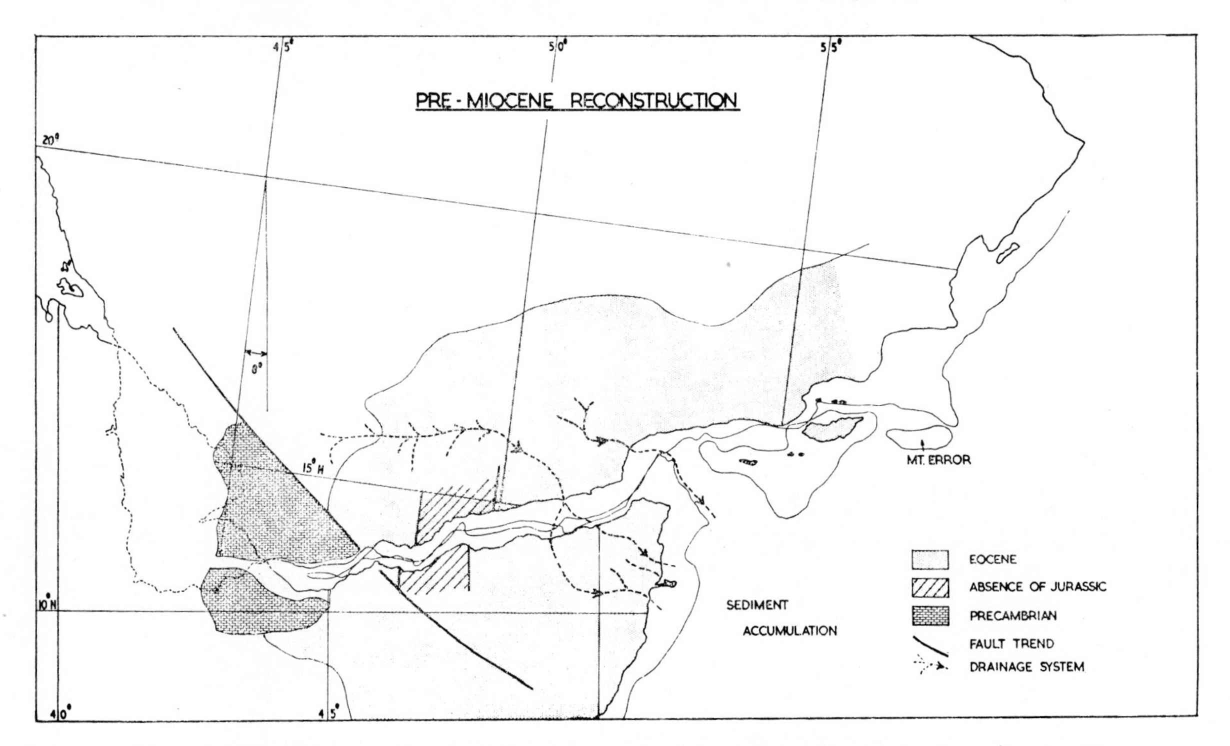

Figure 15. Laughton's (1966a) pre-Miocene reconstruction of the Gulf of Aden (reproduced with permission from *Phil. Trans. Roy. Soc. London, A*, vol. 259, p. 163).

The inferred mechanism for event 23, an earthquake in the Alula-Fartak trench, is in excellent agreement with Laughton's interpretations. The fracture zones within the Gulf of Aden appear to be a series of *dextral* transform faults. A similar fracture zone may intersect the ridge near 14°N, 54°E. Nearly all of the earthquakes within the Gulf are confined to the rough central zone or to portions of fracture zones between ridge crests.

OWEN FRACTURE ZONE

In the Arabian Sea (Fig. 14), however, the configuration of the ridge crest and the locations of earthquake foci (Matthews, 1963; 1966; Sykes and Landisman, 1964; Heezen and Tharp, 1965a) suggest that the Owen fracture zone is a *sinistral* transform fault in the region between the two crests of the Carlsberg Ridge (i.e., between 10°N, 57°E and 13°N, 58°E). The mechanism for event 22 is in agreement with this interpretation. Thus, the first motions for event 22 are of opposite polarity to those for event 23 at most of the stations that recorded the two earthquakes.

Although many of the earthquakes on the Owen fracture zone are situated between the crests of the mid-oceanic ridge, several other earthquakes have been observed along the northern and southern extensions of the fracture zone (Matthews, 1963; Sykes and Landisman, 1964; Stover, 1966). Wilson (1965a) suggested that the Carlsberg Ridge is spreading faster on the eastern side of the Owen fracture zone; hence, this differential spreading is being absorbed to the east of the northern termination of the Owen fracture zone in the Hindu Kush and Himalayas mountains. If so, earthquakes along this northern extension of the fracture zone should be characterized by a predominance of *sinistral* strike-slip motion.

The inferred mechanism for an earthquake on this portion of the Owen fracture zone (event 24), however, is characterized by a large component of *dextral* strike-slip motion. Although more data will be required to verify it, the following interpretation is suggested. The rate of spreading in the Gulf of Aden is now greater than the rate of spreading from the Carlsberg Ridge; hence, motion on the northern extension of the Owen fracture zone is similar to that indicated by event 24. Wilson (1965a) suggested that sinistral motion on this portion of the fracture zone may have been the dominant sense of motion for a long period of time prior to the development of the Gulf of Aden.

One piece of evidence suggesting a higher rate of spreading in the Gulf of Aden is the observation that the rate of heat flow is higher in the Gulf than it is on much of the Carlsberg Ridge (Von Herzen, 1963; Sclater, 1966; Von Herzen and Langseth, 1966).

RED SEA AND THE MOVEMENT OF THE ARABIAN PENINSULA

The strike of the inferred faulting for event 25, an earthquake in the Red Sea, is nearly the same as the strike of the major fracture zones in the Gulf of Aden and in much of the Indian Ocean (Heezen and Tharp, 1965a, b). Event 25 was not as well recorded as many of the other earthquakes studied in this paper. Hence, the results must be viewed with some reservation. The mechanism for the earthquake in the Red Sea suggests that this feature is also opening by sea floor spreading from a series of en echelon segments of ridge. The data for this solution, however, are also consistent with normal faulting on a plane approximately parallel to the axis of the Red Sea. The direction of motion on either side of the Red Sea is very similar to that proposed by Girdler (1966) on the basis of strike-slip movements along the Dead Sea—Aqaba rift, paleomagnetic measurements that suggest an anticlockwise rotation of Arabia (Irving and Tarling, 1961), and other geophysical and geological evidence. Compressional features in Iran, the earthquake mechanisms presented here, and other evidence cited by Girdler (1966) and Laughton (1966a, b) are consistent with a translational movement of the Arabian Peninsula to the NNE relative to Africa.

EAST AFRICA

The mechanisms of three earthquakes in East Africa all indicate a large component of normal faulting. The inferred axis of maximum tension for event 13, an earthquake in the western rift of East Africa on March 20, 1966, is nearly perpendicular to the strike of the rift in the vicinity of the epicenter (Fig. 9). Although evidence for transform faults within East Africa was not found in this study, Bloomfield (1966) suggested that a transform fault may connect two portions of the Malawi rift. Mohr (1967) indicated that a series of faults in the Ethiopian rift may be transform faults.

East Pacific Rise and the Gulf of California

EAST PACIFIC RISE

The inferred mechanisms for three earthquakes on the East Pacific Rise and five events in the Gulf of California are illustrated in Figures 16 and 17. Each of the three shocks on the East Pacific Rise was located on a major fracture zone. The mechanisms

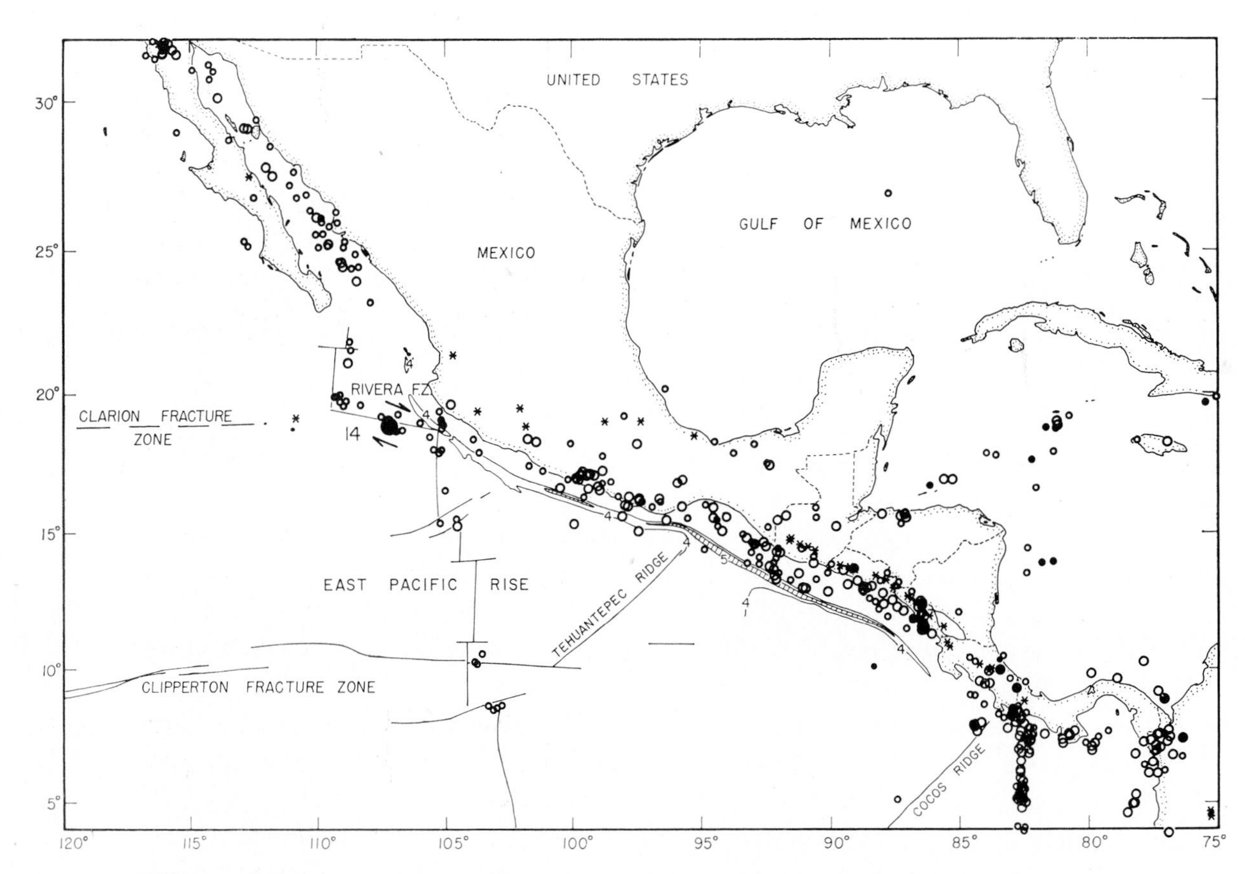

Figure 16. Epicenters of shallow earthquakes (depth less than about 70 km) in Mexico, Central America, and adjacent oceanic areas for the period 1954-1962 (after Sykes, 1967b). Bathymetry of Middle America trench after Fisher (1961); water depths in km. Fracture zones and crests of East Pacific Rise after Menard (1966). Mechanism solution indicated for event 14 on the Rivera fracture zone. Asterisks indicate historically active volcanoes (after Gutenberg and Richter, 1954); circles, epicenters.

of the eight earthquakes are all characterized by a predominance of strike-slip motion. Solutions with a large component of dip slip were not detected on the East Pacific Rise. Event 14 was located off the west coast of Mexico on the Rivera fracture zone (Fig. 16); event 15 occurred on the Easter fracture zone near 27°S, 114°W. In both cases the inferred mechanisms, the bathymetry (Menard, 1966), and the seismicity (Sykes, 1963; Sykes, 1967b) are indicative of dextral transform faults.

Because of its small size and remote location, it was not possible to obtain a unique solution for event 16, an earthquake on the Eltanin fracture zone. However, if strike-slip faulting on a steeply dipping plane with the same strike as the fracture zone is assumed, then the sense of motion can be inferred from the first motions. The bathymetry (Menard, 1964), seismicity (Sykes, 1963), and mechanism are oriented in the correct sense for a sinistral transform fault.

It is interesting that the pattern of epicenters and the inferred mechanism of an earthquake on the Rivera fracture zone are rotated about 25° with respect to the strike of the Clarion zone further west (Fig. 16). Menard (1966) and Chase and Menard (personal communication) deduced a similar strike for the Rivera fracture zone. The configuration of the active seismic zone and the mechanism for event 14 support Menard's (1966) interpretation that the Rivera zone and the Clarion fracture zone may be distinct tectonic features. The data for the Rivera fracture zone and the solutions for earthquakes in the Gulf of California (Fig. 17) support Vine's (1966) contention that the direction of spreading from the East Pacific Rise has changed from east-west to approximately northwest-southeast.

GULF OF CALIFORNIA

The northern and southern extensions of the San Andreas system of faults have long been an enigma because the fault system extends into water-covered areas at either end. Recently a system of mid-oceanic ridges and fracture zones has been recognized off the coasts of British Columbia, Washington, Oregon, and northern California (Menard, 1964; Talwani et al., 1965; Wilson, 1965b; Vine and Wilson, 1965). Although no solutions of earthquake mechanisms were made for shocks in this region, an analysis of future events should be of great value in deciphering the tectonic interaction of these features both with one another and with the San Andreas system.

Seismic activity in the Gulf of California has often been described as marking the southern extension of the San Andreas fault system. Kovach et al. (1962) and Biehler et al. (1964), however, have shown that individual faults near the northern end of the Gulf of California trend more easterly than the axis of the Gulf, and thus they do not appear to continue into the center of the Gulf. Rusnak et al. (1964) reported a series of en echelon faults that trend across the axis of the Gulf. These scarps were interpreted as dextral strike-slip faults arranged en echelon to the San Andreas fault. A series of deep troughs in the Gulf that strike northeast were interpreted as tensional features. Hamilton (1961), Rusnak et al. (1964), Rusnak and Fisher (1964), and others have suggested that the Gulf of California opened by large-scale horizontal motion. In their reconstructions the tip of Baja California is placed in the vicinity of 21°N, 106°W near the indented coastline of the Mexican states of Jalisco and Nayarit.

Crustal structure in the central and southern portions of the Gulf is similar to that reported for the east Pacific, but the structure of the northern Gulf is more nearly continental in character and is similar to that described for the continental borderland off southern California (Phillips, 1964). Thus, with the possible exception of its northern part, the structure of the Gulf does not resemble that of a graben of continental material that has been depressed below sea level. High heat flow in the Gulf (Von Herzen, 1963), the continuity of the belt of earthquake epicenters (Figs. 16 and 17), and the continuity of morphologic and structural features (Menard, 1955, 1964, 1966) indicate that the tectonic setting of the Gulf of California, the San Andreas system, and the ridges north of the Mendocino fracture zone is closely related to the tectonics of the East Pacific Rise. These characteristics and the absence of intermediate and deep-focus earthquakes between central Alaska and central Mexico (Gutenberg and Richter, 1954; Girdler, 1964; Tobin and Sykes, 1966) indicate that the present tectonics of this portion of western North America is distinctly different from the tectonics of island arcs that border much of the western Pacific.

Structural features, earthquake epicenters, and earthquake mechanisms in the Gulf of California are illustrated in Figure 17. The locations of major scarps in the Gulf striking northwest and prominent troughs striking northeast were inferred from the chart "Submarine topography of Gulf of California" (Fisher et al., 1964). The six mechanisms illustrated in Figure 17 are each characterized by a predominance of strike-slip motion. The northwesterly striking plane was chosen as the fault plane in each case. These mechanisms as well as the fact that most earthquakes in this area are confined to the Gulf itself suggest that the faults striking northwest are transform faults. The troughs striking northeast are thought to be the loci of either the growth of new sea floor or crustal material that is being thinned. The interpretation of the northwesterly striking faults as transform faults removes the objection of Biehler et al. (1964) that the en echelon arrangement of faults does not appear to be consistent with right-lateral displacements across the zone defined by the Gulf of California and Salton trough. If the northwesterly trending scarps represent transform faults, their strike suggests that Baja California has moved about 300 to 400 km to the northwest from a position near the indented coastline of the mainland of Mexico. The right-lateral motion detected for event 30 is in agreement with observed displacements near the head of the Gulf of California and in the Imperial Valley (Biehler et al., 1964). The present direction of ocean floor spreading in the northeastern Pacific and in the Gulf of California as inferred from magnetic anomalies (Vine and Wilson, 1965; Vine, 1966) and earthquake mechanisms (Fig. 17) supports Wilson's (1965a, b) contention that much of this spreading is being accommodated by right-lateral motion along the San Andreas.

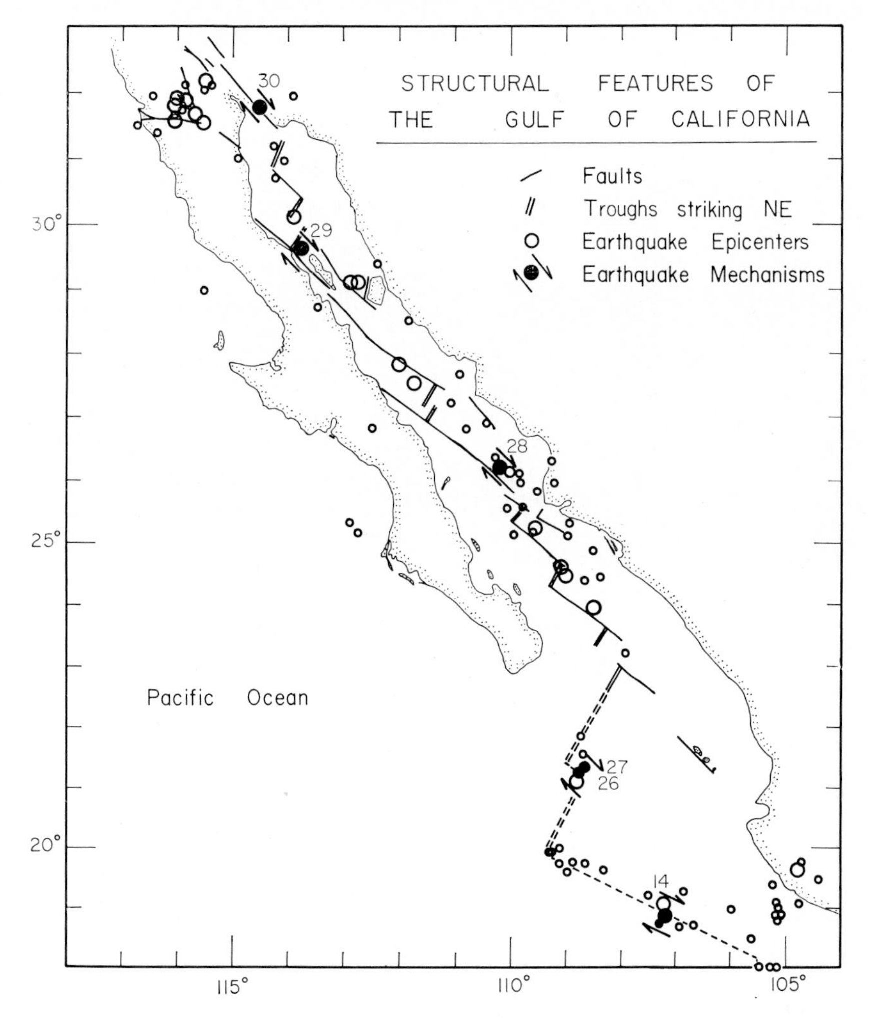

Figure 17. Structural features of the Gulf of California. Earthquake epicenters recomputed for period 1954-1962. Interpretations of major faults and troughs based on chart of submarine topography of the Gulf of California (Fisher et al., 1964). Deep troughs striking northeast are interpreted as tensional features that mark sites of growth of new sea floor.

Macquarie Ridge

In the previous solutions emphasis was placed on examining data for features that are generally recognized as mid-oceanic ridges or the continental extensions of this ridge system. Seismically active features such as the Macquarie, West Chile, and Azores-Gibraltar ridges have not been universally interpreted either as mid-oceanic ridges or as fracture zones (Heezen et al., 1959; Menard, 1965, 1966).

The inferred mechanism for event 17, an earthquake located south of New Zealand on the Macquarie Ridge (Figs. 18 and 19) is unlike any of the other solutions described in this paper. Although interpretations involving greater or lesser amounts of strike-slip motion than that shown in Figure 19 are possible, still a large component of thrust

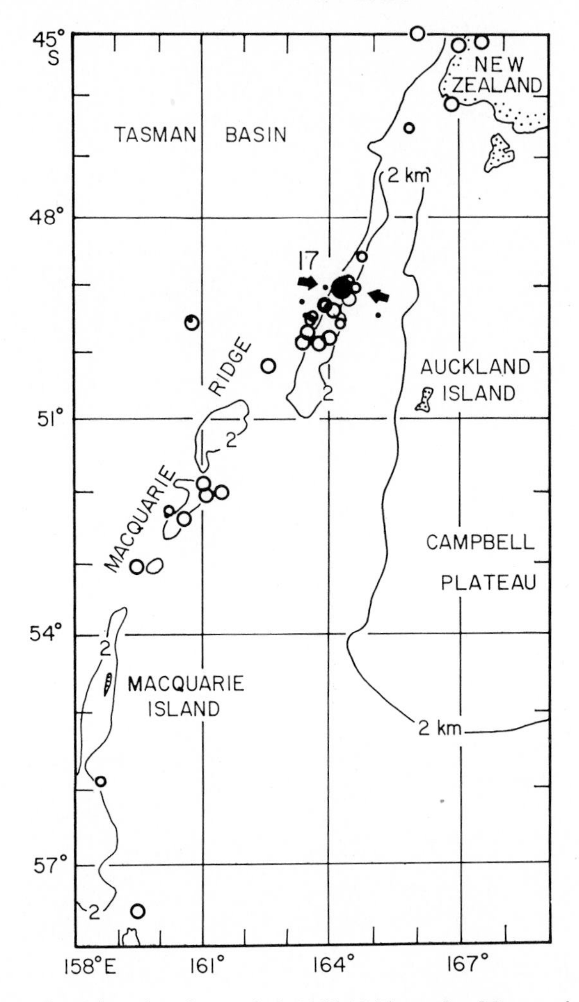

Figure 18. Epicenters of earthquakes for period 1955-1965 on the Macquarie Ridge. Inferred axis of maximum compression shown for event 17. Bathymetry after Brodie and Dawson (1965). Other symbols same as Figure 10.

faulting seems to be demanded by the compressional arrivals at the more distant stations.

The inferred compressional axis for this solution is nearly horizontal and is approximately perpendicular to the trend of the Macquarie Ridge (Brodie and Dawson, 1965) and to the strike of the seismic belt (Sykes, 1963; Cooke, 1966). The orientation of the compressional axis is nearly the same as the direction of principal horizontal stress deduced by Lensen (1960) for southern New Zealand. This suggests that the tectonics of this portion of the Macquarie Ridge may be more similar to that of New Zealand than to the tectonics of the mid-oceanic ridge. This portion of the Macquarie Ridge does not appear to possess a series of parallel, linear magnetic anomalies which seem to be characteristic of spreading ridges in other oceanic areas (Hatherton, 1967).

Conclusions and Discussion

Because of several remarkable improvements in seismic instrumentation and in the availability of this data, mechanism solutions of high precision can now be obtained for

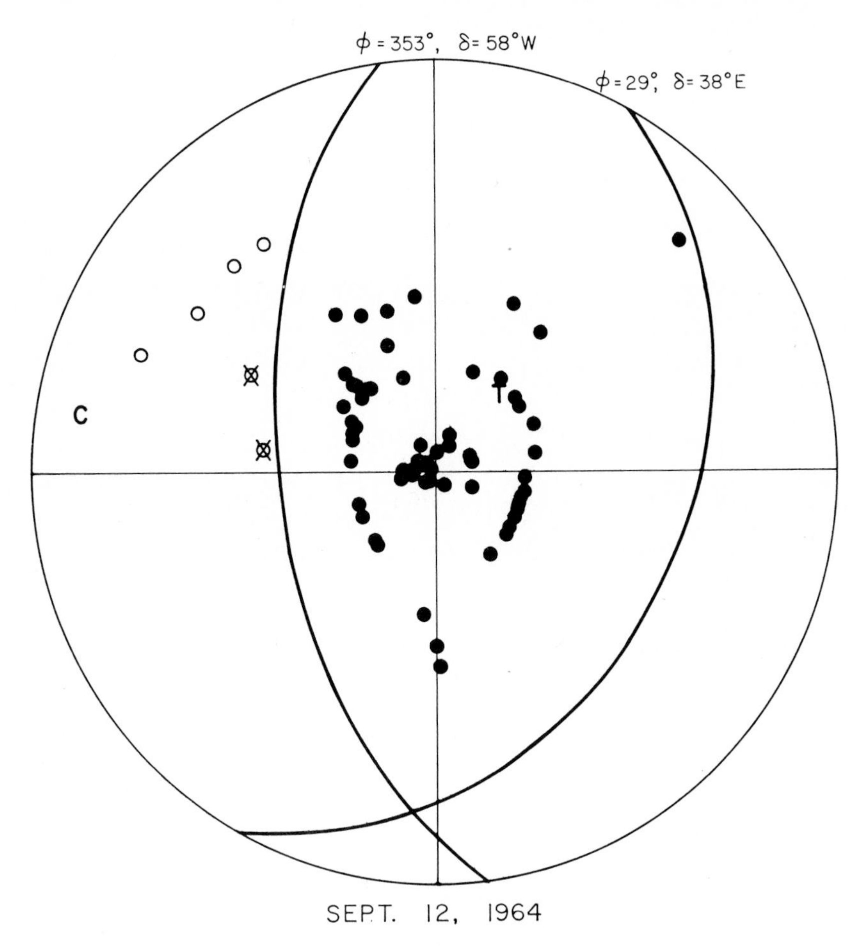

Figure 19. Mechanism solution for event 17 on the Macquarie Ridge (Figures 2 and 18). Symbols same as Figures 3 and 7.

many areas of the world. Earthquakes on the mid-oceanic ridge and the extensions of this ridge system into East Africa, western North America, and the Arctic seem to be characterized by one or the other of two principal mechanisms: (1) a predominance of strike-slip motion on steeply dipping planes or (2) a large component of normal faulting with the inferred axis of tension approximately perpendicular to the ridge.

All of the events in this paper that are characterized by a predominance of strike-slip motion were located on major fracture zones that intersect the crest of the ridge. In each case the inferred sense of displacement is in agreement with that predicted for transform faults; the motion is opposite to that expected for a simple offset of the ridge. Thus, displacements inferred by the use of the assumption of simple offset appear to be incorrect. The events that are characterized by a large component of normal faulting were situated on the ridge system in regions where fracture zones have not been found. Although the East African rift valleys have been cited as examples of compressional, extensional, or strike-slip tectonics, more recent discussions have centered largely on extensional mechanisms (Davies, 1951; Girdler, 1963; Heezen and Ewing, 1963; Holmes, 1965). This interpretation became more prevalent with the realization that the mid-oceanic ridge system existed on a worldwide scale (Ewing and Heezen, 1956).

Seismic activity on fracture zones is confined almost exclusively to the region between the two crests of the ridge. This distribution of activity is a strong argument in favor of the hypothesis of transform faults. Information obtained from earthquake mechanisms and seismicity is, of course, limited to a time scale of a few or a few tens of years for the mid-oceanic ridge. Because a very similar behavior was found for all of the fracture zones and ridge crests investigated in this paper, the results very likely closely approximate the tectonics of the ridge system for much longer periods of time. This view is supported by the similarity of the shape of the ridge crest and the shapes of the continental margins in the Gulf of Aden and in the Gulf of California. These gulfs appear to have opened by a process of sea floor spreading and transform faulting.

As Wilson (1965a) has pointed out, transform faults can only exist if there is crustal displacement. The deduced mechanisms and the distribution of earthquakes both seem to demand a process of sea floor growth at or near the crest of the mid-oceanic ridge system. Transform faults also provide a relatively simple mechanism for continental drift. Thus, the results presented in this paper as well as other evidence for sea floor spreading seem to be indicative of an earth that is much more mobile than many geologists and geophysicists had previously realized.

Unfortunately, neither the seismological evidence nor the magnetic evidence for sea floor growth gives information directly pertinent to the tectonics of the ridge in the third dimension, depth. Seismic activity along the ridge system seems to be confined to the upper 50 or 100 km of the earth. When focal depths can be determined with greater accuracy, these events may prove to be even shallower.

The close spacing of fracture zones in the Gulf of Aden, the Gulf of California, and the equatorial Atlantic suggests that an independent convection current is not present between each set of fracture zones. Likewise, it seems unlikely that the short segments of new ridge in the Gulf of California can be considered to "drive" the long faults that connect these segments. Rather, the complex pattern at the surface of the earth may represent the response of the crust (and perhaps the uppermost part of the mantle) to a less complex process at depth in the mantle. The earthquake mechanisms on the ridge system suggest that the upper layers of the earth are being pulled apart at the ridge crest rather than being pushed apart by the forceful injection of materials. The surficial evidence also indicates that the en echelon arrangement of various segments of ridge in the Gulfs of Aden and California has existed since the initiation of spreading in these areas. In a given region the sense of motion on various transform faults is often of the same sense. This consistency in the arrangement of ridge crests indicates that the positioning of these segments of ridge is not random, but is controlled by some large-scale system of forces in the mantle. The distribution of stresses and displacements in three dimensions remains largely unresolved.

Investigations of the mechanism and distribution of earthquakes in island arcs may provide information about these tectonic processes in the mantle. Although there seems to be a considerable variation in the orientation of earthquake mechanisms in island arcs, one factor that seems to be common among many of the solutions for these regions is that the horizontal component of compression is approximately perpendicular to the strike of the arc (Balakina et al., 1960; Lensen, 1960; Ichikawa, 1961; Honda, 1962; Balakina, 1962; Ritsema, 1964; Isacks and Sykes, in preparation). Thus, results from investigations of the mechanisms of earthquakes seem to be indicative of a system of compressional tectonics in island arcs and of extensional tectonics along the mid-oceanic ridges.

Acknowledgments

I thank Drs. J. Oliver and P. Pomeroy of the Lamont Geological Observatory for critically reading the manuscript. Mr. J. E. Brock and Mr. R. L. Laverty helped assemble the data for analysis and computation. This study would not have been possible without the high-quality data that are now available from the stations of the World-Wide Standardized Seismograph Network of the U.S. Coast and Geodetic Survey, ESSA; the Canadian seismograph network, and the institutions that have cooperated with the Lamont Observatory in the operation of long-period instruments. Dr. P. W. Pomeroy kindly made records available from new stations in Abéché, Chad, and Lamto, Ivory Coast; Dr. B. L. Isacks supplied similar records from a new network of stations in the Fiji-Tonga region of the southwest Pacific; Mr. R. J. Halliday provided microfilm copies of Canadian records. Mr. James Lander of the Coast and Geodetic Survey provided a list of earthquakes in Mexico and Central America that was a valuable aid in selecting the earthquakes to be relocated. Portions of this paper and several of the figures were reproduced from the article "Mechanisms of earthquakes and nature of faulting on the mid-oceanic ridges" (Sykes, 1967a). I am grateful to the *Journal of Geophysical Research* for permission to reproduce that material.

Computing facilities were made available by the NASA Goddard Space Flight Center, Institute for Space Studies, New York. This study was partially supported under Dept. of Commerce Grants C-348-65-(G) and E-22-96-67(G) from the Environmental Science Services Administration and through the Air Force Cambridge Research Laboratories, Office of Aerospace Research, under contract AF19(628)-4082 as part of Project Vela-Uniform of the Advanced Research Projects Agency.

Earth Sciences Laboratories, ESSA
Lamont Geological Observatory
Columbia University

Discussion

DR. ANDERSON: What offset the ridges in the equatorial Atlantic in the first place?

DR. SYKES: I don't know. The only thing I can suggest is that you can have a series of en echelon features which are determined by the stress distribution. There is no requirement that two points on the ridge which are apparently offset by a fracture zone were ever together.

DR. VINE: You are getting into the problem of what is happening in the third dimension. With just surface evidence, the best you can do is say that the pattern observed is simply a superficial accommodation to a change in direction of the underlying presumed convection cell. If you fit the margins of South America and Africa, as we all know, then the locus of this fit, and therefore presumably the locus of initiation of drift and of the mythical underlying convection cell, had this curved form.

DR. DEWEY: Are you justified in taking the nodal planes orthogonal to each other?

DR. SYKES: I have looked into this. It appears that you can fit some of the dip-slip solutions a little better with 70°. For the earthquakes on the fracture zones the east-west plane is very well defined. It is well defined, whether you demand the other plane be orthogonal to it or not. I certainly hope we can get enough data so we can look at this question further and determine whether the angle is 90° or a little less. This would not influence my main arguments.

REFERENCES

Allen, C. R. (1965), Transcurrent faults in continental areas, *Phil. Trans. Roy. Soc. Lond. A* 258:82-89.

Balakina, L. M. (1962), General regularities in the directions of the principal stresses effective in the earthquake foci of the seismic belt of the Pacific Ocean, *Bull. Acad. Sci. U.S.S.R., Geophys. Ser., English Transl.*, No. 11, 918-926.

Balakina, L. M., H. I. Shirokova, and A. V. Vvedenskaya (1960), Study of stresses and ruptures in earthquake foci with the help of dislocation theory, *Publ. Dominion Obs. Ottawa* 24:321-27.

Biehler, Shawn, R. L. Kovack, and C. R. Allen (1964), Geophysical framework of the northern end of the Gulf of California structural province, pp. 126-43 in *Marine Geology of the Gulf of California—A Symposium*, American Association of Petroleum Geologists, Mem. 3.

Bloomfield, K. (1966), A major east-north-east dislocation zone in central Malawi, *Nature* 211: 612-14.

Bodvarsson, G., and G. P. L. Walker (1964), Crustal drift in Iceland, *Geophys. J.* 8:285-300.

Brodie, J. W., and E. W. Dawson (1965), Morphology of North Macquarie ridge, *Nature* 207: 844-45.

Cann, J. R., and F. J. Vine (1966), An area on the crest of the Carlsberg ridge: petrology and magnetic survey, *Phil. Trans. Roy. Soc. Lond. A* 259:198-217.

Cooke, R. J. S. (1966), Some seismological features of the North Macquarie ridge, *Nature* 211: 953-54.

Davies, K. A. (1951), The Uganda section of the Western Rift, *Geol. Mag.* 88:377-85.

Dietz, R. S. (1961), Continent and ocean basin evolution by spreading of the sea floor, *Nature* 190:854-57.

Drake, C. L., N. J. Campbell, G. Sander, and J. E. Nafe (1963), A mid-Labrador Sea ridge, *Nature* 200:1085-86.

Drake, C. L., and R. W. Girdler (1964), A geophysical study of the Red Sea, *Geophys. J.* 8:473-95.

Ewing, M., and B. C. Heezen (1956), Some problems of Antarctic submarine geology, pp. 75-81 in *Antarctica in the International Geophysical Year*, A. Crary et al., eds., Geophysical Monograph 1, American Geophysical Union, Washington, D.C.

Fisher, R. L. (1961), Middle America trench: topography and structure, *Geol. Soc. Amer. Bull.* 72:703-20.

Fisher, R. L., G. A. Rusnak, and F. P. Shepard (1964), Submarine topography of the Gulf of California, chart accompanying *Marine Geology of the Gulf of California*, T. H. Van Andel and G. G. Shor, eds., Amer. Assoc. Petroleum Geologists, Tulsa, Oklahoma.

Girdler, R. W. (1963), Geophysical studies of rift valleys, *Physics Chem. Earth* 5:121-56.

Girdler, R. W. (1964), How genuine is the circum-Pacific belt, *Geophys. J.* 8:537-40.

Girdler, R. W. (1966), The role of translational and rotational movements in the formation of the Red Sea and Gulf of Aden, pp. 65-77 in *the World Rift System*, T. N. Irvine, ed., Geol. Soc. Can., Pap. 66-14.

Gutenberg, B., and C. F. Richter (1954), *Seismicity of the Earth*, 2nd edn., Princeton University Press.

Hamilton, W. (1961), Origin of the Gulf of California, *Geol. Soc. Amer. Bull.* 72:1307-18.

Harding, S. T., and W. Rinehart (1966), Preliminary seismological report, pp. 1-16, in *The Parkfield, California, Earthquake of June 27, 1966*, Coast and Geodetic Survey, ESSA, U.S. Govt. Printing Office, Washington, D.C.

Hatherton, T. (1967), Total magnetic force measurements over the North Macquarie ridge and Solander trough, *N.Z. J. Geol. Geophys.* 10 (3).

Heezen, B. C., E. T. Bunce, J. B. Hersey, and M. Tharp (1964a), Chain and Romanche fracture zones, *Deep-Sea Res.* 11:11-33.

Heezen, B. C., and M. Ewing (1963), The mid-oceanic ridge, pp. 388-410 in *The Sea*, Vol. 3, M. N. Hill, ed., Wiley-Interscience, New York.

Heezen, B. C., R. D. Gerard, and M. Tharp (1964b), The Vema fracture zone in the equatorial Atlantic, *J. Geophys. Res.* 69:733-39.

Heezen, B. C., and M. Tharp (1961), Physiographic diagram of the South Atlantic Ocean, the Caribbean Sea, the Scotia Sea, and the eastern margin of the South Pacific Ocean, Geological Society of America, New York.

Heezen, B. C., and M. Tharp (1965a), Descriptive sheet to accompany physiographic diagram of the Indian Ocean, Geological Society of America, New York.

Heezen, B. C., and M. Tharp (1965b), Tectonic fabric of the Atlantic and Indian oceans and continental drift, *Phil. Trans. Roy. Soc. Lond. A* 258:90-106.

Heezen, B. C., M. Tharp, and M. Ewing (1959), The floors of the oceans: I. the North Atlantic, *Geol. Soc. Amer., Spec. Paper* 65, pp. 1-122.

Heirtzler, J. R., X. LePichon, and J. G. Baron (1966), Magnetic anomalies over the Reykjanes ridge, *Deep-Sea Res.* 13:427-43.

Hess, H. H. (1955), The oceanic crust, *J. Mar. Res.* 14:423-39.

Hess, H. H. (1962), History of ocean basins, pp. 599-620 in *Petrologic Studies: A Volume to Honor A. F. Buddington*, A. E. J. Engel et al., eds., Geological Society of America, New York.

Hodgson, J. H. (1957), Nature of faulting in large earthquakes, *Geol. Soc. Amer. Bull.* 68:611-44.

Hodgson, J. H., and W. M. Adams (1958), A study of inconsistent observations in the fault plane project, *Bull. Seismol. Soc. Amer.* 48:17-31.

Holmes, A. (1965), *Principles of Physical Geology*, 2nd ed., Nelson, London, 1260 pp.

Honda, H. (1962), Earthquake mechanism and seismic waves, *J. Phys. Earth (Tokyo)* 10:1-97.

Ichikawa, M. (1961), On the mechanism of earthquakes in and near Japan during the period from 1950 to 1957, *Geophys. Mag. (Tokyo)* 30:355-403.

Irving, E., and D. H. Tarling (1961), Palaeomagnetism of the Aden volcanics, *J. Geophys. Res.* 66: 549-56.

Isacks, B. L., and L. R. Sykes (in preparation), Focal mechanism of deep and shallow earthquakes in the Tongan and Kermadec island arcs.

Kovach, R. L., C. R. Allen, and F. Press (1962), Geophysical investigations in the Colorado Delta region, *J. Geophys. Res.* 67:2845-71.

Laughton, A. S. (1966a), The Gulf of Aden, *Phil. Trans. Roy. Soc. Lond. A* 259:150-71.

Laughton, A. S. (1966b), The Gulf of Aden in relationship to the Red Sea and the Afar depression of Ethiopia, in *The World Rift System*, T. N. Irvine, ed., Geol. Surv. Can., Pap. 66-14, Ottawa.

Lensen, G. J. (1960), Principal horizontal stress directions as an aid to the study of crustal deformation, *Publ. Dominion Obs. Ottawa* 24:389-97.

Loupekine, I. S. (1966), Uganda, The Toro earthquake of 20 March 1966, pp. 1-38 UNESCO (earthquake reconnaissance mission), Paris.

McEvilly, T. V. (1966), Parkfield earthquakes of June 27-29, 1966, Monterey and San Luis Obispo Counties, California-Preliminary report: Preliminary seismic data, June-July 1966, *Bull. Seismol. Soc. Amer.* 56:967-71.

Matthews, D. H. (1963), A major fault scarp under the Arabian Sea, displacing the Carlsberg Ridge near Socotra, *Nature* 198:950-52.

Matthews, D. H. (1966), The Owen fracture zone and the northern end of the Carlsberg Ridge, *Phil. Trans. Roy. Soc. Lond. A.* 259:172-86.

Menard, H. W. (1955), Deformation of the northeastern Pacific basin and the west coast of North America, *Geol. Soc. Amer. Bull.* 66:1149-98.

Menard, H. W. (1964), *Marine Geology of the Pacific*, McGraw-Hill, New York.

Menard, H. W. (1965), Sea floor relief and mantle convection, *Phys. Chem. Earth* 6:315-64.

Menard, H. W. (1966), Fracture zones and offsets of the East Pacific Rise, *J. Geophys. Res.* 71: 682-85.

Mohr, P. A. (1967), Major volcano-tectonic lineament in the Ethiopian rift system, *Nature*, 664-65.

Phillips, R. P. (1964), Seismic refraction studies in Gulf of California, pp. 90-121 in *Marine Geology of the Gulf of California*, T. H. Van Andel and G. G. Shor, eds., Amer. Assoc. Petroleum Geologists, Tulsa, Oklahoma.

Pitman, W. C., and J. R. Heirtzler (1966), Magnetic anomalies over the Pacific-Antarctic Ridge, *Science* 154:1164-71.

Quon, S. H., and E. G. Ehlers (1963), Rocks of the northern part of the mid-Atlantic ridge, *Geol. Soc. Amer. Bull.* 74:1-8.

Ritsema, A. R. (1964), Some reliable fault plane solutions, *Pure Appl. Geophys.* 59:58-74.
Romney, C. (1957), Seismic waves from the Dixie Valley–Fairview Peak earthquakes, *Bull. Seismol. Soc. Amer.* 47:301-20.
Rothé, J. P. (1954), La zone seismique mediane Indo-Atlantique, *Proc. Roy. Soc. Lond. A* 222: 387-97.
Rusnak, G. A., and R. L. Fisher (1964), Structural history and evolution of Gulf of California, pp. 144-56 in *Marine Geology of the Gulf of California*, T. H. Van Andel and G. G. Shor, eds., Amer. Assoc. Petroleum Geologists, Tulsa, Oklahoma.
Rusnak, G. A., R. L. Fisher, and F. P. Shepard (1964), Bathymetry and faults of Gulf of California, pp. 59-75 in *Marine Geology of the Gulf of California*, T. H. Van Andel and G. G. Shor, eds., Amer. Assoc. Petroleum Geologists, Tulsa, Oklahoma.
Sclater, J. G. (1966), Heat flow in the northwest Indian Ocean and Red Sea, *Phil. Trans. Roy. Soc. Lond. A* 259:271-78.
Shand, S. J. (1949), Rocks of the mid-Atlantic ridge, *J. Geol.* 57:89-92.
Stauder, W., S.J. (1960), The Alaska earthquake of July 10, 1958; seismic studies, *Bull. Seismol. Soc. Amer.* 50:293-322.
Stauder, W., S.J., and G. A. Bollinger (1966), The S-wave project for focal mechanism studies, earthquakes of 1963, *Bull. Seismol. Soc. Amer.* 56:1363-71.
Stover, C. W. (1966), Seismicity of the Indian Ocean, *J. Geophys. Res.* 71:2575-81.
Sykes, L. R. (1963), Seismicity of the South Pacific Ocean, *J. Geophys. Res.* 68:5999-6006.
Sykes, L. R. (1965), The seismicity of the Arctic, *Bull. Seismol. Soc. Amer.* 55:501-18.
Sykes, L. R. (1966), The seismicity and deep structure of island arcs, *J. Geophys. Res.* 71:2981-3006.
Sykes, L. R. (1967a), Mechanism of earthquakes and nature of faulting on the mid-oceanic ridges, *J. Geophys. Res.* 72:2131-53.
Sykes, L. R. (1967b), Seismicity and earth structure in Central America and Mexico (in preparation).
Sykes, L. R., and M. Ewing (1965), The seismicity of the Caribbean region, *J. Geophys. Res.* 70: 5065-74.
Sykes, L. R., and M. Landisman (1964), The seismicity of East Africa, the Gulf of Aden, and the Arabian and Red seas, *Bull. Seismol. Soc. Amer.* 54:1927-40.
Talwani, M. (1964), A review of marine geophysics, *Marine Geol.* 2:29-80.
Talwani, M., X. LePichon, and J. R. Heirtzler (1965), East Pacific rise: the magnetic pattern and the fracture zones, *Science* 150:1109-15.
Tilley, C. E. (1947), The dunite-mylonites of St. Paul's Rocks (Atlantic), *Amer. J. Sci.* 245:483-91.
Tobin, D. G., and L. R. Sykes (1966), Relationship of hypocenters of earthquakes to the geology of Alaska, *J. Geophys. Res.* 71:1659-67.
Tolstoy, I. (1951), Submarine topography in the North Atlantic, *Geol. Soc. Amer. Bull.* 62:441-50.
Tolstoy, I., and M. Ewing (1949), North Atlantic hydrography and the mid-Atlantic ridge, *Geol. Soc. Amer. Bull.* 60:1527-40.
Vacquier, V. (1962), Magnetic evidence for horizontal displacements in the floor of the Pacific Ocean, pp. 135-144 in *Continental Drift*, S. K. Runcorn, ed., Academic Press, New York and London.
Vine, F. J. (1966), Spreading of the ocean floor: new evidence, *Science* 154:1405-15.
Vine, F. J., and D. H. Matthews (1963), Magnetic anomalies over oceanic ridges, *Nature* 199:947-49.
Vine, F. J., and J. T. Wilson (1965), Magnetic anomalies over a young oceanic ridge off Vancouver Island, *Science* 150:485-89.
Von Herzen, R. P. (1963), Geothermal heat flow in the Gulfs of California and Aden, *Science* 140: 1207-08.
Von Herzen, R. P., and M. G. Langseth (1966), Present status of oceanic heat-flow measurements, *Phys. Chem. Earth* 6:365-407.
Washington, H. S. (1930a), The origin of the mid-Atlantic ridge, *J. Maryland Acad. Sci.* 1:20-29.
Washington, H. S. (1930b), The petrology of St. Paul's Rocks (Atlantic), pp. 126-44, in *Report on the Geological Collections made during the voyage of the 'Quest,'* British Museum, London.
Wilson, J. T. (1965a), A new class of faults and their bearing on continental drift, *Nature* 207: 343-47.
Wilson, J. T. (1965b), Transform faults, oceanic ridges and magnetic anomalies southwest of Vancouver Island, *Science* 150: 482-85.
Wilson, J. T. (1965c), Evidence from ocean islands suggesting movement in the Earth, *Phil. Trans. Roy. Soc. Lond. A* 258:145-67.
Wiseman, J. D. H. (1966), St. Paul Rocks and the problem of the upper mantle, *Geophys. J.* 11: 519-25.
Wiseman, J. D. H., and R. B. S. Sewell (1937), The floor of the Arabian Sea, *Geol. Mag.* 74:219-31.
Wohlenberg, J. (1966), Remarks on the Uganda earthquake of March 20, 1966 (Part II), *Chronique de L'IRSAC* (Institut pour la Recherche Scientifique en Afrique Centrale, Lwiro, Rep. Congo) 1(no. 2):7-12.

3. EVIDENCE FROM THE CONTINENTS

REVIEW OF AGE DATA IN WEST AFRICA AND SOUTH AMERICA RELATIVE TO A TEST OF CONTINENTAL DRIFT

P. M. HURLEY J. R. RAND

INTRODUCTION

The hypothesis of continental drift was first studied intensively by geologists who utilized the matching of Phanerozoic geologic ages, ecology, and structures as their evidence. The advent of radiometric age dating techniques extends the range of time over which correlations can be tested, and new interest in the matching of pre-Cambrian age provinces has arisen (Miller, 1965; Almeida et al., 1966; Hurley et al., 1966; Hart et al., 1966).

In this review we have adjusted the coordinates for the separate continents according to a pre-drift reconstruction of Bullard et al. (1965). We are concerned only with the basement age provinces in West Africa, Brazil, the Guianas, and Venezuela, so that information beyond these geographical limits, and in the Phanerozoic, will be ignored. We have also not included lead-alpha and model and chemical lead ages (except in the schematic presentation in Fig. 1) to avoid a mixing of data that becomes difficult to interpret, although it is recognized that these methods gave important early indications of the general age provinces in Africa (Holmes and Cahen, 1957; Cahen, 1961). The combination of mineral ages by the K-Ar and Rb-Sr methods and whole-rock ages by the Rb-Sr method is considered to present the best information on metamorphic assemblages of all grades, and these data are indicated separately in the figures by the notations. All age values are presented on the basis of $\lambda\ (Rb^{87}) = 1.39 \times 10^{-11}\ yr^{-1}$; $\lambda_e\ (K^{40}) = 0.584 \times 10^{-10}\ yr^{-1}$; $\lambda_\beta\ (K^{40}) = 4.72 \times 10^{-10}\ yr^{-1}$.

THE WEST AFRICAN CRATON

The pre-Cambrian Guinean Shield of West Africa is dominated by two principal age provinces, with a sharp boundary between them extending northerly from near Accra through Ghana, Togo, Dahomey, Upper Volta, and Niger (Fig. 1). To the west of this boundary, in Ghana and the Ivory Coast, the basement rocks record a period of orogenesis that occurred mostly in the time interval 1800-2100 m.yr. B.P., as first noted by Bonhomme (1962), who gave the name Eburnean to the orogeny (Fig. 2). Other geochronological investigations represented on Figure 2 include Bonhomme (1961), Vachette (1964a), Hurley et al. (1966 and unpublished), and Kolbe et al. (1966). The schematic age data in Figure 1 include also the measurements of Wilson (1954), Bernazeaud and Grimbert (1956), Bassot et al. (1963), Lay and Ledent (1963), Eber-

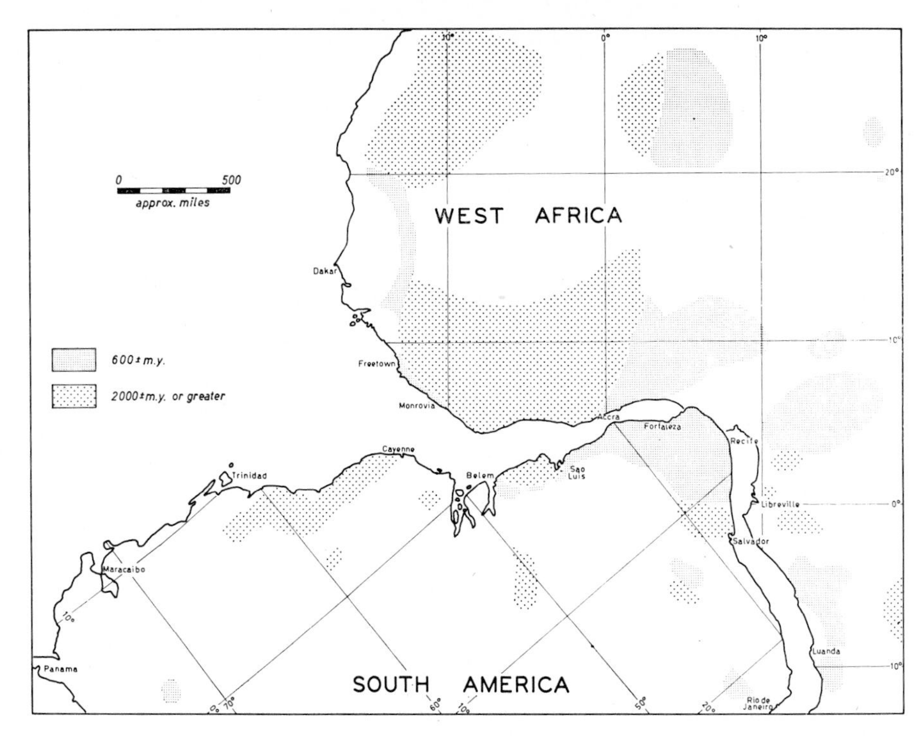

Figure 1. West Africa and northern South America according to Bullard's pre-drift reconstruction, showing regional distribution of 2000 ± m.yr. and 600 ± m.yr. age provinces.

hardt et al. (1964), Ferrara and Gravelle (1966), Laing (1966), Picciotto et al. (1966), Schnetzler et al. (1966), in the region west of the principal age boundary.

It can be seen (Fig. 2) that the 2000(±) m.yr. Eburnean province is bounded on the west by 2700(±) m.yr. rocks in Western Liberia, and apparently also to the north by similar ages in western Ahaggar (Sahara). The West African Craton is therefore a complex of ancient rocks with at least two age groups. In this report we shall tentatively use the term Liberian for the 2700(±) m.yr. orogenic episode, since the first homogeneous province of this age was found in Liberia.

To the east of the principal age boundary (Fig. 1), in Togo, Dahomey, Nigeria, and Cameroun the age values fall abruptly to the range 500-650 m.yr. by both whole-rock Rb-Sr and K-Ar analysis. Occasionally the whole-rock method indicates an ancient relic of a pre-existing basement. Recently it has been found that this age province extends as a great continuous belt of rejuvenated rocks around several of the Archean cratons of Africa, and has thus been called by Kennedy (1964) the Pan-African Orogenic Cycle.

Age measurements in this province (Fig. 3) include the work of Holmes and Cahen (1957), Hurley et al. (1958), reported by Cahen (1961), Snelling (1963), Jacobsen et al. (1963), Vachette (1964b), Lasserre (1964), Bessoles et al. (1956), Boissonnas et al. (1964), Roubault et al. (1965), Bonhomme et al. (1965, 1966), Cahen and Snelling (1966), and Hurley et al. (1966). Younger intrusives were emplaced as late as 100 m.yr., but for the most part the age range of 500-650 m.yr. covers both K-Ar

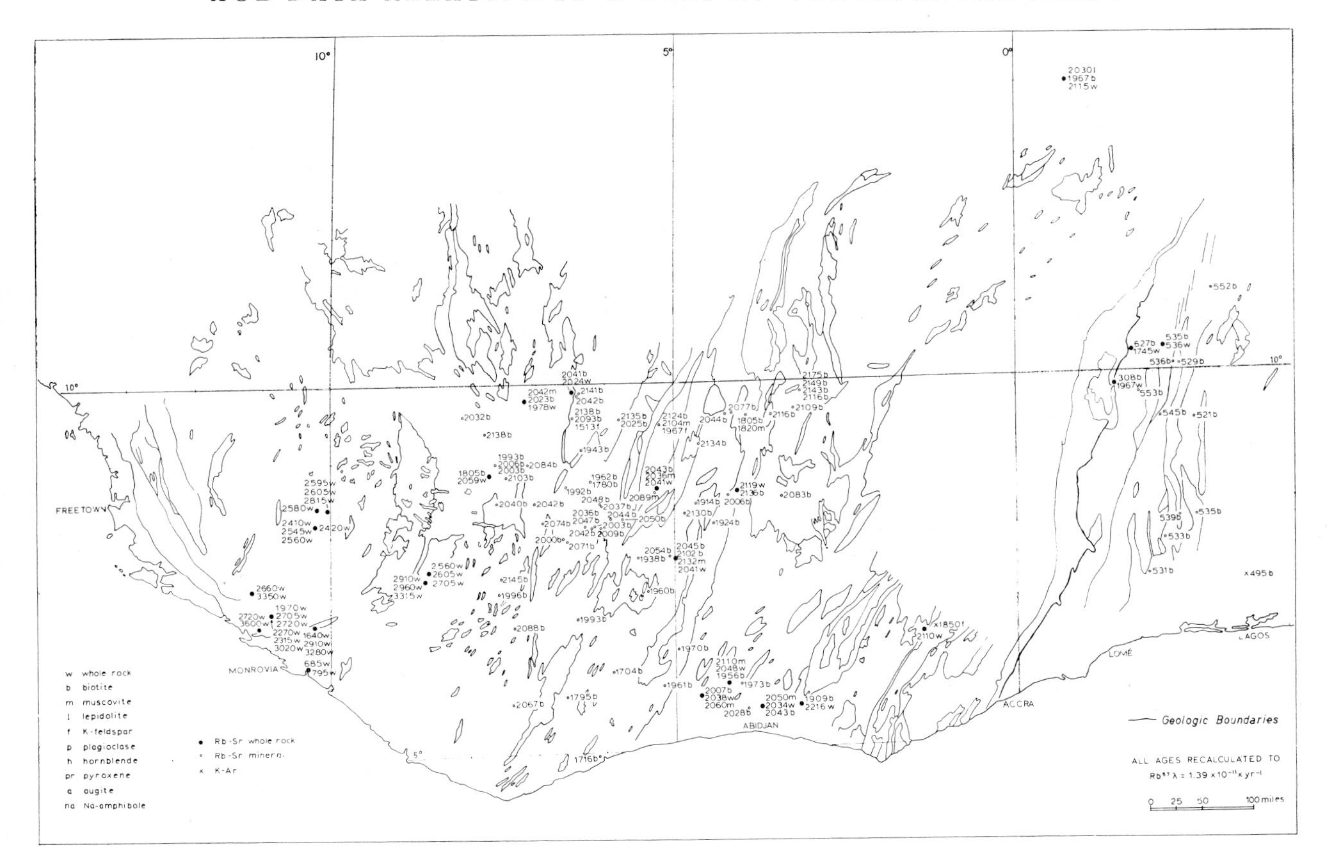

Figure 2. Reported and unpublished radiometric age data and generalized geological trends in the West African craton including Sierra Leone, Liberia, Ivory Coast, Ghana, Togo, and Dahomey.

All ages recalculated to $Rb^{87}\lambda = 1.39 \times 10^{-11} \times yr.^{-1}$. w, whole rock; b, biotite; m, muscovite; l, lepidolite; f, K-feldspar; p, plagioclase; h, hornblende; pr, pyroxene; a, augite; na, Na-amphibote. ●, Rb-Sr whole rock; ○, Rb-Sr mineral; X, K-Ar.

and whole-rock Rb-Sr values, although the Rb-Sr ages tend to be slightly higher than the K-Ar ones. To the south in Gabon, the ages increase again to 2000 ± 200 m.yr. (with scattered regions of ~2700 m.yr.) in the Congo Craton (Lasserre, 1964; Vachette, 1964b).

Basement Ages in South America

Geochronological investigations in Venezuela and the Guianas are represented in Figure 4. Published and unpublished reports include the following: Short and Steenken (1962), McDougall et al. (1963), Snelling (1963, 1962, 1964, 1965), McConnell et al. (1964), Choubert (1964), Chase (1965), Priem et al. (1966), Posadas (1966), Hurley et al. (1966). Two good whole-rock Rb-Sr dates on the gneisses on the Imataca Complex in Venezuela give ages of 2900 and 3050 m.yr., suggesting that the northern part of the Guayana Shield (or the western part, in the pre-drift orientation) includes an ancient province similar to that in the West African Craton. The coincidence of iron ranges in the Imataca Complex of Venezuela and in the Bomi and Nimba ranges of the

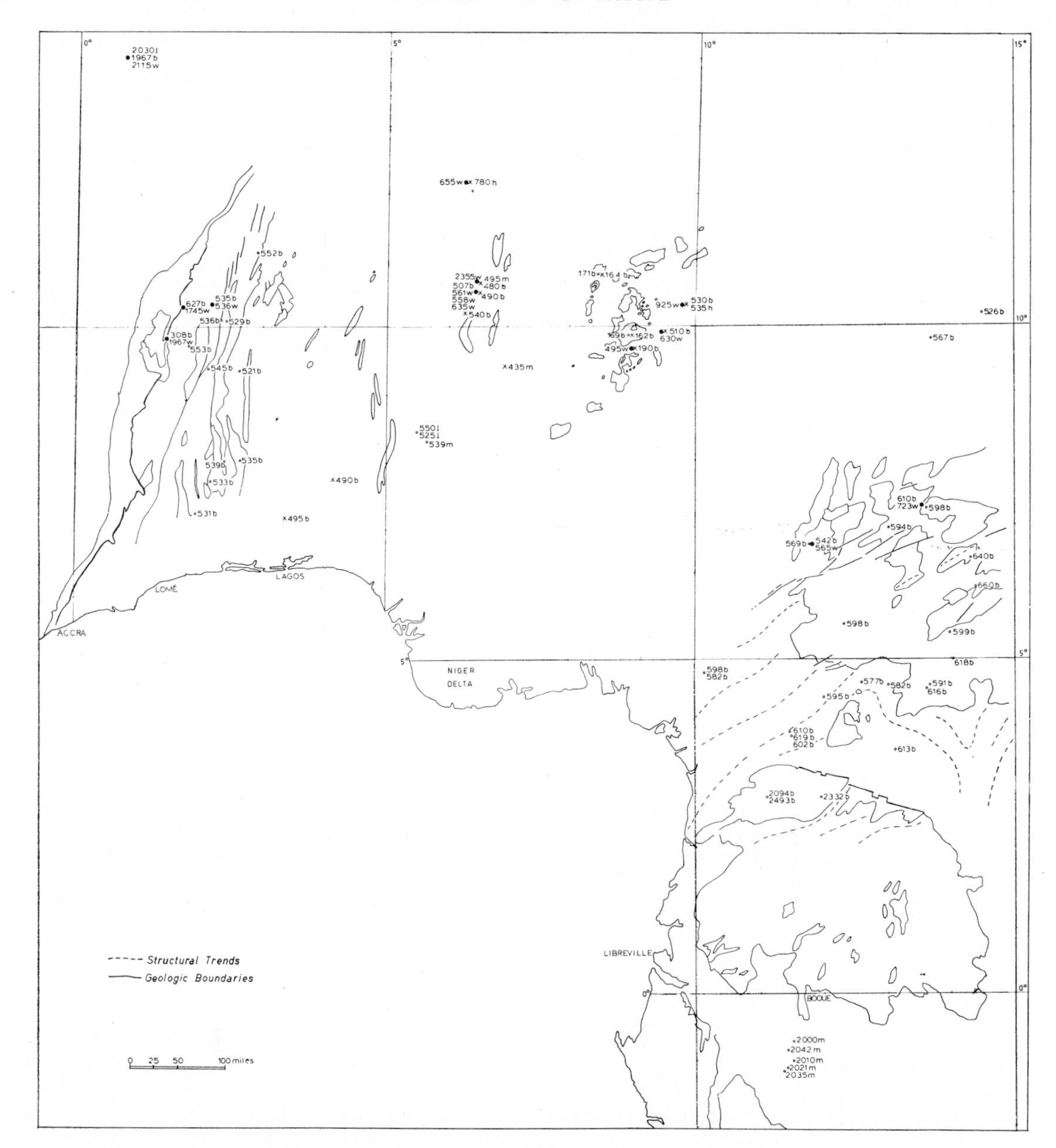

Figure 3. Reported and unpublished radiometric age data in the area of West Africa affected by the Pan African Orogenic Cycle, including Dahomey, Nigeria, Cameroun, and Gabon.

All ages recalculated to $Rb^{87}\lambda = 1.39 \times 10^{-11} \times yr.^{-1}$. w, whole rock; b, biotite; m, muscovite; l, lepidolite; f, K-feldspar; p, plagioclase; h, hornblende; pr, pyroxene; a, augite; na, Na-amphibote. ●, Rb-Sr whole rock; ○, Rb-Sr mineral; X, K-Ar.

same age in Liberia is mentioned in passing. The remainder of the Eastern Guayana Shield appears to be dominated by the 2000(±) m.yr. age value, similar to that found for the Eburnean orogeny in West Africa.

In Figure 1 it appears that the principal age boundary between the Eburnean and Pan-African age provinces in West Africa would enter Brazil close to the town of São

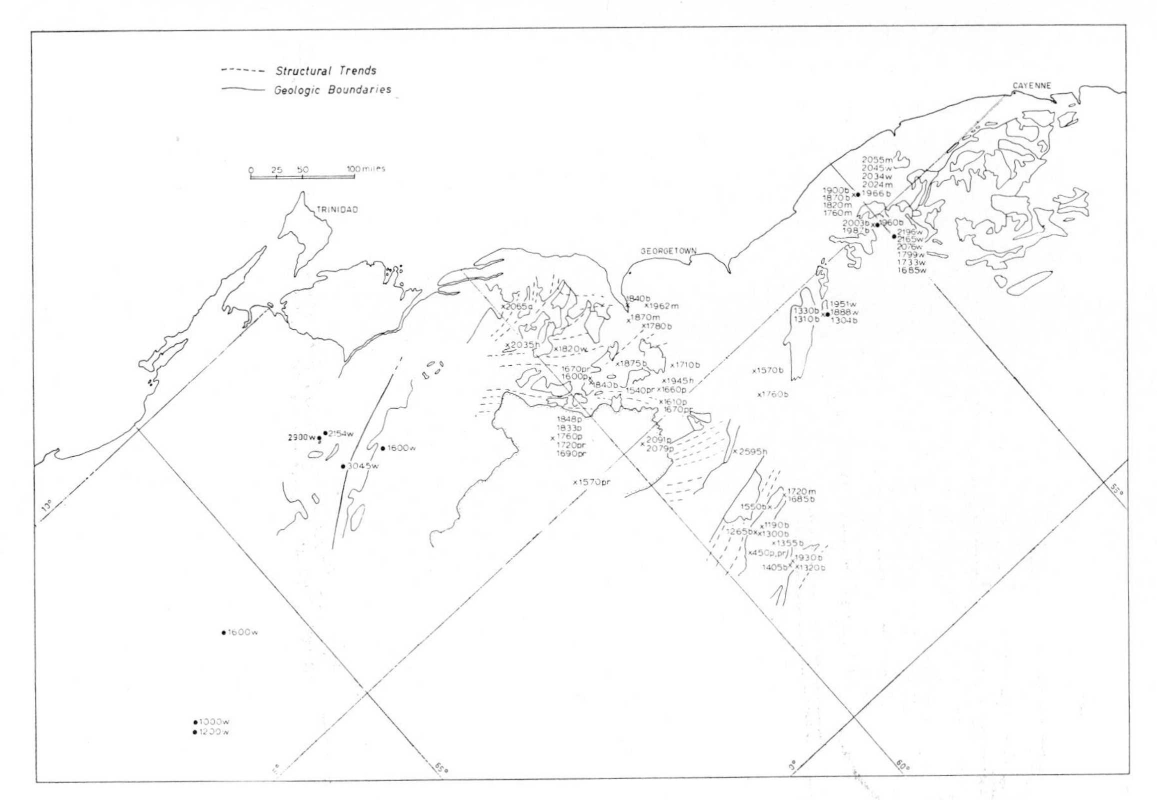

Figure 4. Reported and unpublished radiometric age data in the Guayana Shield of South America. All ages recalculated to $Rb^{87}\lambda = 1.39 \times 10^{-11} \times yr.^{-1}$. w, whole rock; b, biotite; m, muscovite; l, lepidolite; f, K-feldspar; p, plagioclase; h, hornblende; pr, pyroxene; a, augite; na, Na-amphibole. ●, Rb-Sr whole rock; ○, Rb-Sr mineral; X, K-Ar.

Luis if the continents had been together at that time. Recent investigations by the geochronology laboratories at the University of São Paulo and the Massachusetts Institute of Technology have verified this age boundary in this location (Almeida et al., 1966; Hurley et al., 1966). A combined report by these two groups has just been published (Hurley et al., 1967). A schematic presentation representing about 150 age measurements in northeastern Brazil by the K-Ar and whole-rock Rb-Sr methods is given in Figure 1.

In the vicinity of São Luis the age results showed two groups. Samples to the east of São Luis fell in a young age group similar to the Pan-African age province in West Africa. This younger province appears to extend to the eastern coast of Brazil; it has been called the Caririan Orogenic Episode. Ages are mostly in the range 450-650 m.yr. by both methods.

Samples in São Luis and westward to Amapá generally showed ages of about 2000(±) m.yr., with some exceptions that are being investigated actively at present.[1] This 2000(±)

[1] Since this presentation was given there is a suggestion from recent whole-rock determinations that a narow belt of Caririan activity swings westward south of São Luis and turns northerly just east of Belém. If so, this would match a similar Pan-African belt of rejuvenated rocks that appears to occur in coastal Liberia near Monrovia and to extend through Sierra Leone to enter the West African fold belt in Senegal.

m.yr. province appears to extend westward (with the possible gap footnoted above) into the Guianas and Venezuela and matches the 2000(±) m.yr. Eburnean age province directly opposite in West Africa.

North of Salvador, near the east coast of Brazil, the rocks affected and intruded by the 600(±) m.yr. Caririan orogeny become bounded by an undisturbed basement complex referred to as the São Francisco Craton. This cratonic area underlies the larger part of northern Bahia. A few measurements to date indicate that the age of gneisses in the São Francisco Craton is 2000(±) m.yr. or greater, similar to the findings of Vachette (1964b) for the Congo Craton directly opposite on the African side.

Summary

A reconstruction of the position of South America adjacent to West Africa (Bullard et al., 1965) and a comparison of reported age data on basement rocks leads to the following observations:

1. The age boundary between the 2000(±) m.yr. Eburnean and 600(±) m.yr. Pan-African age provinces in central West Africa appears to continue on line into Brazil.
2. The distribution of age values by the K-Ar and whole-rock Rb-Sr methods on both sides of this boundary in Brazil are identical to those in Africa within limits of present knowledge.
3. The extent and age range of the region affected by the Caririan orogeny in Brazil matches those of the Pan-African orogeny in Dahomey, Nigeria, and Cameroun.
4. The 2700(±) m.yr. Liberian age province could be correlated with the Imataca complex in Venezuela of about the same age, insofar as present indications of structural trends will permit.
5. There is a suggestion of another correlation between the Congo Craton in Gabon and the São Francisco Craton directly opposite in eastern Brazil, with similar 2000(±) m.yr. or greater ages.
6. The present limited data give a small indication that the mobile fold belt extending along the coastal region of West Africa through Mauritania, Senegal, and Guinea angles through southwestern Liberia and is matched by a belt of similar age entering Brazil near Belém.

Acknowledgments

This work has been supported by the Division of Research, U.S. Atomic Energy Commission. Although in this review we are not reporting the actual age data in Brazil (prior publication expected in *Science*), we gratefully acknowledge our partnership in the Brazilian effort with F. F. de Almeida, G. C. Melcher, U. G. Cordani, K. Kawashita, and P. Vandoros of the University of São Paulo, and our colleagues at M.I.T., W. H. Pinson, Jr., and H. W. Fairbairn.

Department of Geology and Geophysics
Massachusetts Institute of Technology

REFERENCES

Almeida, F. F. M. de, G. C. Melcher, U. G. Cordani, K. Kawashita, and P. Vandoros (1966), Absolute age determinations from Northern Brazil (Abstr.), *Geol. Soc. Amer. Program, 1966 Ann. Meetings*, p. 3.

Bassot, J. P., M. Bonhomme, M. Roques, and M. Vachette (1963), Mesures d'âges absolus sur les series précambriennes et paléozoiques du Sénégal Oriental, *Bull. Soc. Geol. France* (7th ser.) 5:401-405.

Bernazeaud, J., and A. Grimbert (1956), Conditions de gisement et âge de l'uraninite du Bas-Cavally (Côte d'Ivoire), *Compt. Rend. Acad. Sci.* 242:2744-46.

Bessoles, B., J. Cosson, J. Grassaud, and M. Roques (1956), Âge apparent de la diorite quartzique des Saras dans le Mayombe (A.E.F.), *Compt. Rend. Soc. Geol. France* 7:86-87.

Boissonnas, J., L. Duplan, J. Maisonneuve, M. Vachette, and Y. Vialette (1964), Étude géologique et géochronologique de roches du Compartiment Suggarien du Hoggar Central, Algérie, *Ann. Fac. Sci. Univ. Clermont, Ser. Geol.* 8 (Études Géochronologiques I):73-90.

Bonhomme, M. (1961), Absolute ages in the Bouna granite massif in Upper Volta and in the Windene granite in Ivory Coast, *Compt. Rend. Acad. Sci. Paris* 252 (No. 25):4016-17.

Bonhomme, M. (1962), Contribution à l'étude géochronologique de la plateforme de l'Ouest Africain, *Ann. Fac. Sci. Univ. Clermont, Ser. Geol.* 5.

Bonhomme, M., F. Weber, and R. Favre-Mercuret (1965), Âge par la méthode rubidium-strontium des sédiments du bassin de Franceville (République Gabonaise), *Bull. Serv. Carte Geol. Alsace Lorraine* 18(4):243-52.

Bonhomme, M., J. Lucas, and G. Millot (1966), Signification des déterminations isotopiques dans la géochronologie des sédiments, *Colloq. Intern. Centre Nat. Rech. Sci. Paris*, No. 151, pp. 335-59.

Bullard, E. C., J. E. Everett, and A. G. Smith (1965), The fit of the continents around the Atlantic, in A symposium on continental drift, *Phil. Trans. Roy. Soc.* 1088:41-51.

Cahen, L. (1961), Review of geochronological knowledge in middle and northern Africa, *Ann. N.Y. Acad. Sci.* 91:535-567.

Cahen, L., and N. J. Snelling (1966), *The geochronology of Equatorial Africa*, North-Holland Publ. Co., Amsterdam.

Chase, R. L. (1965), El complejo de Imataca, la anfibolita de Panamoy la tronjemita de Guri: rocas Precambricas del cuadrilatero de Las Adjuntas-Panamo, Edo. Bolivar, Venezuela, *Bol. Geol.* 7:105.

Choubert, B. (1964), Âges absolus du précambrien guyanais, *Compt. Rend. Acad. Sci. Paris* 258: 631-34.

Eberhardt, P., G. Ferrara, L. Glangeaud, M. Gravelle, and E. Tongiorgi (1964), Sur l'âge absolu des series metamorpiques de l'Ahaggar occidental dans la region de Silet-Tibehaouine (Sahara Central), *Compt. Rend. Acad. Sci. Paris* 256:1126-28.

Ferrara, G., and M. Gravelle (1966), Radiometric ages from Western Ahaggar (Sahara) suggesting an eastern limit for the West African craton, *Earth Planet. Sci. Letters* 1:319-24.

Hart, S. R., T. E. Krogh, G. L. Davis, L. T. Aldrich, and F. Munizaga (1966), A geochronological approach to the continental drift hypothesis, *Ann. Rep. Dept. Terrestrial Magnetism*, Washington, D.C., pp. 57-59.

Holmes, A., and L. Cahen (1957), Géochronologie africaine; resultats acquis au 1 Juillet 1956, *Acad. Roy. Sci., Colonials (Brussels), Classe Sci. Nat. Med.*, 5 Fasc. 1.

Hurley, P. M., with L. S. Cahen (1958), Age program in the Lower Congo, *M.I.T. Age Studies*, NYO-3939, *6th Ann. Progress Report for 1958*, USAEC Contract AT(30-1)-1381.

Hurley, P. M., H. W. Fairbairn, W. H. Pinson, Jr. (1966a), Continental drift investigations, *M.I.T. Ann. Progress Report*, 1381-14, USAEC, pp. 3-15.

Hurley, P. M., G. C. Melcher, J. R. Rand, and H. W. Fairbairn (1966b), Rb-Sr whole rock analyses in Northern Brazil correlated with ages in West Africa (Abstr.), *Geol. Soc. Amer., 1966 Ann. Meetings Program*, pp. 100-101.

Hurley, P. M., F. F. M. de Almeida, G. C. Melcher, U. G. Cordani, J. R. Rand, K. Kawashita, P. Vandoros, W. H. Pinson, Jr., and H. W. Fairbairn (1967), Test of continental drift by comparison of radiometric ages, *Science* 157:495-500.

Jacobson, R. R. E., N. J. Snelling, and J. F. Truswell (1963), Age determinations in the geology of Nigeria, with special reference to the Older and Younger Granites, *Overseas Geol. Min. Res.* 9: 168-82.

Kennedy, W. Q. (1964), The structural differentiation of Africa in the Pan-African ($\pm$500 m.y.) tectonic episode, *Res. Inst. African Geol. (Leeds, U.K.), 8th Ann. Rept.*, p. 48.

Kolbe, P., W. H. Pinson, Jr., J. M. Saul, and E. Miller (1967), Rb-Sr study on country rocks of the Bosumtwi Crater, Ghana, *Geochim. Cosmochim. Acta* 31:869-75.

Laing, E. M. (1966), Private communication to P. M. Hurley on 2 K-Ar dates in Sierra Leone, 1966.

Lasserre, M. (1964), Étude géochronologique par la méthode strontium-rubidium de quelques échantillons en provenance du Cameroun, *Ann. Fac. Sci. Univ. Clermont*, No. 25, Ser. Geol. 8 (Études Géochronologiques I):53-68.

Lay, C., and D. Ledent (1963), Mesures d'âges absolus de minéraux et de roches du Hoggar (Sahara-Central), *Compt. Rend. Acad. Sci. Paris* 257 (No. 21):3188-3191.
McConnell, R. B., E. Williams, R. T. Cannon, and N. J. Snelling (1964), A new interpretation of the geology of British Guiana, *Nature* 204:115-18.
McDougall, I., W. Compston, and D. D. Hawkes (1963), Leakage of radiogenic argon and strontium from minerals in Proterozoic dolerites from British Guiana, *Nature* 198:564-67.
Miller, J. A. (1965), Geochronology and continental drift—the North Atlantic, in A symposium on continental drift, *Phil. Trans. Roy. Soc.* 1088:180-91.
Picciotto, E., D. Ledent, and C. Lay (1966), Étude géochronologique de quelques roches du socle cristallophyllien du Hoggar (Sahara Central), *Colloq. Intern. Centre Natl. Rech. Sci.*, No. 151, pp. 277-89.
Posadas, V. G. (1966), Rb-Sr whole rock age in the Imataca Complex, Venezuela, M.Sc. Thesis, Dept. of Geol. and Geophys., M.I.T., Cambridge, Mass.
Priem, H. N. A., N. A. I. M. Boelrijk, R. H. Vershure, and E. H. Hebeda (1966), Isotopic age determinations on Surinam rocks, *Geologie en Mijnbouw*, 45:16-19.
Roubault, M., R. Delafosse, F. Leutwein, and J. Sonet (1965), Géochronologie-Premières données géochronologiques sur les formations granitiques et cristallophylliennes de la Republique Centre-Africaine, *Compt. Rend. Acad. Sci. Paris* 260:4787-92.
Schnetzler, C. C., W. H. Pinson, Jr., and P. M. Hurley (1966), Rubidium-strontium age of the Bosumtwi crater area, Ghana, compared with the age of the Ivory Coast tektites, *Science* 151: 817-19.
Short, K. C., and W. E. Steenken (1962), A reconnaissance of the Guayana Shield from Guasipati to the Rio Aro, Venezuela, *A.V.G.M.P. Bol. Informativeo (Caracas)* 5:189.
Snelling, N. J. (1963), Age of the Roraima formation, British Guiana, *Nature* 198:1079.
Snelling, N. J. (1962-1965), Age determination unit, *Repts. Overseas Geol. Surveys, 1960-1964, London.*
Vachette, M. (1964a), Essai de synthèse des déterminations d'âges radiométriques de formations cristallines de l'Ouest Africain (Côte d'Ivoire, Mauritanie, Niger), *Ann. Fac. Sci. Univ. Clermont*, No. 25, Ser. Geol. 8 (Études Geochronologiques I):7-30.
Vachette, M. (1964b), Âges radiometriques de formations cristallines d'Afrique Equatoriale, *Ann. Fac. Sci. Univ. Clermont*, No. 25, Ser. Geol. 8 (Études Géochronologiques I):31-38.
Wilson, J. T. (1954), Estimates of the age for some African minerals, *Nature* 174:1006.

APPALACHIAN AND CALEDONIAN EVIDENCE FOR DRIFT IN THE NORTH ATLANTIC

JOHN DEWEY MARSHALL KAY

Introduction

It is apposite, in the light of the large amount of information arising from recent studies in the Maritime Appalachians of Canada and in the British Caledonides, to set out concisely cogent facts bearing on the structural-stratigraphic fit of the two orogenic belts. Interpretations of drift in the North Atlantic expressed in recent reconstructions based on bathymetry (Bullard, Everett, and Smith, 1965) and tectonic belts (Kay, 1965) are essentially alike and resemble that made long ago by Taylor (1910). They imply separation of the continental shelf edges by more than 1,000 miles, with a vector of dextral displacement. The composite width of the continental shelves of northeast Newfoundland and Britain is approximately 350 miles, and the geology is barely known. Thus one is confronted with a gap of several hundred miles, so that it is only realistic to compare broad structural and stratigraphic units, belts, and chronologies in the North Atlantic. Extensions of geologic belts, accessible to direct examination, to the margins of the continental shelves will be possible, as geophysical methods develop and are applied, to enable direct matching of presumed opposing continental margins as is now more feasible for such South American and African coasts that have very narrow continental shelves.

The authors describe the main characteristics of each of several distinct belts in Newfoundland and compare them with elements of the British Caledonides. Newfoundland is characterized by three belts of pre-Carboniferous age (Williams, 1964; Kay, 1966) that contrast in their stratigraphy and structure and are sharply separated by lines drawn from White Bay through the Codroy Valley and from the west side of Bonavista Bay to the north side of Fortune Bay (Fig. 1).

1. The Western Belt has a basement of pre-Cambrian metamorphic rocks unconformably below an autochthonous platform sequence of sandstones, carbonates, and argillites of Cambrian and Early Ordovician age. This sequence is structurally overlain by allochthonous plates (thrust sheets) of Cambrian and Lower Ordovician clastic rocks, volcanics, and basic intrusions, thought to have come from the east. The thrust sheets are overlain by a neo-autochthonous Upper Ordovician to Devonian sequence (Rodgers, 1965).

2. The Central Mobile or Volcanic Belt has Ordovician, Silurian, and Devonian sedimentary and volcanic rocks with older metamorphic rocks to the northwest. Two major serpentinite bands intimately associated with Ordovician basic volcanics cross

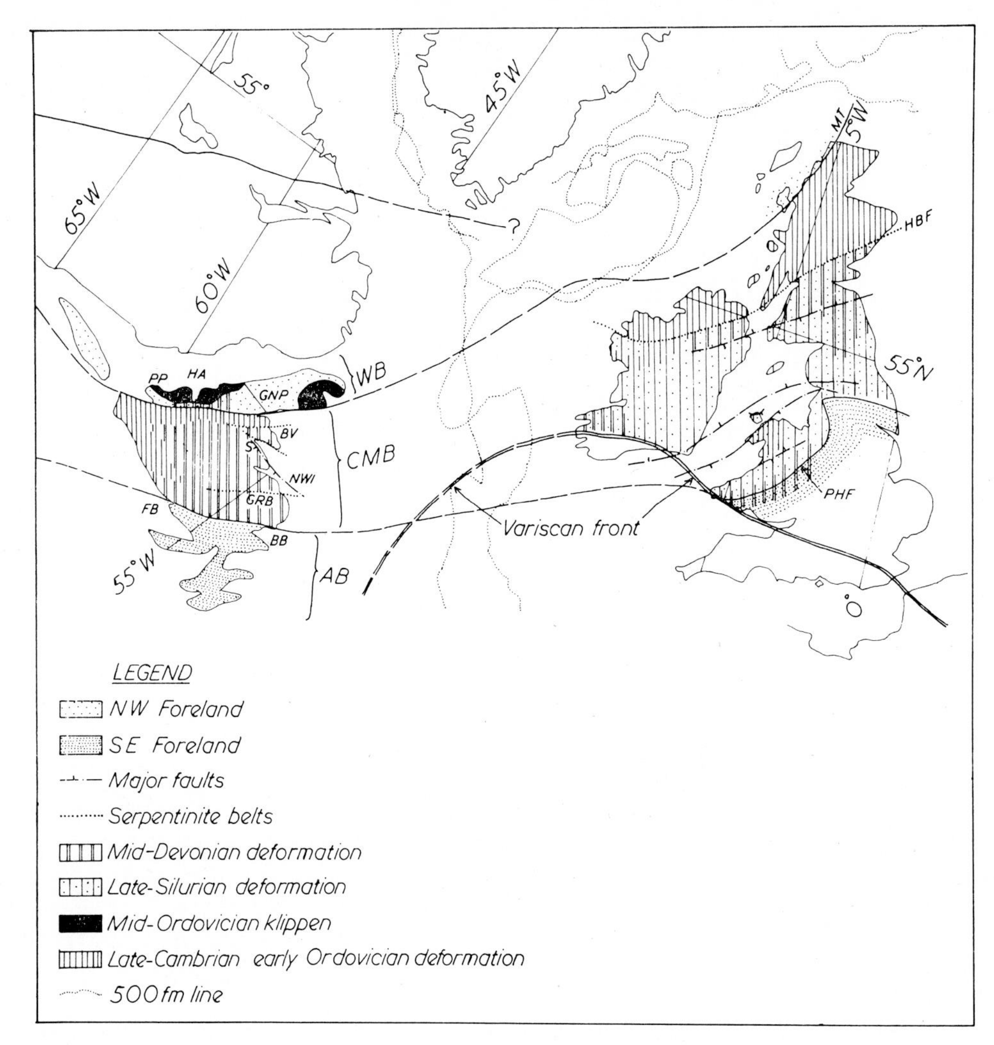

Figure 1. Generalized tectonic map of the Newfoundland Appalachians and the British Caledonides, with the continents and other structural blocks restored to their pre-drift positions, using the construction of Bullard et al. (1965).

the central belt, which is also characterized by numerous granitic and granodioritic bathyliths.

3. The southeastern part of the island, the Avalon Belt, has a basement of folded Proterozoic volcanic and sedimentary rocks overlain unconformably by a thin terrigenous sequence of Cambrian and Early Ordovician age. The pre-Cambrian basement rocks are found in shoals on the Grand Banks, but to the southeast they are covered by thousands of feet of Mesozoic and Cenozoic sediments.

The Western Belt

Pre-Cambrian rocks are exposed in the core of the Great Northern Peninsula (Clifford and Baird, 1962) and in several inliers north of Stephenville (Riley, 1962). These gneisses have been little studied and are regarded as Grenville equivalents on the basis of a few potassium-argon ages from 830 to 960 m.yr.

The autochthonous Cambro-Ordovician succession unconformably overlying the pre-Cambrian basement in the Great Northern Peninsula consists of sandstones, limestones, and argillites a few thousand feet thick (Lochman, 1937; Schuchert and Dunbar, 1934; Whittington and Kindle, 1963; Whittington, 1965). To the south around the Lower Humber River the succession is thicker, considerably more deformed, and in parautochthonous relationship to the pre-Cambrian (Lilly, 1965). Cambro-Ordovician trilobite faunas are essentially similar to western Appalachian faunas of Quebec and the Champlain Valley. Cambro-Ordovician clastics, volcanics, and limestone breccias lie structurally above the autochthonous sequence in two klippen, one extending along the west coast from Port au Port Bay to Daniels Harbour and the other in the north around Canada Bay (Kay, 1945; Rodgers and Neale, 1963). These allochthonous sheets are interpreted as having moved westward as gravity slides from welts in central Newfoundland in Middle Ordovician times, since the west Newfoundland klippe is unconformably overlain by a sequence of later Ordovician, Silurian, and Devonian sediments on the Port au Port Peninsula (Rodgers, 1965). Both klippen include large masses of basic and ultrabasic intrusives (Cooper, 1954) which are regarded as allochthonous, like the sediments and volcanics. The west Newfoundland klippe clearly consists of several slices (Brueckner, 1966) whose surfaces are characterized by chaotic mélange units. The strata of the allochthon are strongly deformed, possessing a regional flat schistosity axial planar to west-facing recumbent folds commonly deformed by later steeper cleavages.

Carboniferous terrigenous rocks overlie the Paleozoic (Baird, 1966; Bell, 1948; Hayes and Johnson, 1938) and are cut by high-angle, northeasterly trending faults. Similar faults in the Maritime Provinces bounded tilt blocks during Carboniferous sedimentation (Belt, 1965; Howie and Cumming, 1963), faulting continuing after Triassic times.

COMPARISONS

The analogies between stratigraphic and structural sequences of the rocks of western Newfoundland, southeastern Quebec, and western New England are evident (Rodgers and Neale, 1963).

West of the Moine Thrust in the northwest foreland of the British Caledonides, autochthonous Cambrian and Lower Ordovician orthoquartzites and limestones of the Durness sequence lie unconformably upon the pre-Cambrian. Unlike Newfoundland the pre-Cambrian consists not of Grenville basement, but of older Lewisian rocks (Sutton and Watson, 1951), yielding ages from 2500 m.yr. to 1400 m.yr. A late pre-Cambrian sedimentary sequence, the Torridonian Series, rest unconformably upon the Lewisian and itself is overlain unconformably by the Cambrian. No rocks of Lewisian or Torridonian facies or age are known in the western belt of Newfoundland; and it therefore seems likely that the Grenville "front" (Fig. 1) is cut out by the Caledonian front somewhere southeast of Greenland.

Central Mobile Belt

This belt is dominated by Ordovician and Silurian sedimentary and volcanic rocks and mid-Devonian granite bathyliths. An older terrain of metamorphic rocks, the Fleur de Lys Group (Neale and Nash, 1963) showing a polyphase deformation history, extends along the northwestern margin of the belt in the Burlington Peninsula. The oldest fossil-dated rocks are Lower Ordovician volcanics in eastern Notre Dame Bay (Williams, 1963; Kay, 1966) and graptolite-bearing argillite and radiolarian chert in the eastern part of the Burlington Peninsula.*

The Ordovician and Silurian sequences show great thickness and facies variation, volcanism being prevalent in Early Ordovician, and widespread in Silurian, times. The distribution of thousands of feet of volcanic and plutonic boulder-bearing conglomerates of Late Ordovician and Silurian age suggests derivation from fault blocks raised during the late Ordovician (Kay, 1965, 1966). Regional unconformity below the Silurian in eastern Notre Dame Bay and granites radiometrically dated as Late Ordovician emphasize the importance of Late Ordovician movements.

Two major serpentinite belts with associated basic intrusives, the Baie Verte and Gander River belts, are intimately associated with Ordovician volcanic rocks and are regarded by Williams (1964) as bounding a major Ordovician rift.

Great volumes of acid plutonic rock in central Newfoundland intrude Ordovician, Silurian rocks and a limited outcrop of Lower Devonian rocks on the south coast (Cooper, 1954). Potassium-argon ages on these rocks range from 450 to 335 m.yr. (Lowden, 1963; Wanless, 1965), and the granites are attributed to the mid-Devonian Acadian orogeny.

The Acadian deformation of the lower Paleozoic rocks is very variable in intensity in the central belt. In Notre Dame Bay deformation is polyphase, and recumbent folds were developed before the intrusion of the granites, whereas around Springdale Silurian rocks are disposed in broad flexures and are only strongly deformed against fault blocks. Regionally, however, the earlier deformed Fleur de Lys possesses flat penetrative cleavages, whereas the Ordovician and Silurian rocks show steep cleavages. The total history of the central mobile belt seems to have been dominated by the effects of steeply inclined fundamental faults that produced the Ordovician and Silurian tilt-blocks that were effective on paleogeography until the Acadian orogenic episode.

COMPARISONS

Direct and specific spatial and temporal comparisons with sequences and structures of the Caledonian mobile belt in Britain are of little value since both belts show extreme variations in stratigraphy within very short distances, exemplified by the contrast in such nearby Ordovician sections as those at Snowdon and Bala in North Wales, or the Silurian sequences in the different fault blocks on New World Island, Newfoundland. The resemblances and differences are but the records of belts of similar history, with mobile crusts,

* (In 1967, low Middle Cambrian trilobites were found interbedded in volcanic rocks in central Newfoundland [Kay and Eldredge, 1968].)

deeply subsiding in some places, rising in others, generally in fault-bounded blocks.

In the Caledonian mobile belt, opposing margins of quite different character reflect the asymmetry of the belt and its divisibility into an early northern orthotectonic belt involving late Proterozoic and Cambrian strata deformed during Late Cambrian or Early Ordovician times and a later southern paratectonic belt in which Cambrian, Ordovician, and Silurian sequences suffered a climactic deformation in Late Silurian times. The northern orthotectonic belt shows great strike persistence in stratigraphic sequence, particularly in the Dalradian, which may be compared with the Fleur de Lys Group of Newfoundland.

A major belt of fundamental faulting with associated serpentinites, the Highland Boundary Fault Belt, postdates the intense deformation and metamorphism of the Dalradian rocks and had a long and complex history of movement influencing Ordovician sedimentation in the west of Ireland, and Devonian sedimentation in Scotland. The position of the belt and its relations to the Dalradian and Ordovician rocks invites a direct correlation with the Baie Verte serpentinite belt in Newfoundland.

However, in spite of the similar relationships between a northern orthotectonic, and a southern paratectonic, belt in both Newfoundland and the Caledonides, the timing of the deformation and metamorphism of the paratectonic belts was somewhat different. The paratectonic belt of Newfoundland suffered its climactic deformation in the Middle Devonian, whereas in Britain climactic deformation was during the Late Silurian, although Middle Devonian movements were locally important.

The Avalon Belt

This belt comprises that part of Newfoundland southeast of a line from northwestern Bonavista Bay to northern Fortune Bay. The oldest rocks in the Avalon Peninsula are an assemblage of thick volcanic rocks and are overlain by thousands of feet of sometimes red, terrigenous clastics. These sequences of late Proterozoic age are unconformably overlain by a thin Cambrian and Lower Ordovician succession consisting of basal orthoquartzites passing up into shales, algal limestones, manganese-bearing sediments, and, in the Lower Ordovician, oölitic iron ores. The Lower Paleozoic rocks are little deformed but show some thickness variation, and locally the Cambrian contains pillow lavas. The Avalon Belt during Cambrian and Ordovician times formed a broad stable platform, and thus the occurrence of pillow lavas is unusual as is also the presence of mid-Devonian granite bathyliths (Lowden, 1963) in the southwestern part of the belt.

COMPARISONS

The Avalon Belt was within a mobile belt of widespread limits during the late pre-Cambrian. It became more stable during the Early Paleozoic, forming the Avalon Platform to the southeast of the deeply subsiding Central Mobile Belt. Southeast of the Pontesford Hill fault line in the Welsh Borderlands of Britain, thin non-volcanic Cambrian, Ordovician, and Silurian sequences were deposited upon a stable platform of folded pre-Cambrian rocks. The pre-Cambrian consists partly of scattered inliers of gneisses and schists of doubtful age (Malvern Gneiss, Rushton Schists, Primrose Hill Gneiss) but also of presumed late Proterozoic volcanics (Uriconian) overlain by terrigenous clastic sediments (Longmyndian). The Uriconian and Longmyndian rocks are very similar to the volcanic and clastic sequences of the Avalon Belt and may both represent Grenville molasse sequences. Whereas the Cambro-Ordovician sequences of northwest Scotland and Newfoundland contain a Pacific trilobite, cephalopod faunas, the Avalon Cambrian contains a distinctive fauna similar to that of the Welsh Cambrian rocks (Hutchinson, 1962; Whittington, 1966).

Summary

In this brief résumé it is argued that in the Newfoundland Appalachians three belts of contrasting structural and stratigraphic sequence are matched in the British Caledonides by three similar belts, and it is suggested that these belts were once contiguous.

Acknowledgment

This study was supported by grants to Columbia University by the National Science Foundation. A volume of papers comparing geology along the North Atlantic is being published (Kay, 1968).

Sedgwick Museum, Cambridge, England (J. D.)
and
Columbia University (M. K.)

References

Baird, D. M. (1966), Carboniferous rocks of the Conche-Groais Island area, Newfoundland, *Can. J. Earth Sci.* 3:247-57.

Bell, W. A. (1948), Early Carboniferous strata of St. Georges Bay area, Newfoundland, *Geol. Surv. Can., Bull.* 10, 45 pp.

Belt, E. S. (1965), Stratigraphy and paleogeography of Mabou Group and related middle Carboniferous facies, Nova Scotia, Canada, *Geol. Soc. Amer. Bull.* 76:777-802.

Brueckner, W. D. (1966), Stratigraphy and structure of west-central Newfoundland, pp. 137-51 in *Geology of Parts of Atlantic Provinces*, W. H. Poole, ed., Guidebook Geol. Assoc. Canada and Mineral. Assoc. Canada, Halifax.

Bullard, E. C., J. C. Everett, and A. G. Smith (1965), The fit of the continents around the Atlantic, *Phil. Trans., A* 258:41-51.

Clifford, P. M., and D. M. Baird (1962), Great Northern Peninsula of Newfoundland—Grenville Inlier, *Bull. Can. Inst. Mining Metal.* 55:150-57.

Cooper, J. R. (1954), Lapoile-Cinq-Cerf map area, *Geol. Surv. Can., Mem.* 276, 62 pp.

Hayes, A. D. and Helgi Johnson (1938), Geology of the St. Georges Bay Carboniferous area, *Geol. Surv. Newfoundland, Bull.* 12.

Howie, R. D., and L. M. Cumming (1963), Basement features of the Canadian Appalachians, *Geol. Surv. Can., Bull.* 89, 18 pp.

Hutchinson, R. D. (1962), Cambrian stratigraphy and trilobite faunas of southeastern Newfoundland, *Geol. Surv. Can., Bull.* 88, 156 pp.

Kay, Marshall (1945), Paleogeographic and palinspastic maps, *Bull. Amer. Assoc. Petrol. Geol.* 29:426-50.

Kay, Marshall (1966), Comparison of the lower Paleozoic volcanic and non-volcanic geosynclinal belts in Nevada and Newfoundland, *Bull. Can. Petrol. Geol.* 14:579-99.

Kay, Marshall (1967), Stratigraphy and structure of northeastern Newfoundland bearing on drift in the North Atlantic, *Bull. Amer. Assoc. Petrol. Geol.* 51:579-600.

Kay, Marshall (Editor) (1968), North Atlantic geology and continental drift, *Amer. Assoc. Petrol. Geol.* Mem. 12 (in press).

Kay, Marshall, and E. H. Colbert (1965), *Stratigraphy and Life History*, John Wiley, New York, 736 pp.
Kay, Marshall, and Niles Eldredge (1968), Cambrian trilobites in Central Newfoundland Volcanic Belt, *Geol. Mag.* 105 (in press).
Lilly, H. D. (1963), Geology of Hughes Brook—Goose Arm area, west Newfoundland, *Memorial Univ. Newfoundland, Geol. Rep.* 2.
Lochman, Christina (1938), Middle and Upper Cambrian faunas from western Newfoundland, *J. Paleont.* 12:461-77.
Lowdon, J. A. (1963), Age determinations by the Geological Survey of Canada, Isotopic ages—Report 3, *Geol. Surv. Can., Pap.* 62-17, part 1, 122 pp.
Neale, E. R. W., and W. A. Nash (1963), Sandy Lake (east half) map-area, Newfoundland, *Geol. Surv. Can. Pap.* 62-28, 40 pp.
Riley, G. C. (1962), Stephenville area, Newfoundland, *Geol. Surv. Can. Mem.* 323, 72 pp.
Rodgers, J. (1965), Long Point and Clam Bank formations, western Newfoundland, *Geol. Assoc. Canada Proc.* 12:83-94.
Rodgers, J., and E. R. W. Neale (1963), Possible "Taconic" klippen in western Newfoundland, *Amer. J. Sci.* 261:713-30.
Schuchert, Charles, and C. O. Dunbar (1934), Stratigraphy of western Newfoundland, *Geol. Soc. Amer. Mem.* 1, 123 pp.
Sutton, John, and Janet Watson (1951), The pre-Torridonian metamorphic history of the Loch Torridon and Scoine areas in the north-west Highlands, and its bearing on the chronological classification of the Lewisian, *Quart. J. Geol. Soc. London* 106:241-307.
Taylor, F. B. (1910), Bearing of the Tertiary mountain belt on the origin of the earth's plan, *Geol. Soc. Amer. Bull.* 21:179-226.
Wanless, R. K., R. D. Stevens, G. R. Lachance, and R. Y. J. Rimsaite (1965), Age determinations and geological studies, Part 1—Isotopic ages, Report 5, *Geol. Surv. Can. Pap.* 64-17 (Part 1), 126 pp.
Whittington, H. B. (1965), Trilobites of the Ordovician Table Head Formation, western Newfoundland, *Bull. Mus. Comp. Zool. Harvard Univ.* 132:275-442.
Whittington, H. B. (1966), Phylogeny and distribution of Ordovician trilobites, *J. Paleont.* 40: 696-739.
Whittington, H. B., and C. H. Kindle (1963), Middle Ordovician Table Head Formation, Western Newfoundland, *Geol. Soc. Amer. Bull.* 74:745-58.
Williams, Harold (1963), Twillingate map-area, Newfoundland, *Geol. Surv. Can. Pap.* 63-36, 30 pp.
Williams, Harold (1964), The Appalachians in northeastern Newfoundland—a two-sided symmetrical system, *Amer. J. Sci.* 262:1137-58.

THE DISTRIBUTION OF CONTINENTAL CRUST AND ITS RELATION TO ICE AGES

E. IRVING W. A. ROBERTSON

Discussions of paleomagnetism and its application to the hypothesis of continental drift have been concerned largely with "Wegenerian" drift, that is, the hypothesis that deals with the fragmentation and dispersal of Laurasia and Gondwanaland (or a single Pangaea) during the Mesozoic and Cenozoic (Wegener, 1924). Paleomagnetic work from continental regions has confirmed many aspects of this hypothesis, and has also introduced new elements; in particular, it has provided evidence that continental drift has not been confined to the last few percent of the earth's history but is a recurrent feature of earlier geological time, as would be more in keeping with a uniformitarian view. Discussions based on paleomagnetism have centered about five main points:

1. The possibility of relative motion of continents during the Carboniferous. The apparent polar-wandering path observed paleomagnetically from Australia shows rapid changes in the Carboniferous. This feature is not seen in the paths observed from Europe and North America, and a Carboniferous phase of drift has therefore been postulated (Irving and Green, 1958; Irving, 1966). (Wegener had previously suggested that polar wandering [strict sense] occurred in the late Paleozoic in order to explain the palaeoclimatic evidence, but he did not suggest Paleozoic relative displacements of the continents.) Further evidence for drift in the Carboniferous has been found in the U.S.S.R. (see for example Khramov et al., 1965) where the paleomagnetic pole positions obtained from Permian rocks in the Russian and Siberian platforms agree, whereas results from the Devonian and Lower Paleozoic rocks often do not. It is therefore possible that these two crustal blocks were much further apart during the early Paleozoic and Devonian than in the Permian, by which time they had achieved roughly their present relative positions.

2. The possibility of continental drift at the Paleozoic-Mesozoic boundary. In Europe (Creer et al., 1957), North America (Collinson and Runcorn, 1960) and Africa (Gough et al., 1964) the paleomagnetic directions of the Permian differ by about 30° from those of the Triassic, but in Australia (Irving et al., 1963; Irving, 1966) the directions in the Permian pass upward into the Mesozoic without significant change other than reversal. This indicates that continental drift occurred at the end of the Paleozoic, and one aspect of this postulated episode is the initiation of the break-up of Gondwanaland (Creer, 1965).

3. The possibility of shear between Laurasia and Gondwanaland. Studies of magnetic inclinations in the Tethys and adjacent borderlands have lead de Boer (1963, 1965) and van Hilten (1964) to suggest that east-west relative motions occurred between Laurasia and the northernmost components of Gondwanaland during the Mesozoic and

Lower Tertiary. They suggest that the Alpine orogenic zone is the shear zone between them.

4. The formation of the Atlantic. The longitudinal difference between the paths of apparent polar wandering observed for Permian and later time from Africa (Gough et al., 1964) and South America (Creer, 1965), and from Europe and North America (Creer et al., 1957) suggest that the Old World and New World have moved apart to form the Atlantic between them.

5. The general northerly trend in the latitudes of continents since the Paleozoic (Blackett et al., 1960). The evidence for this is found in the changes in the inclination observed in Permian and younger rocks. This process appears to have occurred intermittently since the Paleozoic and may be a reflection not of continental drift but of polar wandering (strict sense).

It is possible that the events 1, 2, 3, and 4 occurred in that order and are to be regarded as phases in a time sequence of continental drift, and one of the most interesting prospects for paleomagnetic work on continental rocks therefore lies in the study of these early drift phases and their relation to geologically recorded events: orogenies, transgressions, regressions, and glaciations. This paper is concerned primarily with late Paleozoic events (points 1 and 2 above) and their possible relation to glaciation in the southern hemisphere. The paleomagnetic observations are incomplete, and of necessity this discussion is of an exploratory nature.

Ewing and Donn (1956, 1966) have proposed a geographical theory of glaciation. Their idea is that circumpolar glaciation is more likely to develop if the pole is situated in a "thermally isolated" environment, such as a large land mass or a small enclosed sea, than if it were in an open ocean where a free circulation of water would tend to even out temperatures. The ebb and flow of glaciation during the present ice age is attributed by them to special geographical arrangements around the Arctic Basin which do not involve latitude changes and so cannot be studied paleomagnetically. Their general proposition, however, concerns the occurrence of ice ages themselves, not the climatic fluctuations within them, and can be tested. If Ewing and Donn are correct, then a high concentration of land would be expected in high latitudes during ice ages. The latitude distribution of continental crust can be calculated from the paleomagnetic results. The method is applicable to the late Paleozoic and the present ice ages, because they were confined to circumpolar regions when both cold and warm climates occurred on the earth at the same time. Its application to the infra-Cambrian glaciation is less clear because it is possible, as Harland (1964a) contends, that this was of exceptional severity and affected all latitudes.

The relation of glacial occurrence to pole position and land in the southern hemisphere during the late Paleozoic is set out in Figure 1. Although there is much paleomagnetic evidence in support of the idea that the southern continents were close together in the late Paleozoic, a reliable reconstruction for the whole of this time cannot yet be made from paleomagnetism, and a reconstruction based on geology is therefore adopted.

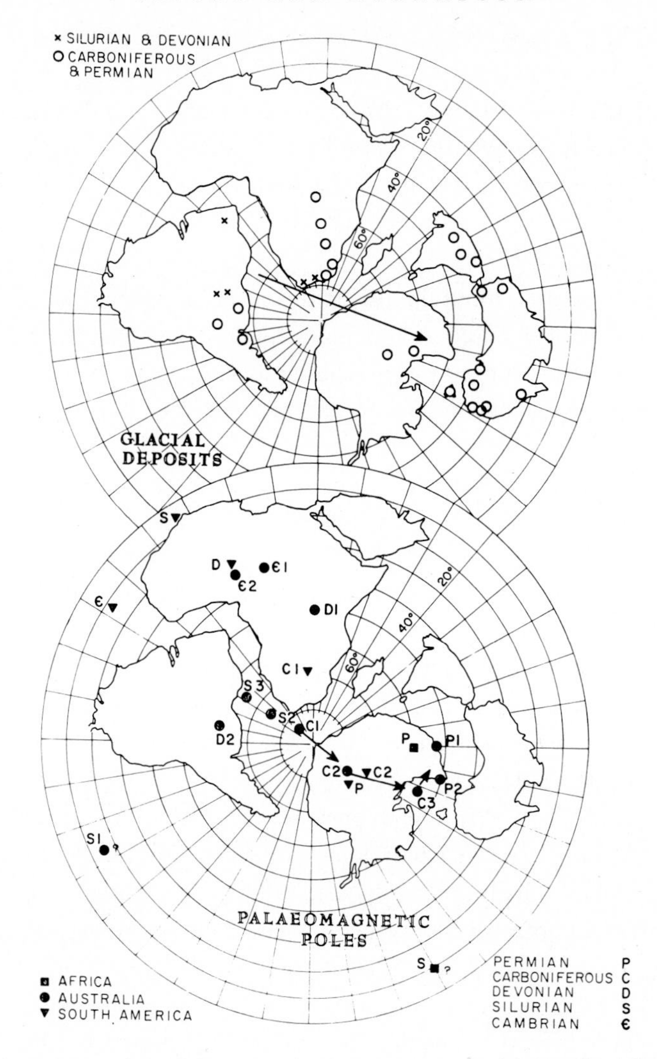

Figure 1. King's reconstruction of Gondwanaland upon which glacial deposits (above) and Paleozoic paleomagnetic poles have been plotted. This is a geological not a paleomagnetic reconstruction. The paleomagnetic poles have been compiled from the following sources: South America (Creer, 1965), Africa (Opdyke, 1964), Australia (Irving and Green, 1958; Irving and Parry, 1963; Green, 1959; Briden, 1966; Irving, 1966).

King's map of Gondwanaland is used (King, 1962). He assumes that Gondwanaland was a single continent and that the gaps between the present land masses were then occupied by material now comprising the ocean ridges. Other workers would place the continents closer together and regard the ocean ridges as a more recent development, but this is a matter of detail insofar as this discussion is concerned; the argument would equally well apply to any other drift reconstruction of Gondwanaland. Glacial occurrences (Maack, 1960) are found in the Silurian and late Paleozoic. The oldest occurrences (Silurian and Devonian) are in South America and Africa, and the youngest (late Carboniferous to Late Permian) are in Australia.

In the lower part of Figure 1 the Paleozoic paleomagnetic poles are plotted with respect to the observing localities. Poles observed from different continents are distinguished by different symbols. One Cambrian pole and one Silurian pole are near the margin of Gondwanaland, and two Silurian poles are placed well away; none of these poles is based on a substantial number of observations. All other Paleozoic poles, including many based on detailed demagnetization studies, fall within Gondwanaland. There is a general trend with time, all Late Carboniferous and Permian poles being located in eastern Gondwanaland. In Australia this polar movement (indicated by arrows in Fig. 1) has been found to occur during a comparatively short period of time within the Carboniferous. Therefore all the paleomagnetic poles for the late Paleozoic of Gondwanaland fall over continental crust, and the migration with time of the glacial beds is mirrored by a migration of poles.

So far we have had to rely on geological reconstruction, but for the Permian and later time more paleomagnetic data are available, and we can begin to make reconstructions from them. In Figure 2 a reconstruction for the Permian is given. Only one period of

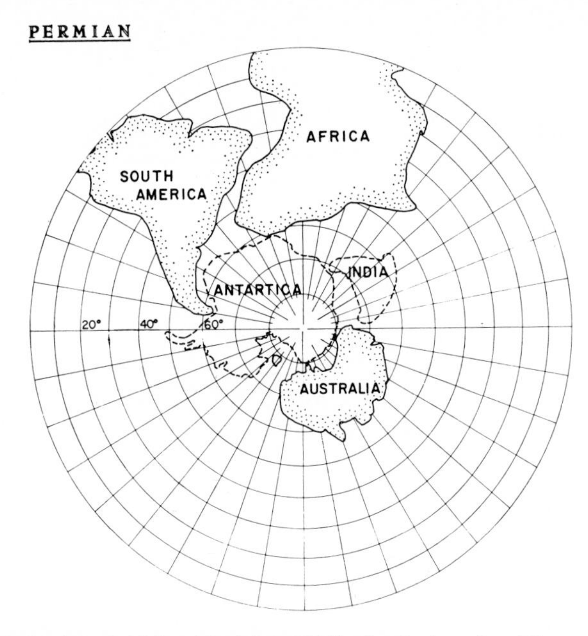

Figure 2. Permian latitudes of Africa (Opdyke, 1964), South America (Creer, 1965), and Australia (Irving and Parry, 1963).

time is considered, and there is therefore an uncertainty in longitude. South America, Africa, and Australia are near the pole. There are no Permian results from Antarctica, so that its latitude is not yet known, but as the dotted outline shows, it fits well between east and west Gondwanaland, indicating that one large land mass surrounded the south pole. If an agnostic view is taken and no place is assigned to Antarctica, the south pole at that time would then have centered in a comparatively small sea or have been marginal to a large continent; it does not appear to have been in the middle of a large ocean.

In Figure 3 the latitude spread of continental crust for the Permian is plotted, including information from the northern hemisphere. This is calculated as the length of crust

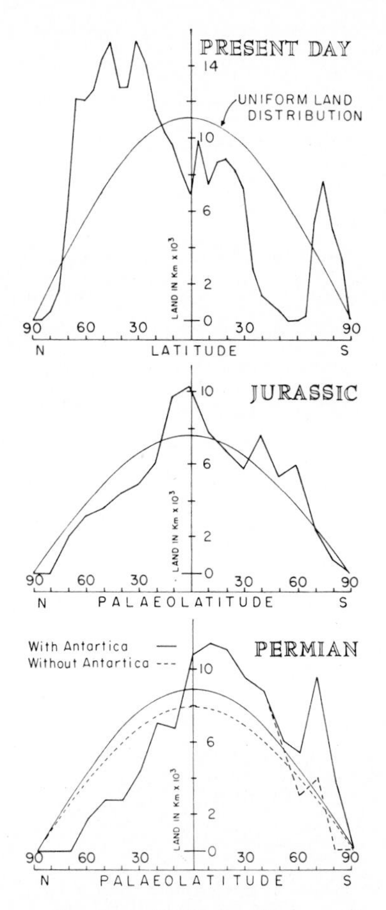

Figure 3. Distribution of land as a function of latitude. The smooth curve is the theoretical expectation for a uniform distribution of land through all latitudes.

in small circles of latitude. In these calculations the continental crust is assumed to be bounded by the present coastlines. It would probably have been more realistic to have used the edge of the continental shelf, but if this had been done, subjective choices would have had to be made with regard to certain large shelf areas (for example in the East Indies). The effect of using the present coastlines is to underestimate the extent of continental crust, but the general shape of the distribution is unlikely to be deformed. In the Permian there was no continental crust at the north pole so far as our present observations go, and most of it was in the southern hemisphere. The distribution is roughly the mirror image of that today.

In Figures 4 and 5 the southern continents are placed in their Jurassic latitudes in two possible longitudinal arrangements. Australia remains near the pole, but South Africa and South America have moved away since the Permian. Whatever the longitudinal arrangement adopted, it does not seem possible to place a large mass of land over the south pole. The histograms for the Jurassic (Fig. 3) show a uniform spread of continental crust. The Triassic data are unfortunately insufficient to allow maps to be drawn for all the southern continents, but the Triassic paleomagnetic poles from Australia and Africa

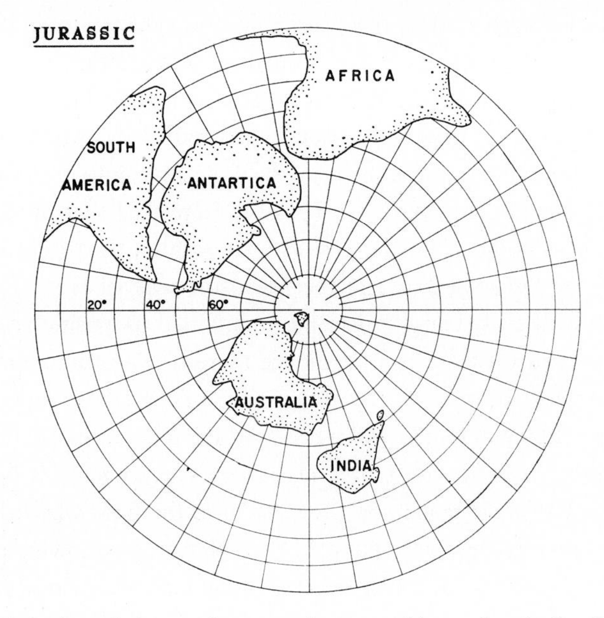

Figure 4. Jurassic latitudes of the southern continents—arbitrary longitude. Calculated for South America from the data of Creer (1965), for Africa from the data of Gough et al. (1964), and for India, Antarctica, and Australia redrawn from base maps in Irving (1964).

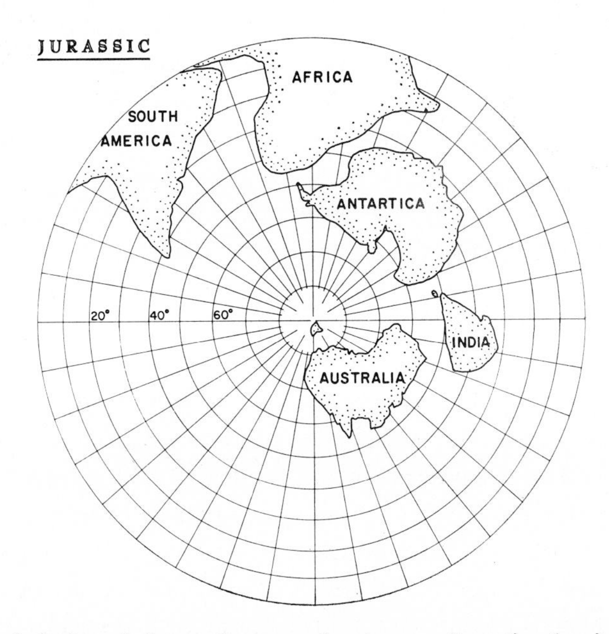

Figure 5. Jurassic latitudes of the southern continents—an alternative longitudinal arrangement.

are indistinguishable from those from the Jurassic. Hence the movement apart of Africa as a representative of western Gondwanaland, from Australia, as a representative of eastern Gondwanaland, occurred between Permian and Triassic time; it is these observations from Africa and Australia which suggest that the disruption of Gondwanaland began between the Permian and the Triassic.

These comparisons suggest that the late Paleozoic glaciation in Gondwanaland is closely associated with the south pole determined paleomagnetically. The last records of this glaciation are at the end of the Paleozoic, at which time the paleomagnetic results suggest that the development of a large ocean occurred at the south pole. It is possible that glaciation was continuous through from the Paleozoic into the Mesozoic, and that it has not been recorded in the Mesozoic because there was then no land at the poles. Alternatively, it is possible that the creation of a substantial ocean destroyed the circumpolar glaciation of the Paleozoic, so that the Mesozoic was ice-free. The first possibility is considered the less likely because Australia maintained its position very near the pole during the Mesozoic and yet there is no evidence there of glaciation, and the temperature measurements that have been made by oxygen isotope methods indicate a mild (14-16°C) but not a cold climate (Lowenstam and Epstein, 1954). The paleomagnetic results are therefore consistent with Ewing and Donn's hypothesis; it is possible that the Silurian and Upper Paleozoic glaciation grew because the pole was located over a land mass and the dissolution of that land mass led to deglaciation, allowing the comparatively mild climate of the Mesozoic to develop. The ending of glaciation would likely affect the climate over the whole earth so that continental drift and consequent deglaciation may have "triggered" the large changes in life at the end of the Paleozoic, just as the deglaciation on the infra-Cambrian may have "triggered" the evolutionary changes at its beginning (Harland, 1964a; Rudwick, 1964). If this argument is correct, continental drift at the end of the Paleozoic is seen to be the essential environmental cause of the geological and paleontological changes which mark the Paleozoic-Mesozoic boundary.

For the reasons already given, the Ewing-Donn hypothesis may not apply to the infra-Cambrian glaciation, and this problem will not become clear until paleomagnetic observations are available from these levels in many parts of the world, a task which is proving to be technically difficult (Harland, 1964b; Mumme, 1963; Briden, 1964). Regarding the current ice age, objections have been made to the Ewing-Donn hypothesis on the grounds that the paleomagnetic results indicate that the pole entered the Arctic basin in the early Cenozoic, yet glaciation did not develop until the end of the Tertiary, so that there was too large a time lapse between the development of an apparently suitable geographical environment and the initiation of glaciation. This argument assumes that there has been no relative movement of the continents. The relevant procedure is to draw latitude maps of the continents for successive periods of time from Cretaceous to the present. Unfortunately, this can at present only be done for the Cretaceous and Upper Tertiary (Figs. 6 and 7). In the Cretaceous, Alaska and the tip of Siberia are near together, and the Atlantic appears on this evidence to have been about half closed. The western Cordillera of North America and the ranges of eastern Siberia are of Cretaceous or later age, and the stippled area is likely to have been open sea. The pole need not therefore have been within an embryo Arctic ocean. In Upper Tertiary time the paleomagnetic results suggest relations similar to those at present, but the ages of most of the rocks on which Figure 7 is based are not, in general, accurately known, and it is therefore uncertain to what part of the Upper Tertiary this map refers. Obviously no case in

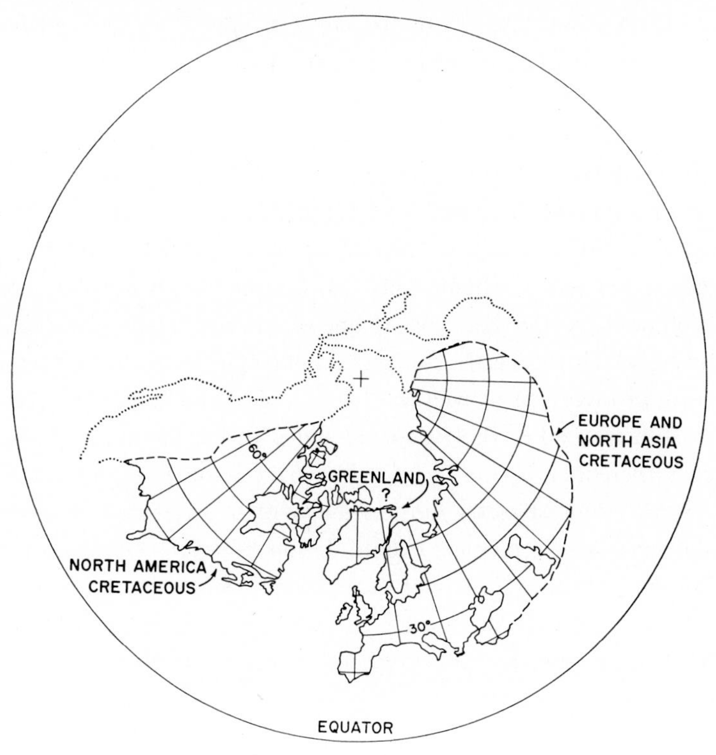

Figure 6. Cretaceous paleolatitudes. Redrawn from base maps in Irving (1964).

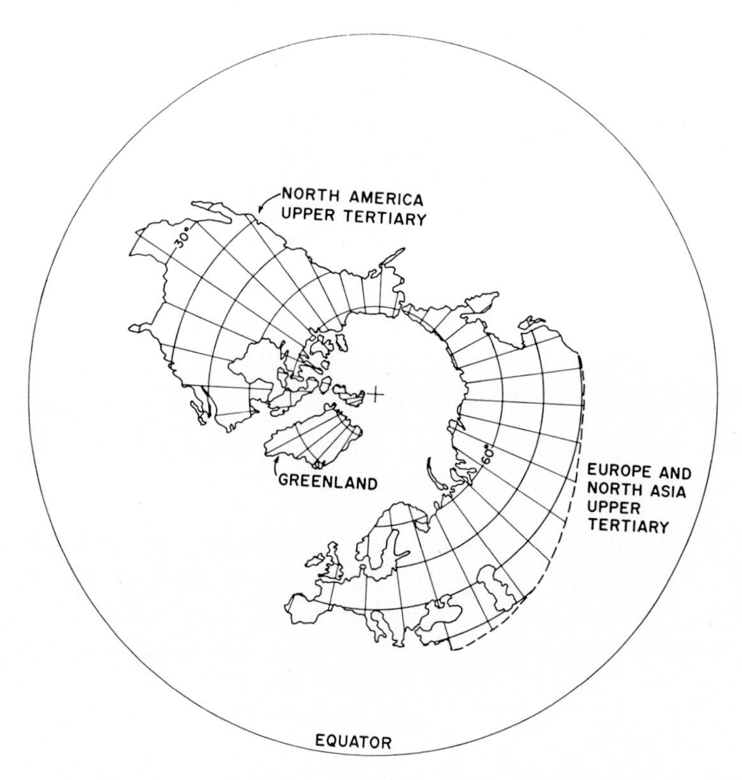

Figure 7. Upper Tertiary latitudes. Redrawn from base maps in Irving (1964).

support of the Ewing-Donn hypothesis for the initiation of the current ice age can be made on the basis of the present incomplete data, but it is clear that the paleomagnetic observations are not necessarily inconsistent with it as may appear at first sight.

Finally, some mention is made of the possible relation of drift phases (and their associated geological expressions) to changes in the frequency of reversal of the geomagnetic field. In the Lower Carboniferous frequent reversals have been observed (e.g., Wilson and Everitt, 1963). The apparent movement of the pole (and glaciation) across Gondwanaland in the late Carboniferous, as seen in the Australian observations, is immediately followed by the Kiaman Magnetic Interval (late Carboniferous to late Permian), during which the earth's magnetic field appears to have been consistent in polarity (positive or reversed) as well as direction. The end of this interval is correlated approximately with the end of the Paleozoic and with the initiation of the break-up of Gondwanaland. Thereafter in the Triassic the field is observed to reverse frequently (e.g. Picard, 1964). This suggests the possibility of a correlation, in time, of changes of reversal frequency with geological events. If this is substantiated by future observations it would represent an extraordinary coincidence, and some mechanism connecting processes in the core and on the earth's surface would be required. This mechanism would need to have a time constant very much less than those commonly associated with convection in the mantle.

University of Leeds (E. I.)
and
Dominion Observatory (W. A. R.)

Discussion

DR. PRESS: Is there a problem of reliability of the inclination in sediments?

DR. IRVING: Yes, there is a problem, and if a substantial systematic inclination error is present it would produce errors in the latitude maps. In most of the maps we have shown there is good "igneous" control of inclinations, and the statistical comparison of directions in igneous rocks and sediments (Creer, 1967; Irving, 1967) shows that no systematic errors are observable in real sediments, although they are known to occur under certain laboratory conditions (King, 1955).

DR. ANDERSON: Did you suggest that the maximum drift of continents took place just when the field of the earth was reversing polarity?

DR. IRVING: This is a puzzle. It does appear in these two cases that there are recognizable drift episodes which occur at the change of reversal frequency in the late Carboniferous, and at the end of the Paleozoic. Why this is so, I really don't know.

REFERENCES

Blackett, P.M.S., J. A. Clegg, and P. H. S. Stubbs (1960), An analysis of rock magnetic data, *Proc. Roy. Soc. A.* 256:291-322.

Boer, de J. (1963), The geology of the Vicentinian Alps (N.E. Italy) with special reference to their palaeomagnetic history, *Geol. Ultriectina* 11:1-178.

Boer, de J. (1965), Paleomagnetic indications of megatectonic movements in the Tethys, *J. Geophys. Res.* 70:931-44.

Briden, J. C. (1964), Thesis, Australian National Univ.

Briden, J. C. (1966), Estimates of direction and intensity of the palaeomagnetic field from the Mugga Mugga Porphyry, Australia, *Geophys. J. (Roy. Astron. Soc.)* 11:267-78.

Collinson, P. W., and S. K. Runcorn (1960), Polar wandering and continental drift: evidence of paleomagnetic observations in the United States, *Geol. Soc. Amer. Bull.* 71:915-58.

Creer, K. M. (1965), Paleomagnetic data from the Gondwanic continents, *Phil. Trans. Roy. Soc.* 258:27-40.

Creer, K. M. (1967), Systematic errors in paleomagnetic inclination in sedimentary rocks, *Nature* 213:282-83.

Creer, K. M., E. Irving, and S. K. Runcorn (1957), Geophysical interpretation of paleomagnetic directions from Great Britain, *Phil. Trans. Roy. Soc. A.* 250:144-56.

Ewing, M., and W. L. Donn (1956), A theory of ice ages, *Science* 123:1061-66.

Ewing, M., and W. L. Donn (1966), A theory of ice ages III, *Science* 152:1706-12.

Gough, D. I., N. D. Opdyke, and M. W. McElhinny (1964), The significance of paleomagnetic results from Africa, *J. Geophys. Res.* 69:2509-19.

Green, R. (1959), Ph.D. Thesis, Australian National Univ.

Harland, W. B. (1964a), Critical evidence for a great infra-Cambrian glaciation, *Geol. Rundschau* 54:45-61.

Harland, W. B. (1964b), Evidence of Late Precambrian glaciation and its significance, pp. 119-149 in *Problems in Palaeoclimatology*, A. E. M. Nairn, ed., Interscience-Wiley, New York.

Hilten, D. van (1964), Evaluation of some geotectonic hypotheses by paleomagnetism, *Tectonophysics* 1:3-71.

Irving, E. (1964), *Paleomagnetism*, Wiley, New York, 399 pp.

Irving, E. (1966), Paleomagnetism of some Carboniferous rocks from New South Wales and its relation to geological events, *J. Geophys. Res.* 71:24.

Irving, E. (1967), Evidence for paleomagnetic inclination error in sediments, *Nature* 213:283-84.

Irving, E., and R. Green (1958), Polar movement relative to Australia, *Geophys. J.* 1:64.

Irving, E., and L. G. Parry (1963), The magnetism of some Permian rocks from New South Wales, *Geophys. J.* 7:395-411.

Irving, E., W. A. Robertson, and P. M. Stott (1963), The significance of the paleomagnetic results from Mesozoic rocks of eastern Australia, *J. Geophys. Res.* 68:2281-87.

Khramov, A. N., V. P. Rodionov, and R. A. Komissarova (1965), New data on the Paleozoic history of the geomagnetic field in the U.S.S.R., pp. 206-213 in *Present and Past Geomagnetic Field*, Nauka Press, Moscow.

King, L. C. (1962), *The Morphology of the Earth*, Oliver and Boyd, Edinburgh.

King, R. F. (1955), The remanent magnetism of artificially deposited sediments, *Monthly Notices Roy. Astron. Soc., Geophys. Suppl.* 7:115-34.

Lowenstam, H. A., and S. Epstein (1954), Paleotemperatures of the post-Aptian Cretaceous as determined by the oxygen isotope method, *J. Geol.* 62:207-48.

Maack, R. (1960), The paleogeography of Gondwanaland, *Proc. 21st Intern. Geol. Congr.* 12:35-55.

Mumme, W. G. (1963), Ph.D. Thesis, Univ. Adelaide, Australia.

Opdyke, N. D. (1964), The paleomagnetism of the Permian Red Beds of Southwest Tanganyika, *J. Geophys. Res.* 69:12.

Picard, M. D. (1964), Paleomagnetic correlation of units within Chugwater (Triassic) Formation, West Central Wyoming, *Bull. Amer. Assoc. Petrol. Geol.* 48:269-91.

Rudwick, M. J. S. (1964), The infra-Cambrian glaciation and the origin of the Cambrian fauna, pp. 150-155 in *Problems in Paleoclimatology*, A. E. M. Nairn, ed., Interscience.

Wegener, A. (1924), *The Origin of Continents and Oceans* (English translation by J. G. A. Skerl), Methuen, London, 212 pp.

Wilson, R. L., and C. W. F. Everitt (1963), Thermal demagnetization of some Carboniferous lavas for palaeomagnetic purposes, *Geophys. J.* 8:149-64.

PALEOCLIMATIC EVIDENCE OF A GEOCENTRIC AXIAL DIPOLE FIELD

JAMES C. BRIDEN

In a symposium devoted to aspects of the history of the earth's crust the place of a contribution on climates (essentially an atmospheric phenomenon) and the earth's magnetic field (ultimately a deep internal phenomenon) may not be immediately obvious. The link between climates and the magnetic field as observed at the earth's surface is that each shows variation with latitude. It may be supposed that some sort of variation of climate and field direction with latitude has persisted through geological time. If this is so, then studies of past climate and the ancient magnetic field both have a bearing on ancient latitudes and hence on the problem of continental drift—the crustal phenomenon.

As far as we know, the factors controlling climate are independent of those controlling the earth's magnetic field, so comparison of paleoclimatic and paleomagnetic data provides an extremely powerful test of whether the interpretation of both these sets of data is sound. The particular question to be discussed here is whether the model on which paleomagnetic data are commonly interpreted is sound.

The methods of paleoclimatology and of paleomagnetism are the classical geological ones, of extrapolating from the present into the past. The first step, then, is to try to ascertain which of the features of climatic distribution and of the geomagnetic field have persisted through geological time. In other words, we need paleoclimatic and paleomagnetic models, based on the modern situation, and which are applicable to the greatest possible extent in space and time to the geological record. We also need paleomagnetic and paleoclimatic data to apply to these models, and then the purpose is to study whether both models and both kinds of data are all compatible with the same paleogeography—the same distribution of continents and oceans—at each interval of geological time.

Taking the paleoclimatic aspects first, ideally one would look for "fossil climates" of a particular age in all latitudes (and preferably in all longitudes as well) and compare them with the present distribution. Unfortunately there is no such thing as a fossil climate, so it is necessary to find "paleoclimatic indicators"—geological records which reflect climate, or some element of climate. These records may be found in environments where climate is likely to have been a major factor, namely land areas, shallow seas, and the surface layers of oceans. Three types of paleoclimatic indicator may be recognized: faunas, floras, and sedimentary rocks. The distribution of many of these is latitude-dependent now, and is likely to have been so in the past. To deny the persistence of this latitude-dependence is to deny the operation of present-day physical and biological processes in the past. In particular, it would be a denial of the planetary temperature gradient (Fig. 1). The density distribution of very many climatic indicators at the present time shows a maximum at the equator and either a polar minimum or a whole high-

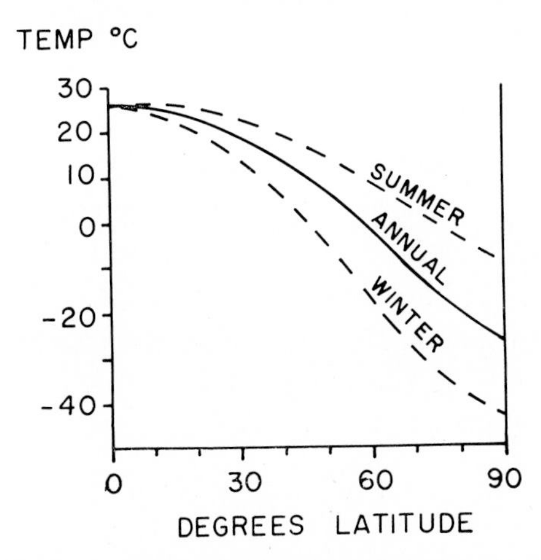

Figure 1. Variation of mean temperature with latitude, after Blackett (1961).

latitude zone from which the indicator is absent (for example reefs, evaporites, and carbonates). A less common distribution, seen today in the distribution of glacial phenomena and some deciduous trees for example, has a maximum in polar or intermediate latitudes.

The paleoclimatic model on which ancient distributions are to be interpreted must be as surely founded as possible, and hence must depend on the minimum of assumptions; it must also be widely and easily applicable to as much of the geological record as possible, and preferably in a quantitative fashion. The simplest model is derived from the fact that the net solar flux reaching the earth's surface has a maximum at the equator and minima at the poles. Temperature follows the same pattern (Fig. 1), and it is difficult to see how it could have been otherwise on an earth as we know it. Even in periods of glaciation when ice caps may have been more pronounced around one pole than the other, and the symmetry of climatic zonation was temporarily disturbed, the general form of the temperature gradient would not have been disrupted. This simple model does *not* require that the actual value of equatorial temperature or the latitudinal temperature gradient was constant through geological time, but it does insist that maxima and minima have not been reversed or destroyed. Plumstead (1965) puts this another way, with reference to plants: "The argument that ancient and extinct plants may have behaved differently (from their modern analogues) would not be supported by modern biologists since light and temperature controls of plant life are fundamental principles." However, this does not mean that the same *tolerance* of a range of environments was exhibited by plants in the past. This tolerance is only one of many potential variables which could alter distributions and have no relevance to latitude change. Others are *barriers*, which may be oceans, zones of excessive rainfall or aridity, exceptionally saline water, or mountain ranges. There are also *distorting* influences affecting distribution, which could be ocean currents, or an overall change in the planetary temperature gradient between glacial and non-glacial times, or a change in the total amount or radiation reaching the earth. These and many other "noise" factors preclude the numerical estimation of paleolatitude from individual occurrences of paleoclimatic indicators.

The method of applying the paleomagnetic model is similar, but it is simpler because we do have more or less direct fossil preservation of the paleomagnetic field direction, the correct determination of which is the purpose of many paleomagnetic techniques. Three possible models for their interpretation come to mind: (1) The field has not been dipolar. This is not a useful model because it does not predict anything which can be tested, nor is it justified by the present situation. It would only be adopted as a model if evidence of a particular non-dipole field configuration came to hand. (2) The field has been predominantly that of a geocentric dipole. This is the present situation, and it is supported by the consistency of paleomagnetic results within single crustal units. This consistency is both lateral at any one time, and sequential at any one place. It has been confirmed at least since Permian time by many paleomagnetic studies, which have been reviewed many times, for example by Irving (1964), and there is no need to repeat the details here. (3) The third model is that of a geocentric axial dipole. The most important property of this model is that paleo*magnetic* poles, latitudes, and equator become paleo*geographic* elements as well. The purely paleomagnetic evidence of this is confined to the last 20 million years. Paleomagnetic poles determined from Plio-Pleistocene rocks from each continent are grouped around the geographic (rather than the geomagnetic) pole. There seems to be further support from inclination measurements on deep ocean sediments (Opdyke et al., 1966; Opdyke, this volume). If continental drift has occurred, it is difficult to imagine how paleomagnetic data *alone* could be used to test the model for pre-Neogene time. The principal alternative line of evidence is from paleoclimatology.

One method of testing the geocentric axial dipole field model is to draw paleogeographic maps using paleoclimatic evidence alone, from which paleolatitudes might be estimated. Paleomagnetic data could then be superimposed; this would then illustrate the ancient magnetic field configuration. This and other methods which depend primarily on the interpretation of paleoclimatic data have failed so far because of the difficulties, mentioned earlier, of making numerical estimates of paleolatitude from paleoclimatic evidence alone. Until this problem is overcome tests must begin with the assumption of the geocentric axial dipole model and the application of paleomagnetic data. Paleolatitudes of occurrences of paleoclimatic indicators can then be estimated. The compatibility of these estimates with the paleoclimatic model is then the test of the validity of the paleomagnetic model.

In applying paleomagnetic data it is important not to exaggerate the range of its validity either in distance (across orogenic belts or on to other continents from the area studied) or in time (over periods in which the sampling area could have moved significantly). A complete paleomagnetic record from all continents would facilitate reconstructions of the relative positions of the continents for all instants of geological time. Due to the inadequacy of the available data, only approximate reconstructions can be made for some geological ages, and in general the data do not give an accurate fix on longitude. They do, however, define paleolatitudes well for many regions and for large portions of Phanerozoic time, and it is on paleolatitudes that the rest of this discussion is concentrated.

The first step in calculating the paleolatitudes of a crustal unit in which a local paleomagnetic study has been made is to compute the paleomagnetic pole. The ancient meridian through the sampling area is given by the declination or bearing D of the mean direction of stable remanent magnetization. The arc distance from the sampling site, along the ancient meridian, to the paleomagnetic pole is given by the simple relation between paleolatitude λ and the mean inclination or dip I of the stable remanence: $\tan I = 2 \tan \lambda$. This locates the pole. Paleolatitude lines can now be drawn across the crustal unit which was studied; they are arcs of circles centered on the paleomagnetic pole.

Paleolatitudes and the distribution of paleoclimatic indicators may be plotted on four distinct types of diagram: (1) Time variations of paleolatitude at any single place may be compared with paleoclimatic evidence from that place. A diagram of this type was drawn by Wegener (1924, p. 110) and was first used with paleomagnetic data by Irving (1956). Figure 2 shows paleolatitude variations for Grafton, Australia, together with estimates of development of reefs in southeast Australia. The Carboniferous and earlier records come from the Paleozoic basins mainly in central New South Wales. There are no later occurrences until the beginning of the Great Barrier Reef in the mid-Tertiary. The absence of reefs during the period when Australia was in high latitudes, and their

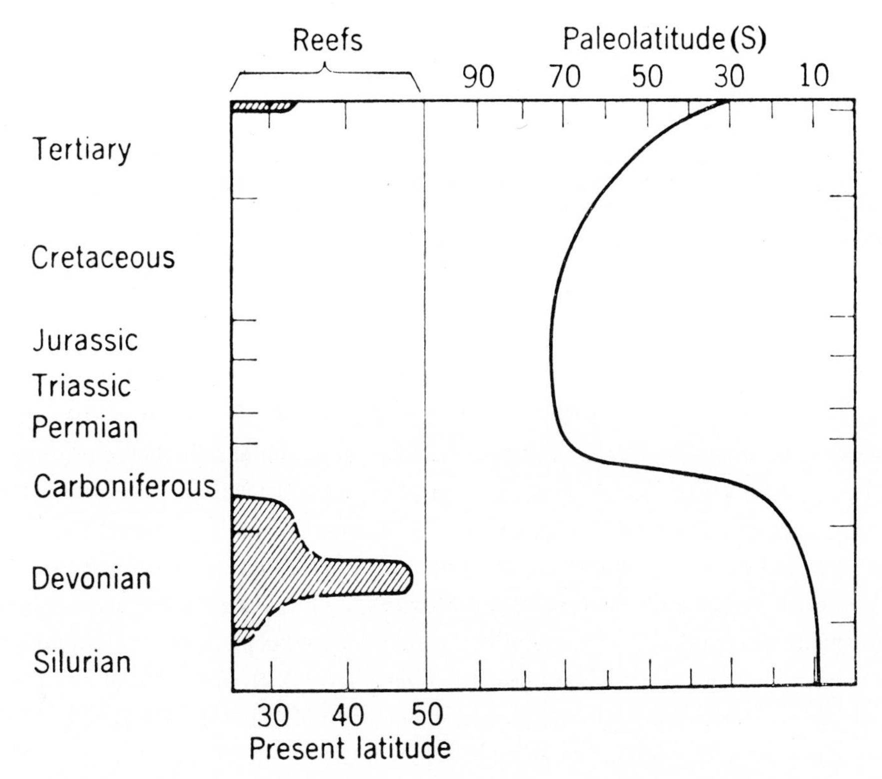

Figure 2. Comparison of reef development and paleolatitude in eastern Australia, reproduced from Irving (1964). The reef development is plotted as a function of present latitude on the left, from Teichert (1952); paleolatitude variation for Grafton, New South Wales, is on the right.

presence when the continent was in low to intermediate latitudes correlates well with the presumed affinity of reef builders with warm environment. (2) Paleolatitudes and the paleoclimatic indicators may be plotted on the same map (Creer et al., 1959; Opdyke, 1959, 1962; Runcorn, 1961; Irving and Gaskell, 1962). In the example shown in Figure 3 most limestones, evaporites and redbeds are in low paleolatitudes, while

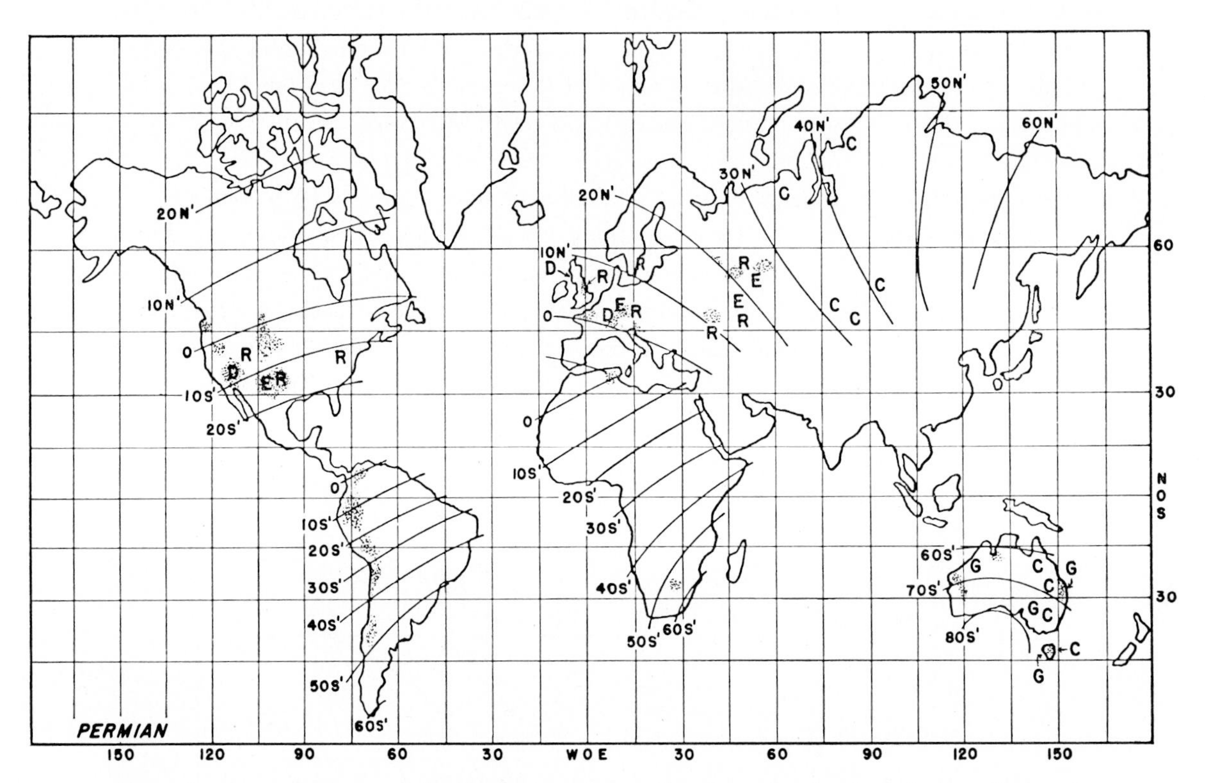

Figure 3. Generalized map of the distribution of some Permian paleoclimatic indicators. Carbonates are denoted by shading, evaporites E, redbeds R, desert sandstones D, glacial deposits G, and coals C. Redrawn, with additions, from Briden and Irving (1964). The paleolatitudes are compiled from Irving's (1964) A-group data plus results from Africa (Opdyke, 1964) and South America (Creer, 1964).

glacial deposits are much nearer the pole. With floral and faunal data, it may be possible to contour such maps to show abundance or diversity variations in these indicators, as Stehli et al. (1967) have done for recent organisms. (3) The latitude (or paleolatitude) range of the main occurrences of paleoclimatic indicators may be plotted against geological time. Figure 4 is of this type; it is highly schematic due to the limitations of the data. The full utilization of this type of representation will have to await nearly complete paleomagnetic coverage. (4) Quantitative estimates may be made of the abundance of paleoclimatic indicators at all latitudes (and paleolatitudes) and plotted in diagrams like Figures 5-7. This approach has been discussed in detail before (Briden and Irving, 1964), and here a few examples will suffice to illustrate the pitfalls involved and the attempts to overcome them. The paleomagnetic coverage is summarized in Table 1 in the form of pole positions and the times and areas to which each is considered applica-

Table 1. Average paleomagnetic poles used in the compilations of Figures 5-7. Irving's (1964) A-group data.

Period	Europe and North Asia	Japan	India	Africa	Australia	North America	South America	Greenland	Antarctica
Cambrian	3N 18W				35N 159W	12N 9W			
Ordovician	15N 148E					20N 153E			
Silurian	25N 132E			50N 11W	32N 97E	26N 139E			
Devonian	31N 157E				65N 165W				
				(Lower)	84N 134W				
Carboniferous	30N 150E					36N 132E			
				(Upper)	48N 46W				
Permian	43N 167E				45N 52W	38N 105E		38N 163E	
Triassic	50N 147E			53N 44W	53N 28W	59N 106E		68N 160E	
Jurassic	57N 107E		13N 69W	71N 85W	47N 34W		83N 112W		52N 40E
Cretaceous	72N 168E	42N 151W		68N 168W[1]	54N 33W	67N 171W			
L. Tertiary	76N 150E		38N 83W		65N 50W				
U. Tertiary	83N 141W	84N 62W				88N 117W			

[1] Madagascar

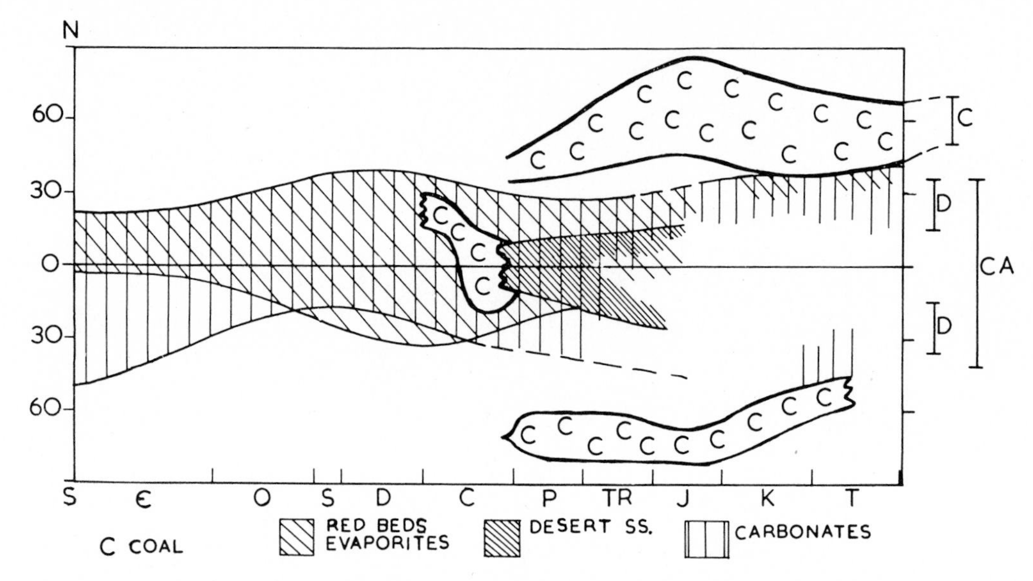

Figure 4. Generalized plot of variation of paleolatitude range of major paleoclimatic indicators with time. Latitude is plotted against Kulp's (1961) time scale, and geological periods are marked. On the right the range of present-day counterparts is shown: desert sands D, shallow marine carbonates CA, and peat C. From Briden and Irving (1964).

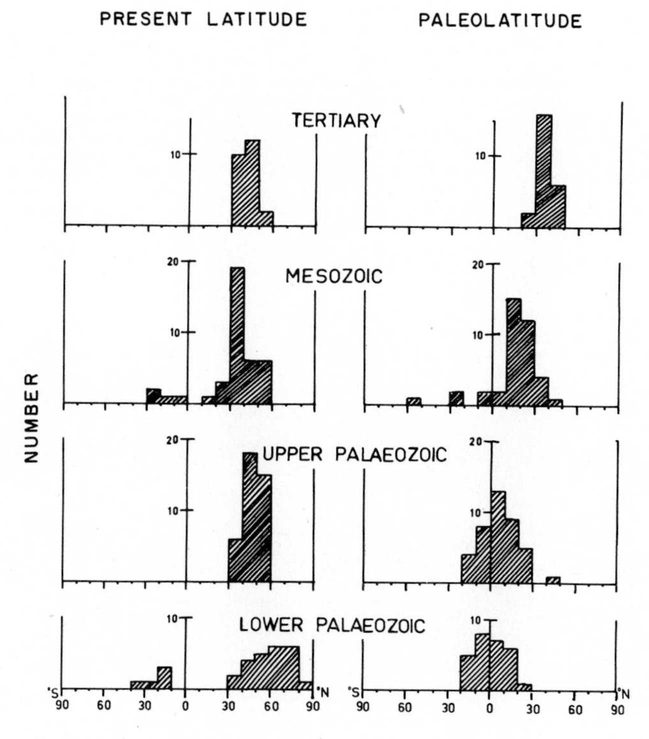

Figure 5. Equal-angle frequency histograms of evaporites, showing time variations. From Briden and Irving (1964), where the sources of evaporite data are given.

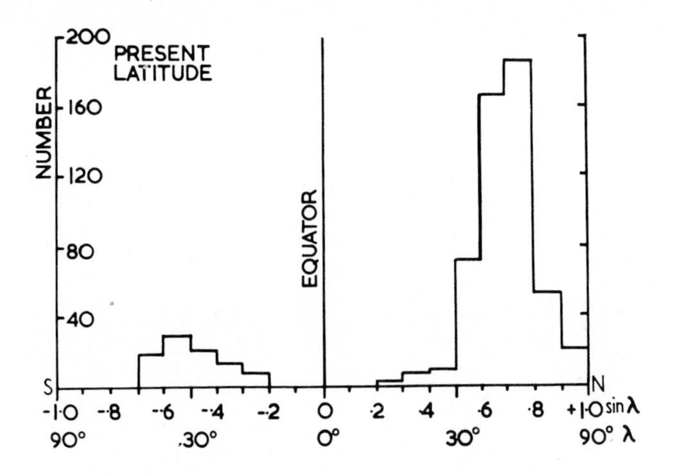

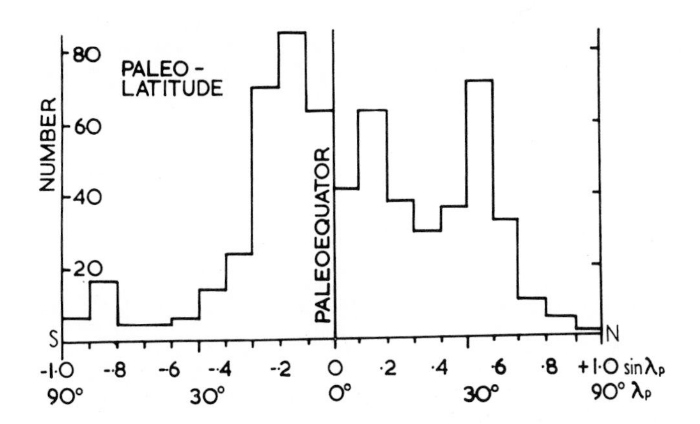

Figure 6. Equal-area frequency histograms for carbonates, from Briden and Irving (1964), where the sources of carbonate data are given.

ble. Figure 5 summarizes a compilation of evaporites of which paleolatitudes can be estimated. One entry is made per geological period for each 300-mile-diameter circular area in which evaporites of that geological age occur. The present latitude spectrum is generally similar to that of present-day land distribution; in other words the sample is fairly uniform with respect to present latitude. The paleolatitudes are generally close to the paleoequator. This is especially true for the Paleozoic; a northward shift of the spectrum in the Mesozoic and Tertiary is interpreted in terms of greater abundance of lagoonal environments in the northern hemisphere. Figure 6 summarizes a similar study of carbonates, with similar results, though breakdown of these results into successive time intervals would show that the symmetry is preserved through the Mesozoic. The weighting in this case was achieved by compiling an extensive list from standard geological literature and again limiting the number of entries per locality to one per geological

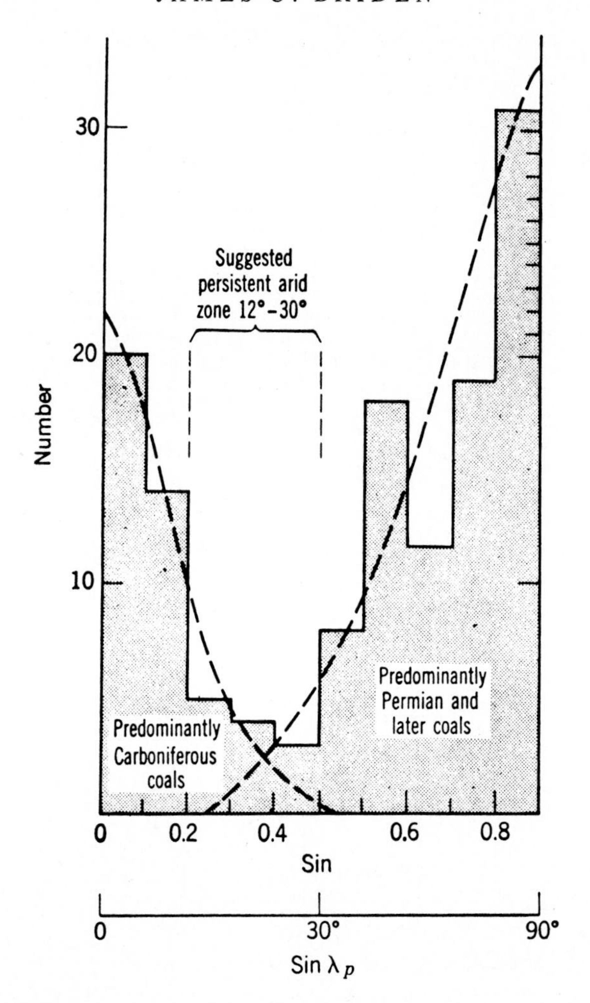

Figure 7. Equal-area frequency histogram of paleolatitudes of coalfields, with both hemispheres plotted together. The dashed line shows the tentative division into two groups, centered one about the equator, the other near the poles. Reproduced from Irving (1964) using data of Briden and Irving (1964).

period. In this histogram the latitude scale is linear in sin λ; each division thus corresponds to an equal area on the earth's surface, and this type of plot is referred to as an equal-area histogram.

The only example available which relates to plant distributions is that of coalfields. In Figure 7 their paleolatitudes are plotted irrespective of which hemisphere they occur in (the histograms for each hemisphere are similar), on an "equal-area" histogram. The occurrences fall into two groups: those in low latitudes are predominantly the Carboniferous coals of western Europe and North America. The high paleolatitude group are mainly the Permian and younger coals from Canada, Siberia, and the southern continents. These two groups are also distinct in the fossil floras they contain.

An example of the paleolatitude distribution of a faunal group is that of Irving and Brown (1964, 1966), on labyrinthodont reptiles. This was an attempt to show latitude variation in taxonomic diversity. The present latitudes show a group in each hemisphere, the northern group being much the more numerous. The paleolatitude distribution (Figure 8) shows fair symmetry about the paleoequator. Discussion of this result (Stehli,

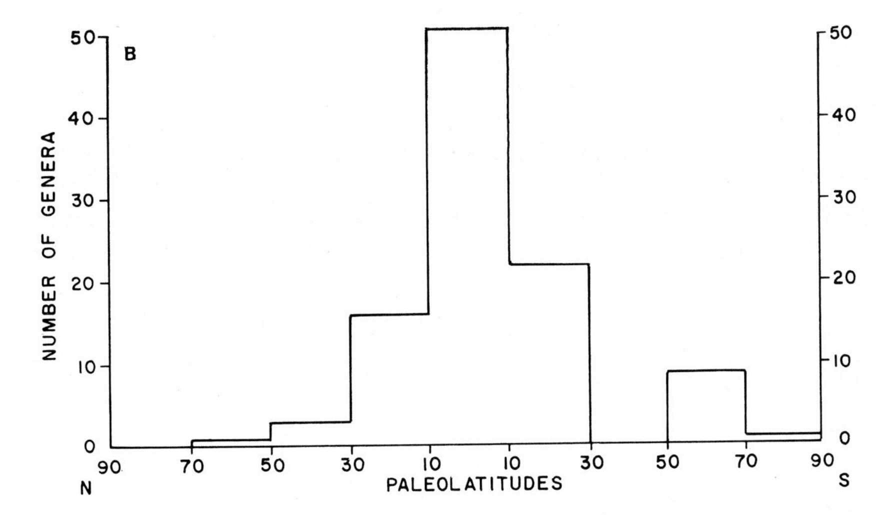

Figure 8. Generic frequency versus paleolatitude for Upper Carboniferous and Permian labyrinthodonts, modified from Irving and Brown (1966).

1966; Irving and Brown, 1966) illustrated some of the decisions which have to be made in analyzing data of this sort. How large an area should be treated as a single station, meriting a single entry in the compilation? What taxonomic level should be treated? Definitions of genera and species can be somewhat subjective and variable from one paleontologist to another. A corollary to this is how one determines taxonomic diversity at each station.

The examples illustrate the three major problems in interpreting these attempts at quantitative analysis. (1) Is the time coverage adequate and (2) is the areal coverage adequate? These problems relate chiefly to the incompleteness of the paleomagnetic record (Table 1). A possible bias here is a consequence of the scarcity of paleomagnetic evidence of high latitude. Briden and Irving (1964) attempted to overcome this by pointing out the rarity or absence of "low paleolatitude indicators" in the three large sedimentary basins which were at high paleolatitude—western Canada (Cretaceous), central Australia (late Carboniferous to early Tertiary), and Siberia (Permian to Cretaceous). To these can now be added South America (Carboniferous and Permian) and Southern Africa (Carboniferous to Cretaceous). In these areas clastic sediments predominate in the marine basins at high paleolatitudes, while within the same continents marine sequences at low to mid-southerly paleolatitudes contain abundant carbonates and occasional evaporites. (3) Is there a bias in the quantitative estimates of geological abundance? The incompleteness of the geological record prohibits a perfect weighting system; the systems used here would not be expected to bias the *range* of the paleolatitudes. The limitation on the number of entries where indicators are abundant tends to flatten out the peaks, yet maxima are still observed and are likely to be genuine. The sources of geological information listed by the original compilers of these histograms are certainly not completely comprehensive, but the authors were of the opinion that they were adequate and that neither the acquisition of new data nor the adoption of different weighting schemes would alter their major conclusions. In support of this view, the new paleomagnetic results from South America and Africa do not alter the paleolatitude histograms very much.

The histograms also respond rather sensitively to the use of invalid models—invalid sets of latitude estimates—in that north-south shifts will result in the shift of an equatorial maximum; rotational movements cause a spreading of the peaks. Longitudinal movement is not detected. One obviously invalid framework (Fig. 9) is one in which a

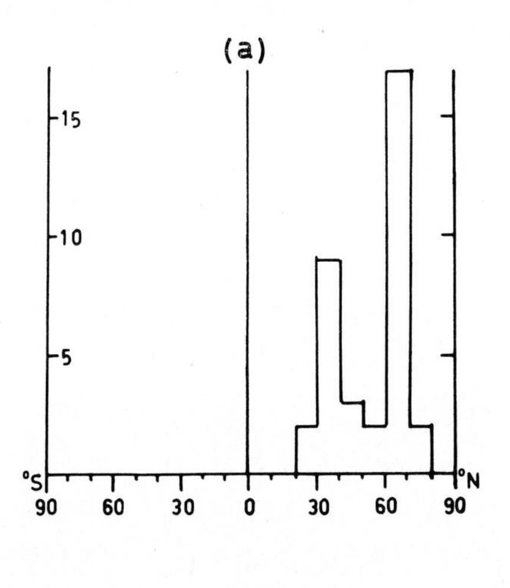

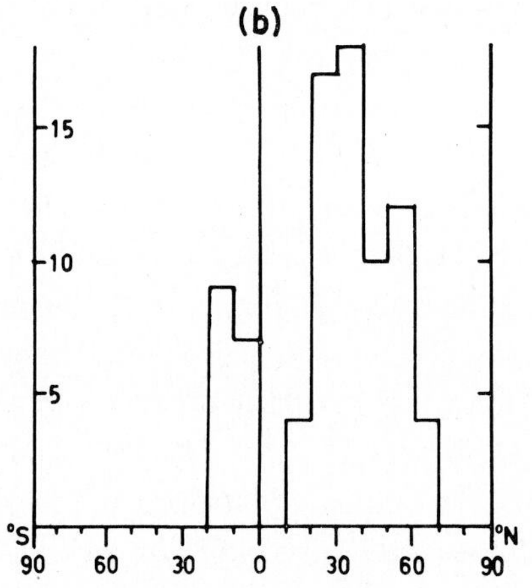

Figure 9. Equal-angle frequency histograms of paleolatitude of (a) Triassic and (b) Devonian carbonates from North America, Eurasia, and Australia calculated from Australian paleomagnetic data only, assuming no subsequent relative movement between continents. From Briden and Irving (1964).

single set of paleomagnetic data has been applied worldwide. In this example the Australian data have been used to calculate paleolatitudes of Triassic and Devonian carbonates from North America and Eurasia as well as Australia. The two histograms are different from each other and from other spectrums of past and present latitude. They are meaningless.

The evidence presented here suggests that the geological and paleomagnetic coverages are adequate for a judgment to be made between the validity of the two paleographic frameworks—one the present distribution of continents and oceans (summarized by the present latitude spectrums) and the other suggested by paleomagnetism. This judgment must be made on the consistency of each set of spectrums for each interval of geological time, and on which of the two frameworks is consistent with the paleoclimatic model outlined at the beginning of this paper. The present latitude spectrums vary through geological time, and none of them shows maxima or minima in the latitudes expected on the paleoclimatic model. The paleolatitude spectrums, by contrast, are in general consistent with each other and with the model. Equatorial maxima are seen in most of them, and the deviations from this situation can be explained in terms of slight biases in the distribution of appropriate environments. The data extend back to the Cambrian, but most of them refer to the Devonian onward. The data strongly favor the view that for that part of geological history, if the earth's magnetic field has not been close to that of a geocentric axial dipole, it has not been far from it.

ACKNOWLEDGMENTS

I am indebted to John Wiley and Sons and Dr. E. Irving for permission to reproduce Figures 2-7 and 9; and the proprietors of the American Journal of Science, and Drs. E. Irving and D. A. Brown for permission to reproduce Figure 8.

Geology Department
University of Birmingham

DISCUSSION

DR. BULLARD: I think one really good example is worth a lot of rather weak ones. One that has always struck me is the existence of the *Glossopteris* flora at 86°S, within a few hundred miles of the south pole. I find it difficult to believe that there is a plant which will grow where there are six months day and six months night.

DR. IRVING: The paleolatitudes of *Glossopteris* extend to 75°S.

DR. BOUCOT: Willow trees grow as far north as land goes in the northern hemisphere (82°N). In northern Greenland there are small birches and willows which experience about two continuous months respectively of both day and night. Coal is forming in the swamp environment there. You do not need twelve months of light to grow trees.

DR. OPDYKE: I agree. The most important feature of the *Glossopteris* distribution is that the same species are found in Antarctica as in Tanganyika (10°S) and in India (15°N), Australia, and South America.

DR. BOUCOT: This is more of a taxonomic than a climatic problem. We have endless genera of plants and mammals growing in arctic conditions which are present in temperate and nearly tropical environments. To asume that our taxonomy of the Permian is so perfect that we can tell without any doubt one species from another and be sure that environmental tolerances are within a few degrees centigrade, is really pushing it a bit far.

DR. STEHLI: This distribution could be explained by a very mild climate over a very broad area, in the same way that you now find these willows that grow in northern Greenland growing in alpine regions of the United States or even Mexico.

DR. OPDYKE: Yes, but this is the argument Axelrod (1963) uses when he says that the low-altitude flora in Africa was selectively destroyed and the high-altitude (*Glossopteris*) flora was selectively preserved. I think this is nonsense.

DR. BOUCOT: It seems pretty obvious that these coal beds at 86°S are not of a high-altitude variety. They give you the impression of being much nearer to sea level.

DR. OPDYKE: . . . As do the African occurrences.

DR. BOUCOT: So if you have a more equable climate, with everything closer to sea level, the fact that you get similar things would not surprise me.

DR. BRIDEN: You find them very close to glacial deposits.

DR. BOUCOT: This is the problem of the glacials in the Horlick, which are not very well dated. Below them you have a disconformity and lower Devonian, and some distance above them you have plants. Since you don't know the times implied, you can't conclude anything about the relation between the plants and the glacials.

DR. KAY: It is of interest to note that the most thorough assemblage of data on the world distribution of carbonates is that by Ronov. I am not judging whether he had a firm basis, but the point is that he came to the conclusion that the present latitudes work perfectly well. Others have come to other conclusions. I suppose that simply means that you have to know which authority you want to get a given conclusion.

DR. BRIDEN: [Written response to Dr. Kay's comment]. I have taken up Dr. Kay's suggestion because Ronov's maps are more likely to give evenly distributed source data than that used in compiling Figures 5-7. It is worth developing further because it leads to a powerful method of presentation and also enables graphical illustration of weaknesses in the sampling coverage. As an example I have analyzed Ronov and Khain's (1957) lithofacies map of the Middle Devonian. They distinguish six sedimentary facies as well as areas of erosion and sub-aerial and marine volcanic facies. I have translated this into quantitative form by tracing along latitude (and paleolatitude) lines at 5°-intervals, using the paleomagnetic data in Table 2, and measuring the arc distances occupied by each facies. Figure 10 shows the present latitude spectrum of continents,

Table 2. Average paleomagnetic poles for the Middle Devonian. The dates of some of the formations studied within the Devonian is not certain. The poles given here are the most reliable ones for each region, using *paleomagnetic* criteria for reliability (but the Chinese data should be treated with reserve), and are considered to be applicable to the Middle Devonian. The data are more up to date than those used in the main body of the paper.

Region	Pole	Formations	References
South America	9N 22W	Red beds	Creer (1964)
North America	25N 119E	Perry Lavas	Black (1964); Phillips and Heroy (1966)
Europe	10N 44W	Old Red Sandstone	Chamalaun and Creer (1963), sediments Stubbs (1958), lavas
North Asia	4N 24W	Sediments, Yenesei River	Popova (1963)
China	61N 49W	"Hematite Rocks"	Wang et al. (1960)
Australia	65N 165W	Canberra igneous rocks Murrumbridge Series Yalwal Stage basalts	Green (1961)

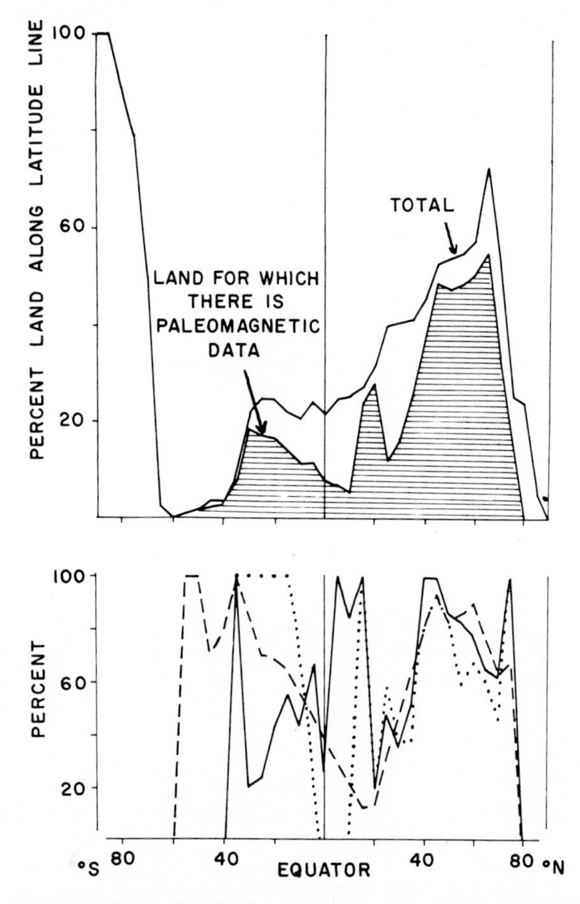

Figure 10. (above) Present latitude distribution of land, and percentage of paleomagnetic coverage in the Middle Devonian. (below) Percentage of paleomagnetic coverage in the Middle Devonian as a function of present latitude for all sedimentary areas (solid line), areas of carbonate and carbonate-mudstone sequences (dotted line), and all present land areas (dashes).

and the proportion of paleomagnetic coverage at each present latitude. Figures 11-13 show the relative proportions of sedimentary facies, the volcanic and erosion areas having been extracted. Figure 11 shows the present latitude distribution for all regions. Figure 12 is similar but refers only to areas for which there is paleomagnetic coverage. The similarity of these two plots is taken to indicate that the paleomagnetic coverage is likely to be adequate for the compilation of a paleolatitude spectrum. They show that

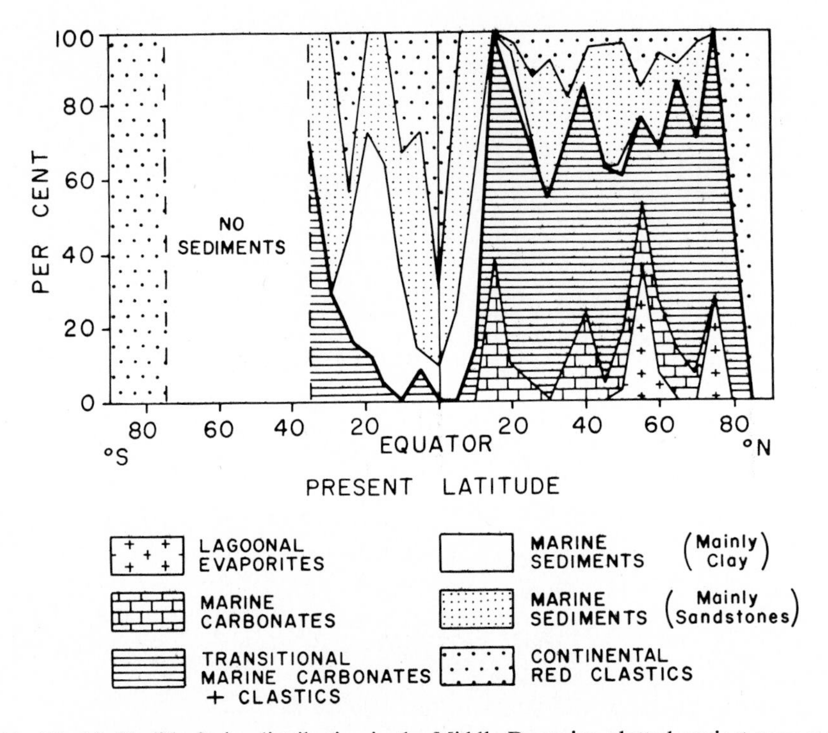

Figure 11. Worldwide lithofacies distribution in the Middle Devonian plotted against present latitude. The thick line divides predominantly chemical from predominantly clastic facies.

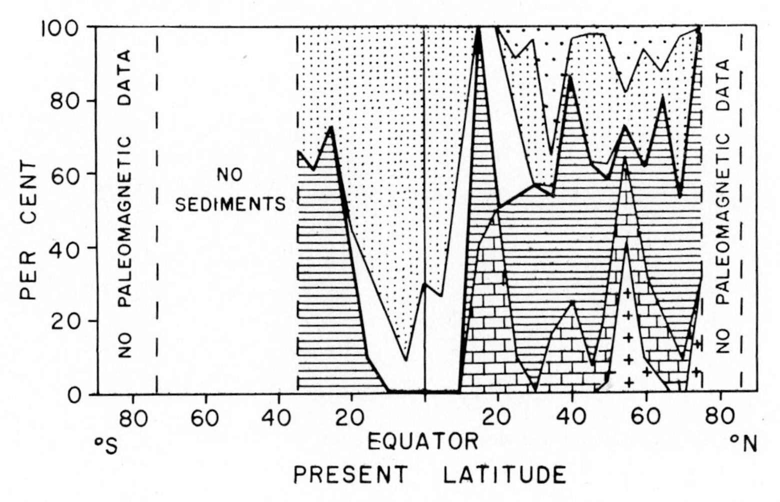

Figure 12. As for Figure 11: areas for which there is paleomagnetic coverage for the Middle Devonian only.

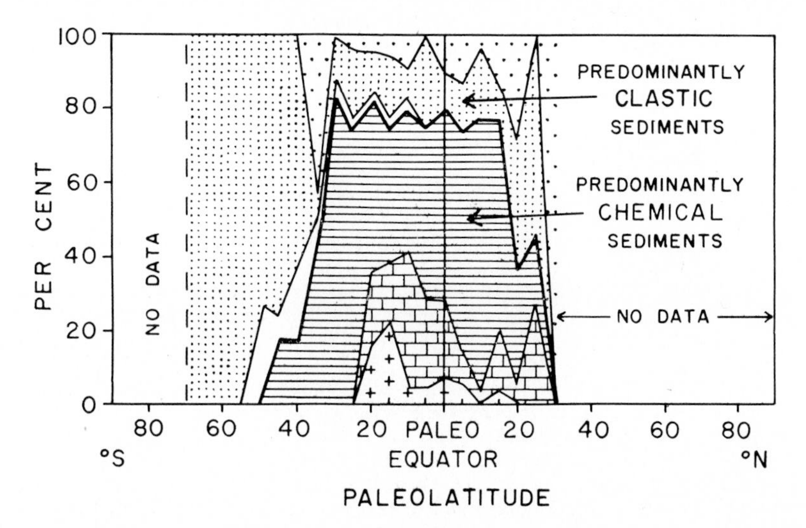

Figure 13. Lithofacies distribution in the Middle Devonian plotted against paleolatitude.

sequences containing a large proportion of chemical sediment all lie farther than 15° from the present equator. This is not what would be expected from the present climatic model. The paleolatitude distribution (Fig. 13), on the other hand, shows a maximum in the relative abundance of chemical sediments near the paleoequator, and increasing proportions of clastics in higher paleolatitudes. This not only agrees with the present climatic model but is consistent with the conclusions presented in the paper. It reinforces the conclusion of Briden and Irving (1964) that their data, though unevenly distributed, were adequate to describe the worldwide situation, and that whatever source of data is used, the same overall conclusion is reached.

REFERENCES

Axelrod, D. I. (1963), Fossil floras suggest stable, not drifting, continents, *J. Geophys. Res.* 68:3257-63.

Black, R. F. (1964), Palaeomagnetic support of the theory of rotation of the western part of the island of Newfoundland, *Nature* 202:945-48.

Blackett, P. M. S. (1961), Comparison of ancient climates with the ancient latitudes deduced from rock magnetic measurements, *Proc. Roy. Soc., Lond. A* 263:1-30.

Briden, J. C., and E. Irving (1964), Palaeolatitude spectra of sedimentary palaeoclimatic indicators, pp. 199-224 in *Problems in Palaeoclimatology*, A. E. M. Nairn, ed., Interscience, New York.

Chamalaun, F. H., and K. M. Creer (1963), A revised Devonian pole for Britain, *Nature* 198:375.

Creer, K. M. (1964), A reconstruction of the continents for the Upper Palaeozoic from palaeomagnetic data, *Nature* 203:1115-20.

Creer, K. M., E. Irving, and A. E. M. Nairn (1959), Palaeomagnetism of the Great Whin Sill, *Geophys. J.* 2:306-23.

Green, R. (1961), Palaeomagnetism of some Devonian rock formations in Australia, *Tellus* 13: 119-24.

Irving, E. (1956), Palaeomagnetic and palaeoclimatological aspects of polar wandering, *Geofis. Pura Appl.* 33:23-41.

Irving, E. (1964), *Paleomagnetism and Its Application to Geological and Geophysical Problems*, Wiley and Sons, New York, 399 pp.

Irving, E., and D. A. Brown (1964), Abundance and diversity of the labyrinthodonts as a function of paleolatitude, *Amer. J. Sci.* 262:687-708.

Irving, E., and D. A. Brown (1966), Reply to Stehli's discussion of Labyrinthodont abundance and diversity. *Amer. J. Sci.* 264:488-96.

Irving, E., and T. F. Gaskell (1962), The palaeogeographic latitude of oilfields, *Geophys. J.* 7:54-63.

Kulp, J. L. (1961), Geologic time scale, *Science* 133:1105-14.
Opdyke, N. D. (1959), The impact of paleomagnetism on paleoclimate studies, *Intern. J. Bioclim. Biomet.* 3:1-6.
Opdyke, N. D. (1962), Palaeoclimatology and continental drift, pp. 41-65 in *Continental Drift*, S. K. Runcorn, ed., Academic Press, New York.
Opdyke, N. D. (1964), The paleomagnetism of the Permian red beds of southwest Tanganyika, *J. Geophys. Res.* 69:2477-87.
Opdyke, N. D., B. Glass, J. D. Hays, and J. Foster (1966), A paleomagnetic study of Antarctic deep sea cores, *Science* 154:349-57.
Phillips, J. D., and P. B. Heroy (1966), Paleomagnetic results from the Devonian Perry lavas near Eastport, Maine, *Trans. Amer. Geophys. Un.* 47:80 (abstr.).
Plumstead, E. P. (1965), Evidence of vast lateral movements of the earth's crust provided by fossil floras, *Intern. Un. Geol. Sci. Upper Mantle Symposium*, New Delhi, 1964, pp. 65-74, Copenhagen.
Popova, A. V. (1963), Paleomagnetic investigations of Paleozoic sedimentary rocks in Siberia, *Akad. Nauk SSSR Izv. Geofiz. Ser.*, pp. 444-50.
Ronov, A. B., and V. E. Khain (1957), Istoriya osadkonakopleniya v srednem i verkhnem paleozoe v svyazi s gertsinskim etapon tektonicheskogo razvitiya zemnoi kory [History of sedimentation during the middle and upper Paleozoic in relation to the stages of the Hercynian tectonic cycle], *Intern. Geol. Congr., 20th, Mexico*, Sect. 5, T1, pp. 113-35.
Runcorn, S. K. (1961), Climatic change through geological time in the light of the palaeomagnetic evidence for polar wandering and continental drift, *Quart. J. Roy. Met. Soc.* 87:282-313.
Stehli, F. G. (1966a), Discussion: Labyrinthodont abundance and diversity, *Amer. J. Sci.* 264:481-87.
Stehli, F. G., A. L. McAlester, and C. E. Helsley (1967), Taxonomic diversity of Recent bivalves and some implications for geology, *Geol. Soc. Amer. Bull.* 78:455-66.
Stubbs, P. H. S. (1958), Continental drift and polar wandering, a palaeomagnetic study of the British and European Trias and the British Old Red Sandstones, Ph.D. thesis, University of London.
Teichert, C. (1952), Fossile Riffe als Klimazeugen in Australien, *Geol. Rundschau* 40:33-38.
Wang, T., H. Teng, C. Li, and S. Yeh (1960), *Acta Geophysica Sinica* 9:125-38.
Wegener, A. (1924), *The Origin of Continents and Oceans* (transl. by J. G. A. Skerl from 3rd German edn.), Methuen, London, 212 pp.

A PALEOCLIMATIC TEST OF THE HYPOTHESIS OF AN AXIAL DIPOLAR MAGNETIC FIELD*

FRANCIS G. STEHLI

Introduction

The design of a test of the axial dipolar nature of the earth's magnetic field is not so simple as has commonly been assumed. Difficulty is encountered primarily because of the problem of obtaining, from either paleomagnetic or paleoclimatic sources, data of a quality and quantity adequate to support a test. The requirements for a test, a consideration of what can actually be tested, and a preliminary test using data for the Permian are presented here.

Requirements for a Test

Assuming that data are available in adequate quantity and quality to support a test, the minimum requirement is that it be possible to fix and compare for the same interval of geologic time the position of some spatial reference point as determined by independent lines of evidence (paleomagnetic and paleoclimatic). If the positions of the reference point coincide, the field is an axial dipole. If they are in substantial disagreement, the field is not an axial dipole.[1]

A simple test would be to compare for some time T_1 the relative goodness of fit of paleoclimatic data on two models of the earth. One of these models should be the paleomagnetic model appropriate for T_1 and the other might be a present-earth model which would be equivalent to a T_1 model, assuming neither continental drift nor polar wandering. For the present test the Permian period is selected to represent T_1. The present-earth model (P.E.M.) is, of course, readily available. It is necessary, however, to examine both the available paleomagnetic and the available paleoclimatic data to see what kind of a test the data will support.

PALEOMAGNETIC DATA

Construction of a Permian paleomagnetic earth model (P.P.E.M.) requires that we invoke three basic assumptions. These assumptions, briefly stated, are the following:

1. Paleomagnetic measurements are valid when extended to Permian rocks.

* Contribution No. 32, Department of Geology, Case Western Reserve University, Cleveland, Ohio 44106.

[1] Since axiality of the field is only one of the assumptions involved in such a test, it is also possible that one or more of the other assumptions could be incorrect. Evidence presented below leads to the conclusion, however, that the assumption of an axial field is the least well substantiated of the assumptions involved.

2. The field in the Permian was dipolar.[2]
3. The field in the Permian was axial.[3]

The result of invoking these assumptions is that the paleomagnetic data then enable one to establish a particular configuration of the earth's continental blocks with respect to their presumed latitudes (longitude is not provided by the measurements as applied to any single time interval). In effect, then, one equates paleomagnetic and paleogeographic latitude. The configuration satisfying the data for the Permian under the assumptions noted is the P.P.E.M. and may be tested for goodness of fit to paleoclimatic data.

The presently available paleomagnetic data for the Permian are extremely limited and are distributed principally in Eurasia west of the Ural mountains and in North America (Fig. 1).[4] It would be desirable, therefore, if a test could be conducted which would utilize one of these regions. Paleomagnetic latitudes have been superimposed on Figure 1 (principally after Briden, this volume) because it is quite clear that with the few measurements available it is not yet possible to make a test involving the entire earth, and it is desirable to seek out for use the regions most likely to provide a critical test. The greater land area and more intensive study which have been accorded to the northern hemisphere suggest that adequate data are most likely to be found there. Figure 1 shows clearly that paleomagnetic measurements for the Permian are concentrated here, and the same is true of potential paleoclimatic indicators. Perhaps the single most critical region is in eastern and southern Asia, where paleomagnetic latitude tends to run almost normal to present latitude. Unfortunately, however, the paleomagnetic latitude lines are not extended into this region by workers in paleomagnetism. Thus, though the area is rich in Permian rocks and paleoclimatic indicators, no test is possible pending further reliable paleomagnetic measurements in the area. The remaining critical area is found in North America and is particularly suitable for a test because in the Permian the paleomagnetic equator ran through the middle of the continent, and it is possible to examine the fit of paleoclimatic data to a spread of latitudes up to about 25°N and 25°S of the presumed equator. From the data for North America given by Irving (1964) and Helsley (1965) a North American pole has been selected which lies at 34°N, 105°E (present-day coordinates). Using this pole, the P.P.E.M. for North America was prepared and is shown in Figure 2. Greenland has been excluded because the virtual pole position from paleomagnetic measurements (38°N, 165°E) suggests it is magnetically unlike the remainder of North America and more akin to Europe.

[2] This assumption requires acceptance of the continental drift hypothesis, for paleomagnetic measurements yield only the position of the magnetic pole with respect to the station at which measurements were made. To obtain a single Permian paleomagnetic pole of each sign, the various continents must be moved into positions which provide coincidence of the paleomagnetic poles as seen from each of them.

[3] An axial field assumes (average) coincidence of the rotational and paleomagnetic poles. Thus, to obtain a unique Permian paleomagnetic pole the single, but otherwise unconstrained, pole made possible by employment of assumption 2 must be brought into coincidence with the rotational pole. This operation may involve acceptance of the polar wandering hypothesis.

[4] The data plotted are those given by Irving (1964) and indicated by him to meet minimum requirements. Data for which no location is given have been omitted.

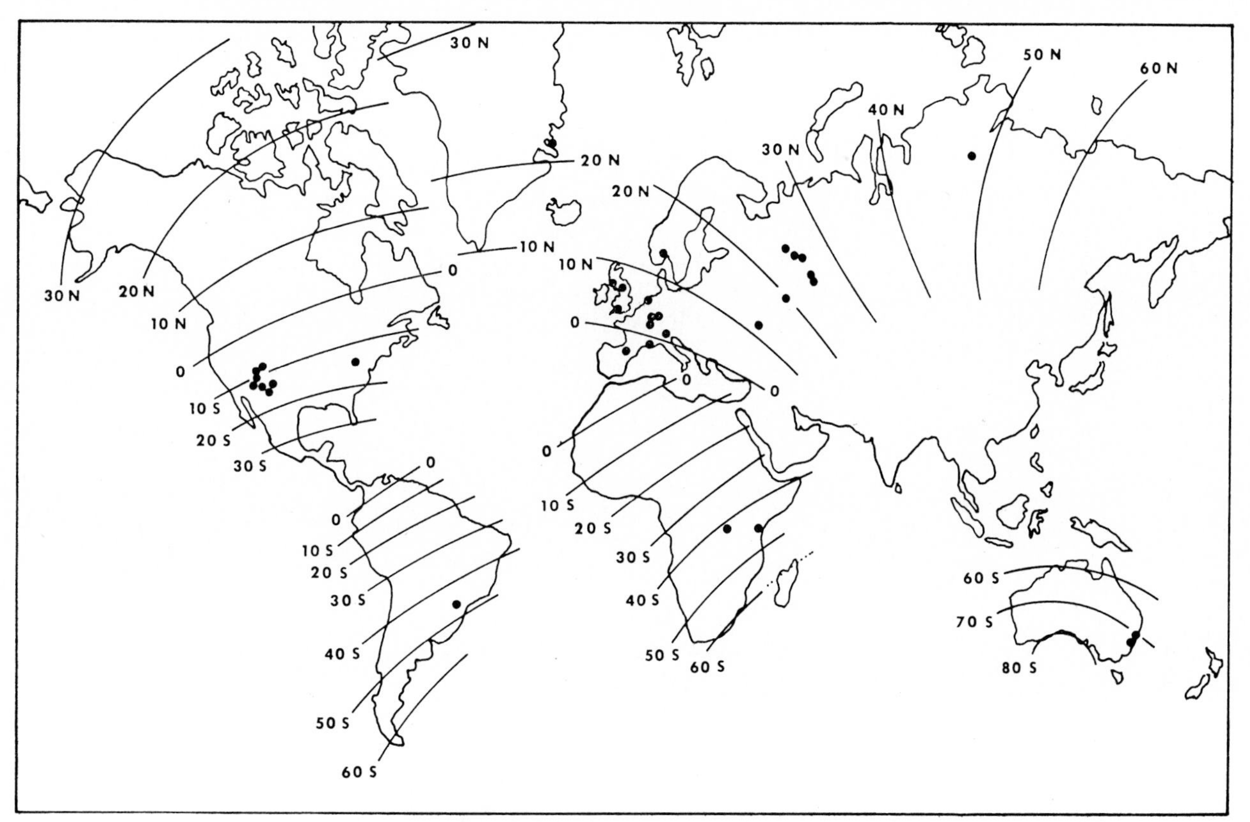

Figure 1. Permian Paleomagnetic Earth Model (P.P.E.M.) principally after Briden (this volume) showing the control stations at which paleomagnetic measurements believed by Irving (1964) to fulfill minimum requirements for reliability have been made. The extreme paucity of control in the southern hemisphere is evident and the same situation prevails over great areas of the northern hemisphere as well. In particular, reliable data are lacking in eastern and southern Asia, an area of unusual interest since paleomagnetic latitude appears to trend normal to present geographic latitude as this area is approached.

PALEOCLIMATIC DATA

The use of paleoclimatic data in a test of the kind proposed requires the employment of two basic assumptions:

1. The paleoclimatic indicator employed provides climate-dependent information when applied to Permian rocks.
2. The Permian earth exhibited temperature and insolation gradients from the equator to the pole, qualitatively similar to the present-day gradients.

By invoking these assumptions one is able to reconstruct, given adequate data, the geographic latitude framework of the Permian. The problem of obtaining suitable sets of data is a difficult one. To be really useful paleoclimatic data should fulfill the following minimum requirements:

1. Be susceptible to unambiguous quantitative treatment.
2. Be directly and simply related to latitude.
3. Be widely enough available to provide adequate data.

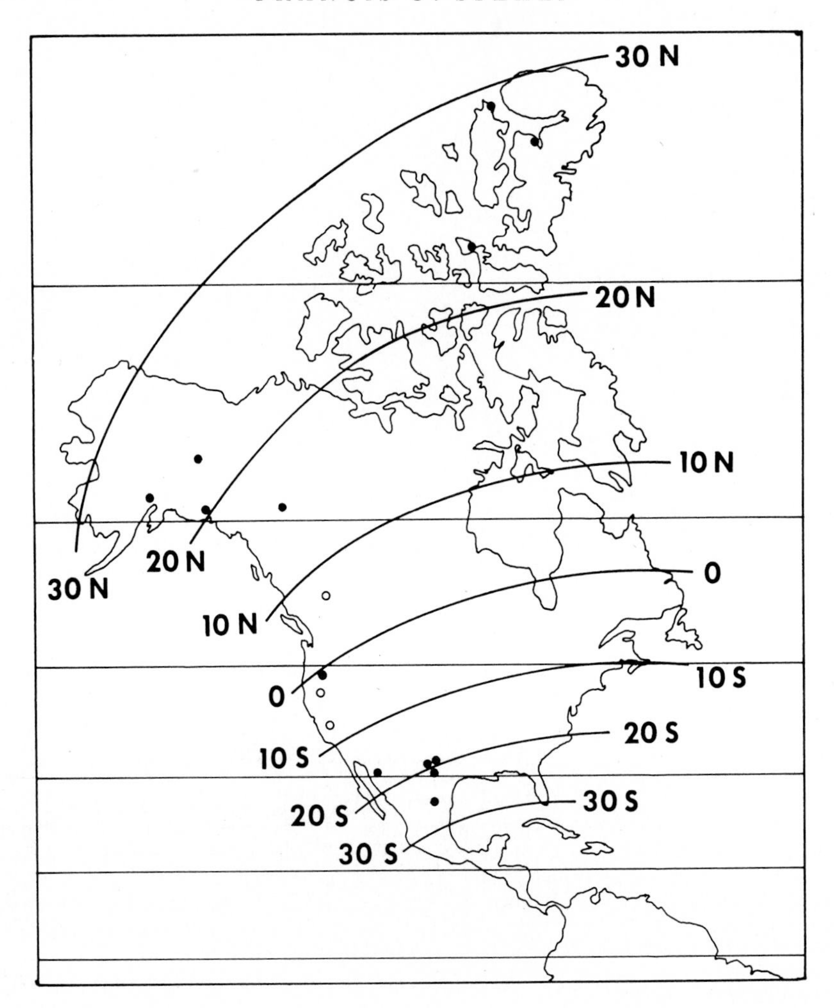

Figure 2. Permian Paleomagnetic Earth Model (P.P.E.M.) for North America assuming a north pole at 34°N 105°E (heavy, curved lines) and Present Earth Model (P.E.M.) light, straight lines. Dots show control areas for the Permian Taxonomic Diversity data used as a paleoclimatic indicator in this paper. Solid black dots are the control stations used in data sets #1 and #2. Open circles refer to control stations available only for data set #1.

Relatively few of the many paleoclimatic parameters that have been used in the past fulfill these minimum requirements. For our purposes it is not, however, necessary to consider in detail all the great variety of parameters that have been employed but only to select for a test the parameter which, at the present state of our knowledge, appears most suitable. Taxonomic diversity has been selected as most reliable, since Stehli (1968) has shown that, using recent distributions taken from the zoological literature, it is possible to locate the rotational pole with considerable accuracy (average pole at 88°N with a cone of confidence at the 95% level having a radius of 3°37′). Two sets of taxonomic diversity data for the Permian of North America have been prepared by the writer with assistance of Dr. G. A. Cooper of the Smithsonian Institution and Dr. R. E. Grant of the U.S. Geological Survey, from published literature, unpublished reports, study of museum collections, and original collecting. The groups involved in each data set and the sources of information used are given in the Appendix. The organisms are calcite-shelled, shallow-water, marine invertebrates, for it is the opinion of the writer that in the fossil record such organisms are most likely to exhibit the superior preservation, widespread occurrence, and breadth of study likely to provide the most reliable diversity data. The areas considered by Cooper, Grant, and the writer to be well enough studied to provide usable data are shown in Figure 2. It is clear from this figure that, while much more will undoubtedly be learned about the Permian faunas in North America, there is at least reasonable coverage through the areas which preserve little metamorphosed normal marine rocks. Figure 2 shows that both of these sets of diversity data cross the paleomagnetic equator and extend from about 25°N to 25°S on the paleomagnetic latitude framework and thus are well suited to a test which would determine the relative goodness of fit of the data to the North American P.E.M. and P.P.E.M.

A Look at Assumptions

A comparison of the fit of the paleoclimatic data applied to the P.E.M. and the P.P.E.M. could lead to any of three possible results. First, the paleoclimatic data could fail to show a superior fit to either the P.E.M. or the P.P.E.M. Secondly, it could show superior fit to the P.P.E.M., and thirdly, it could show superior fit to the P.E.M. If the first result is found, the test is nondiagnostic. If the second result is encountered, paleomagnetic latitudes are the same as paleogeographic latitudes, and the hypothesis of an axial dipolar magnetic field is substantiated. If the third result is encountered, a fundamental incompatibility of the paleomagnetic and paleoclimatic evidence is revealed. The probable explanation of such incompatibility is that a fundamental assumption of one or the other of the two lines of evidence is in error. It is worthwhile to review the necessary assumptions to determine their relative strengths and their importance to the test. If doubt is cast on basic assumptions, it is desirable to know which of them is most likely to be in error.

EXTRAPOLATION OF PALEOMAGNETIC DATA

In order to sustain any kind of a test it is necessary to assume that measurements used to give both paleomagnetic and paleoclimatic data actually measure these parameters, now and in the past. This assumption for both lines of evidence appears extremely strong. There are, however, some boundary conditions that should be considered. Paleomagnetic measurements are made at particular places, and the limit of justifiable extrapolation away from the control stations is not well known. In Eurasia, for instance, measurements east and west of the Ural orogenic belt give essentially the same pole positions, suggesting that measurements may be extrapolated throughout continental blocks, regardless of intervening orogenic belts. On the other hand, in Europe one finds that measurements from stations in the area affected by Alpine deformation give results slightly different from those made farther north. It is not clear whether this is a result of the deformation or relates to differing magnetic realms at the time of deposition. Until enough measurements are made to answer this question, the limits of extrapolation of paleomagnetic measurements must be considered arbitrary. It is here accepted that the paleomagnetic latitude framework indicated by the measurements used is valid throughout North America, exclusive of Greenland.

THE PALEOCLIMATIC MODEL

The paleoclimatic data used are based on diversity gradients. These gradients for large and widely distributed groups of organisms show a maximum at or very near the equator and a decline poleward into each hemisphere (for instance, see Stehli, 1968). None of the major groups of marine invertebrates, so far examined in detail, has shown important departures from the pattern of an equatorial maximum. It is, in fact, extremely unlikely that groups of marine invertebrates which occur in equatorial regions would show a highly asymmetrical diversity distribution in the vicinity of the equator, since few barriers to migration exist in this uniformly warm region. Temperate organisms, especially among fundamentally terrestrial groups, may, on the other hand, show asymmetrical distributions because temperature and other barriers can easily limit their equatorward distribution. Plots of diversity versus latitude for the groups of marine invertebrates studied by Stehli (1968) are shown in Figure 3. The data have been handled in the same manner as those for the two sets of Permian diversity data to be used in the following test. It is clear from examination of the figure that a maximum in the diversity curves, at or close to the equator, is one of their important properties. This relationship of maximum diversity to the equator for widely distributed groups of marine invertebrates is considered equally valid when applied to the Permian because it is based on a response which characterizes virtually all large, broadly distributed groups of present-day marine organisms (and most terrestrial organisms), regardless of relationship, and is thus a fundamental response unlikely to have suffered evolutionary modification.

In summary, for each technique the assumption which states that the measurements are valid for the purposes to which they are here applied is accepted. The reservations are entertained that paleomagnetic measurements may be strictly interpreted only for the local region where the measurements are made; and that special, but unlikely, configurations of land masses may give rise to diversity gradients that are not equatorially centered.

ADDITIONAL ASSUMPTIONS

Beyond the basic assumption of the paleoclimatic method—that its measurements are valid—lies the seemingly unassailable assumption relating to the geometry of the earth-sun system, and this, too, is accepted as valid. Beyond the basic assumption of the paleomagnetic method—that its measurements are valid—lie the assumptions of an axial and a dipolar field. These two assumptions may be examined in an attempt to assess their strength. The earth's field, as viewed today, is dipolar but distinctly non-axial. The north magnetic pole lies at about 70°N, 100°W, and the south magnetic pole lies at

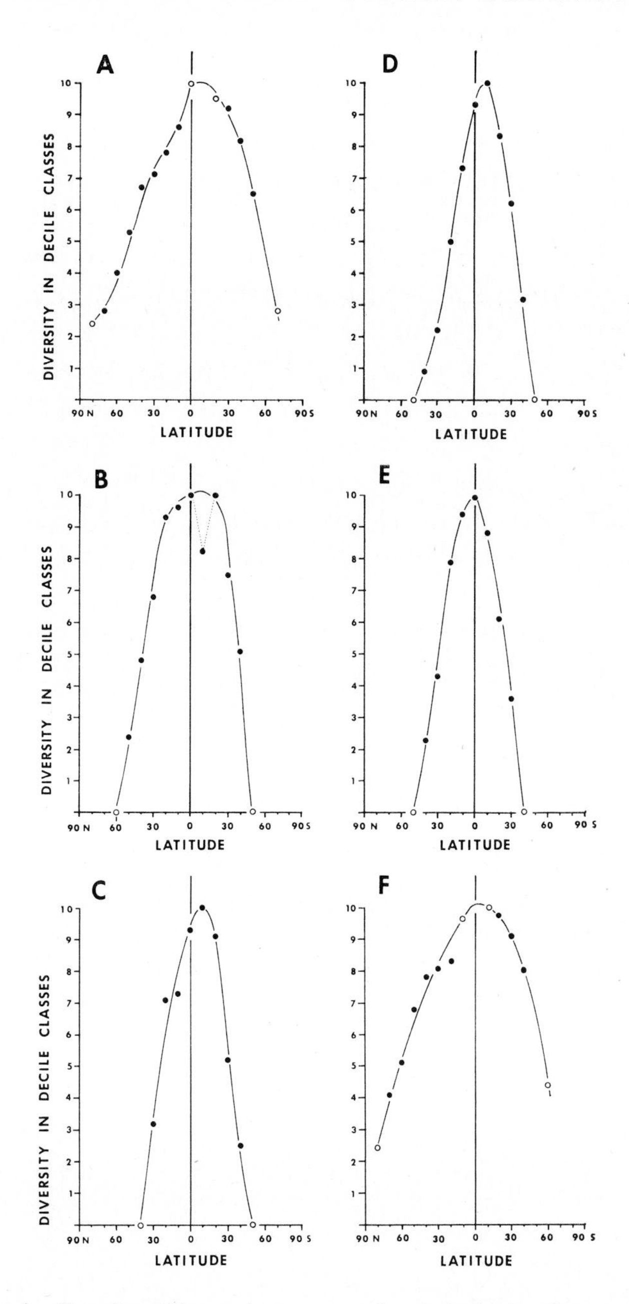

Figure 3. Taxonomic diversity plots against present-day latitude for Recent benthonic marine invertebrates. The number of control points per group varied from 36 to over 100. The raw data have been averaged for 10° latitude classes and a running average of 3 classes has been used in obtaining the control plotted (solid black dots). To normalize the data, it has been plotted on a decile scale. Terminal points and those adjacent to classes without data for which the running average could not be obtained are shown as open circles. Relationship of the diversity maximum to the equator is pronounced though there is a slight tendency for the maximum to occur in the 1st class south of the equator. The data refers to the following groups of organisms, A = genera of pelecypods; B = species of Cypraeid gastropods; C = species of *Strombus* (Gastropoda); D = species of the family Pinnidae (Pelecypoda); E = families of Pelecypoda; F = genera of hermatypic corals.

about 68°S, 143°E. The paleomagnetic argument assumes that over a geologically short period of time secular variation causes the magnetic pole position to average out at the location of the rotational pole (e.g., Kawai et al., 1965). Evidence offered in further support of this assumption is that, through the late Tertiary at least, paleomagnetic pole positions average out near the position of the present rotational pole (at 85°N, 128°W [Irving, 1964, pp. 108-109]). The data as shown by Irving scatter widely, however, and at the 95% level a cone of confidence around the average pole has a radius of 5°, thus allowing the pole to fall anywhere between 80° and 90°N and in the quadrant between 90°W and 180°W. The present rotational pole position thus lies not within but on the periphery of the cone of confidence, and the data seem at least as likely to be indicating a slightly non-axial field as one that is axial. The case for the assumption of an axial field appears not to be a strong one.[5] It could, in fact, be assumed with equal validity that the field is fundamentally non-axial and has migrated through the Mesozoic and Tertiary into its present (and perhaps ephemeral) near-axial configuration. Which of these assumptions is most likely to be correct can be determined by the type of test attempted below.

The dipolar assumption seems more secure since it is demonstrable that the primary field of the present is substantially dipolar. It must be remembered, however, that the mechanism for the generation of the earth's field is not known but only assumed. In assuming that the field must be or has always been dipolar in the absence of a knowledge of the generating mechanism, we are merely falling back on the venerable simplifying assumption of geology—uniformitarianism.

In summary, if we grant the basic assumption of each line of evidence that its measurements validly represent the parameters under investigation and if we grant that adequate data are available for each, we find the secondary assumptions of the two techniques to vary greatly in strength. The assumption most likely to be in error, it would seem, is the assumption that the field is necessarily axial.

A Preliminary Test

It is considered that there is adequate information to allow a confrontation between paleomagnetic and paleoclimatic data for North America. The test is to be regarded as preliminary, however, because the writer is accumulating a large volume of additional data which should, in the near future, allow a far more definitive test to be made and because it is hoped that more paleomagnetic data will accrue to provide a more substantial paleomagnetic latitude framework.

Since neither paleomagnetic data, as used here, nor paleoclimatic data provide longitude information, it will be sufficient to examine latitude alone. Figure 4 presents the Permian diversity data as applied to the North American P.E.M. and P.P.E.M. The diversity data are *compatible* with the use of the P.E.M., although, because all the control is north of the equator on this framework, no maximum at the equator is demonstrated. The diversity data are *incompatible* with the P.P.E.M., for on this framework the diversity

[5] It is preferred by workers in the field because it fits more harmoniously with the assumed generating mechanism for the field.

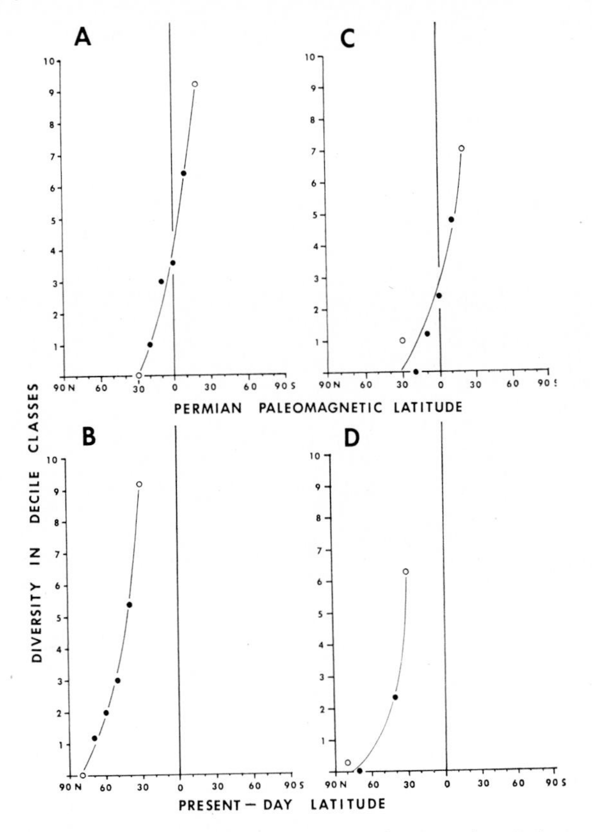

Figure 4. Permian taxonomic diversity data applied to the Permian Paleomagnetic Earth Model (P.P.E.M.) and to the Present Earth Model (P.E.M.). A, shows data set #1 (see appendix for forms involved) applied to the P.P.E.M. C, shows data set #2 (see appendix) applied to the P.P.E.M. It appears quite clear that neither data set is related to the paleomagnetic equator since diversity rises from 0 in the northernmost class to 9 and 7, respectively in the southernmost and fails to show the expected maximum at the equator. B, shows data set #1 applied to the P.E.M. and D shows data set #2 on the same framework. On this framework the data fulfill expectations yielding curves similar to those for the Recent organisms examined. Conventions as in Figure 3.

rises steadily across the paleomagnetic equator to a maximum in the most southerly latitude class. For reasons already discussed, such a diversity distribution among marine invertebrates is very unlikely and, as far as the writer is aware, does not occur among Recent forms.

As a further test, diversity data for Recent pelecypod genera were taken from Stehli et al. (1967) for six stations along the west coast of North America and plotted on the P.P.E.M. framework to see what distortion would occur. The stations and their P.E.M. and P.P.E.M. latitudes are:

Station	No. of Genera	P.E.M. latitude	P.P.E.M. latitude
Point Barrow, Alaska	24	72°N	30°N
Puget Sound, Wash.	63	50°N	5°N
Monterey, Calif.	82	37°N	5°S
Baja Calif., Mexico	89	23°N	20°S
Jalisco, Mexico	97	20°N	30°S
Panama	111	0°N	45°S

The diversity data for genera of Recent pelecypods (normalized to decile values to permit comparison) are shown in Figure 5 plotted against the P.P.E.M. latitude frame-

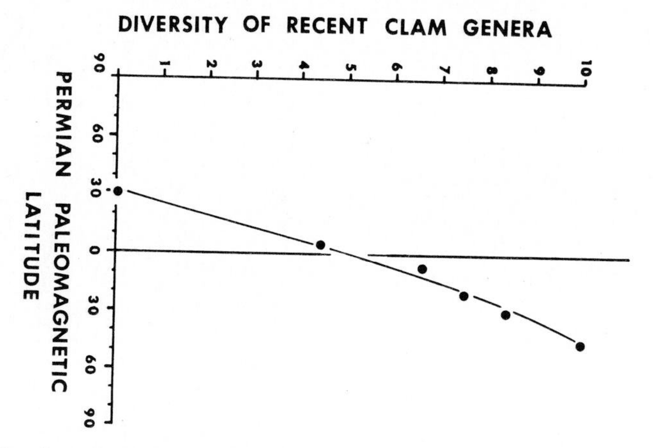

Figure 5. Diversity data for Recent pelecypod genera at six stations along the west coast of North America plotted against a Permian paleomagnetic latitude framework. The distortion introduced by this obviously incorrect procedure produces a curve essentially identical to that seen for Permian diversity data on the Permian paleomagnetic latitude framework. It seems probable, therefore, that the framework is likewise incorrect for the Permian data. (Data not averaged.)

work. It is very clear that on this obviously incorrect framework the diversity distribution (seen against its proper framework in Fig. 3A) is distorted in such a way as to assume a form essentially identical to that seen for Permian diversity data plotted on the P.P.E.M. (Figs. 4A, 4C). One is led to conclude, therefore, that the Permian diversity data applied to the P.P.E.M. is also on an incorrect latitude framework.

Though preliminary and relatively crude, the simple test made above appears adequate to discriminate between the P.E.M. and the P.P.E.M. The available data are not adequate to show that the P.E.M. is necessarily the best obtainable model. The data are sufficient, however, to show that the paleoclimatic data is in substantial disagreement with the P.P.E.M. in that paleomagnetic latitude does not appear to coincide with paleogeographic latitude. It is the axial assumption in the paleomagnetic argument that requires coincidence of paleomagnetic and paleogeographic latitude, and since they are not coincident, suspicion is directed at the axial assumption. As noted earlier, this assumption does not rest on particularly strong grounds. If the axial requirement is dropped and a non-axial field is assumed, it is no longer necessary for paleomagnetic latitude to coincide with geographic latitude, and the incompatibility shown by the North American Permian data disappears.

One can conclude from this preliminary test that, insofar as the data allow a test, the test shows the P.P.E.M. framework to be a very unlikely one. One may also conclude that a great deal more paleomagnetic evidence is needed in order to provide an adequate framework and to answer the unresolved question of how far one is justified in extrapolating paleomagnetic measurements from the control station. It would appear also that considerable caution is necessary in interpreting paleomagnetic data as supporting either polar wandering or continental drift at the present state of our knowledge.

ACKNOWLEDGMENTS

The work reported here was supported by the National Science Foundation (Grant GA 610) and the Petroleum Research Fund of the American Chemical Society (Grant 1614-A2).

Department of Geology
Case Western Reserve University

REFERENCES

Cameron, B. E. B. (1962), Permian fauna from the Yukon Territory, unpublished thesis, Univ. of Alberta.
Cloud, P. E. (1944), Pt. 3, Permian Brachiopods, in R. E. King et al., *Geol. Soc. Amer. Spec. Pap.* 52.
Coogan, A. H. (1960), Stratigraphy and paleontology of the Permian Nosoni and Dekkas formations, *Univ. Calif. Publ. Geol. Sci.* 36:243-316.
Cooper, G. A. (1953), Permian fauna at El Antimonio, Western Sonora, Mexico, *Smithsonian Inst. Misc. Collections* 119, No. 2.
Cooper, G. A. (1957), Permian Brachiopods from Central Oregon: Smithsonian Misc. Coll., V. 134, No. 12.
Crockford, M. B. B., and P. S. Warren (1935), The Cache Creek Series of British Columbia, *Trans. Roy. Soc. Can.* (ser. 3), 29:149-61.
Dutro, J. J., and E. L. Yochelson (1961), New occurrences of *Leptodus* (Brachiopoda) in the Permian of the Western United States, *J. Paleo.* 35:952-54.
Girty, G. H. (1908), The Guadalupian fauna, *U.S. Geol. Surv. Prof. Pap.* 58.
Harker, P., and R. Thorsteinsson (1960), Permian rocks and faunas of Grinnell Peninsula, Arctic Archipelago, *Geol. Surv. Can. Mem.* 309.
Helsley, C. E. (1965), Paleomagnetic results from the Lower Permian Dunkard Series of West Virginia, *J. Geophys. Res.* 70:413-24.
Irving, E. (1964), *Paleomagnetism and its Application to Geological and Geophysical Problems,* John Wiley & Sons, New York, London, Sydney.
Kawai, N., et al. (1965), Counterclockwise rotation of the geomagnetic dipole axis revealed in the world-wide archaeo-secular variations, *Proc. Jap. Acad.* 41:398-403.
Kindle, E. M. (1926), The occurrence of the genus *Leptodus* in the anthracolithic fauna of British Columbia, *Trans. Roy. Soc. Can.* (ser. 3) 20:109-11.
King, R. E. (1930), The geology of the Glass Mountains, Texas. Pt. II. Faunal summary and correlation of the Permian formations with description of Brachiopods, *Univ. Texas Bull.*, No. 3042.
Moffit, F. H. (1938), Geology of the Chitina Valley and adjacent area, Alaska, *U.S. Geol. Surv. Bull.* 894: 1-137.
Moffit, F. H. (1954), Geology of the eastern part of the Alaska Range and adjacent area, *U.S. Geol. Surv. Bull.* 989-D:1-218.
Stehli, F. G. (1954), Lower Leonardian Brachiopoda of the Sierra Diablo, *Bull. Amer. Mus. Nat. Hist.* 105 (art. 3):263-358.
Stehli, F. G. (1968), *Taxonomic Diversity Gradients in Pole Location,* Pt. 1—*The Recent Model,* Peabody Mus. Nat. Hist. Bull., Yale Univ. (in press).

Appendix A

1. Data Set #1 relates to the number of representatives of easily recognized and seldom overlooked groups of "Tethyan" brachiopods found at each of the various control stations. These forms were selected because, even in areas in which collecting has not been extensive, there is a good chance that these forms will have been found and recognized.

The groups included, and the numbers by which reference is made to them, are:

1. *Scacchinella*
2. *Gemmellaroia*
3. Teguliferinidae, Richthofeniinae, Prorichthfeniinae
4. Lyttoniacea
5. Plicated Meekellidae
6. Enteletinae
7. *Composita*

Area	Present latitude	Paleomagnetic latitude	Groups present	Reference
Svartevaeg, Axel Heiberg Island District of Franklin, Canada	82°N	29°N	0	Collections made by Stehli and Helsley, 1964
Cañon Fiord Ellesmere Island District of Franklin, Canada	80°N	27°N	0	Collections made by Stehli and Helsley, 1964
Grinnell Peninsula Devon Island District of Franklin, Canada	77°N	23°N	0	Collections made by Stehli and Helsley, 1964; Harker, P. and Thorsteinsson, R. (1960)
Alaska, U.S.A.	66°N	23°N	7	Grant, R. E. (1965) unpublished U.S. Geol. Survey report on collections made by E. E. Brabb
Alaska, U.S.A.	62°N	25°N	6	Moffit, F. H. (1954)
Alaska, U.S.A.	60°N	21°N	6, 7	Moffit, F. H. (1938)
Yukon Territory, Canada	61°N	16°N	0	Cameron, B. E. B. (1962)
British Columbia, Canada	51°N	7°N	4	Kindle, E. M. (1926); Crockford, M. B. B. and Warren, P. S. (1935)
Central Oregon, U.S.A.	44°N	0°	4, 5, 7	Cooper, G. A. (1957)
Northern California, U.S.A.	41°N	2°S	4, 5, 7	Dutro, J. T. & Yochelson, E. L. (1961); Coogan, A. H. (1960)
Southern Inyo Range, Calif., U.S.A.	36°N	7°S	5, 6	Merriam, C. E. and Hall, W. E. (1956)
Sonora, Mexico	30°N	16°S	7	Cooper, G. A. (1953)
Sierra Diablo, Texas, U.S.A.	31°N	19°S	1, 3, 4, 5, 6, 7	Stehli, F. G. (1954)
Guadalupe Mts., Texas, U.S.A.	32°N	19°S	3, 4, 5, 6, 7	Girty, G. H. (1908) collections by Rigby and Stehli, 1950
Glass Mts., Texas, U.S.A.	30°N	21°S	1, 3, 4, 5, 6, 7	King, R. E. (1930), collections by Cooper, Grant and others
Las Delicias, Coahuila, Mexico	26°N	26°S	3, 4, 5, 6, 7	Cloud, P. E. (1944)

2. Data Set #2 relates to genera of terebratuloid brachiopods. The group was chosen because it has been the subject of a recent study by the writer, in the course of which many of the collections here employed were examined. The genera involved and the numbers by which reference is made to them are:

1. *Dielasma*
2. *Glossothyropsis*
3. *Heterelasma*
4. *Rostranteris*
5. *Oligothyrina*
6. *"Cryptonella"*

Area	Present latitude	Paleomagnetic latitude	Groups present	Reference
Svartevaeg, Axel Heiberg Island District of Franklin, Canada	82°N	29°N	1, 2	Stehli & Helsley, collections made 1964
Cañon Fiord, Ellesmere Island District of Franklin, Canada	80°N	27°N	1	Stehli & Helsley, collections made 1964
Grinnell Peninsula, Devon Island District of Franklin, Canada	77°N	23°N	1	Stehli & Helsley, collections made 1965; Harker, P. & Thorsteinsson, R. (1960)
Alaska, U.S.A.	66°N	23°N	1	Grant, R. E. (1965), unpublished U.S. Geol. Survey Report on collections made by E. E. Brabb
Alaska, U.S.A.	62°N	25°N	1, 3	Moffit, F. H. (1954)
Alaska, U.S.A.	60°N	21°N	1	Moffit, F. H. (1938)
Yukon Territory, Canada	61°N	16°N	1	Cameron, B. E. B. (1962)
Central Oregon, U.S.A.	44°N	0°	1, 4	Cooper, G. A. (1957)
Sonora, Mexico	30°N	16°S	1, 2, 3	Cooper, G. A. (1953)
Sierra Diablo, Texas, U.S.A.	31°N	19°S	1, 2, 3	Stehli, F. G. (1954)
Guadalupe Mts., Texas, U.S.A.	32°N	19°S	1, 2, 3, 5	Girty, G. H. (1908), collections by Rigby and Stehli 1950
Glass Mts., Texas, U.S.A.	30°N	21°S	1, 2, 3, 5, 6	King, R. E. (1930) and collections by Cooper, Grant & others
Las Delicias, Coahuila, Mexico	26°N	26°S	1, 2	Cloud, P. E. (1944)

THE CRUST OF THE EARTH FROM A LOWER PALEOZOIC POINT OF VIEW

ARTHUR J. BOUCOT   W. B. N. BERRY
J. G. JOHNSON

INTRODUCTION

The paleontologist has far more to contribute to the science of geology than the classification of fossils. Throughout his work runs the central theme of increasing understanding of evolution. This increasing understanding of plant and animal evolution depends for its reliability upon a thorough knowledge of morphology, arranged in a reasonable and usable taxonomic scheme. Important to any taxonomic scheme are the biogeographic relationships displayed by the taxa under consideration. These biogeographic relationships in turn are heavily influenced by and partly dependent on reliable paleogeographic and ecologic reconstructions. In other words, any estimate of the potential area in which various taxa have actively interbred during the past will be heavily influenced by whatever is known of paleogeography (including lithofacies), ecology and biogeography. These considerations in turn are heavily dependent upon the available stratigraphic correlations. Ultimately, the value and reliability of the stratigraphic correlations, except in those limited areas where purely physical criteria may be relied upon, depend upon the quality of our knowledge concerning the evolution of the various taxa upon which the correlations are based. In short, we are dealing with a set of interlocking considerations. The paths that evolution has taken, the taxonomy of the groups involved, the biogeography, ecology, paleogeography, and pertinent stratigraphic correlations make up a group of interdependent interpretations of the observed morphology, observed fossil occurrences, and stratigraphic relations. We approach the truth in these matters by a series of continuing and successive approximations. In many ways this is a bootstrap operation. It is never finished. Thus the geologist or geophysicist, dependent in part upon our age determinations, is continually frustrated in his attempt to gain from us the last, the final, the ultimate pronouncement.

We now propose to discuss a series of problems, involving the crust of the earth, dealing with beds of Silurian and Early Devonian age from the point of view of biogeography, paleogeography, ecology, and correlation. These discussions are based on what we know of the evolution of the brachiopods and graptolites of that age, together with what we know of their distribution in time and space, and on the correlation problems and stratigraphic considerations necessary to gain an understanding of their relationships.

Many geologists, fossil collectors, and paleontologists have contributed to this study. Boucot's contribution consists chiefly of taxonomic, stratigraphic, and biogeographic studies begun in 1948 at Harvard under the direction of Dr. Preston E. Cloud, Jr. and continued in collaboration with Dr. J. G. Johnson, California Institute of Technology, since 1962. Beginning in 1960, Dr. W. B. N. Berry, University of California at Berkeley, and Boucot started collaboration upon a new correlation chart, including an extensive text and maps showing the distribution and lithofacies for the North American Silurian based upon information analyzed by Berry for the graptolites and by Boucot for the brachiopods. This correlation chart is now in press, and a similar work for South America, prepared in collaboration with a group of South American paleontologists, is well advanced. In 1965 Berry and Boucot began intensive work on a similar chart, maps, and text for the Old World Silurian in collaboration with a large group of paleontologists and geologists. In 1960 Boucot published a short paper dealing with Early Devonian biogeography, based chiefly on brachiopods; since 1962 this work has been greatly expanded and amplified in collaboration with Johnson and Dr. John Talent, Mines Department of Victoria. In 1965 Dr. Alfred M. Ziegler, University of Chicago, published a short paper describing a method of analyzing a series of marine, chiefly shallow-water, animal communities of Late Llandovery age. This method has universal application, and Ziegler in collaboration with Boucot has since applied it to the large volume of data, amassed by Boucot, to provide a more universal picture of Silurian and Early Devonian animal communities and their distribution and environments.

We do not intend to go into the involved and detailed faunal basis for the zoning scheme for the Silurian and Early Devonian of the world. Our subdivisions are based upon a mixture of old and new work, partly ours, partly that of others, supported by evolutionary studies where possible, but in some cases by less reliable criteria. For the Silurian system the criteria are discussed and summarized in Berry and Boucot (in press). For the Early Devonian the criteria are discussed and summarized in a series of papers of which the following are the most pertinent: Boucot and Johnson (in preparation); Boucot, Johnson, and Talent (in press). For non-marine beds of Early Devonian age we have relied primarily upon the basic work of White (1950) on the evolution of the pteraspid fishes in western Europe; and upon the extension of this work to other parts of the world by a number of authors. The bulk of the faunal basis is made up of information gained from study of brachiopods and graptolites, done on a worldwide scale, but additional, critical data have been gained from work on conodonts, particularly from the basic work of Walliser (1964) and its elaboration in other regions by subsequent studies conducted by a number of workers. The corals and trilobites have not yet been seriously considered in this context on a worldwide basis.

World Silurian Lithofacies and Its Consequences

The stratigraphic history of the earth during the Silurian is a fabric woven tightly from the threads of paleogeography (including lithofacies), biogeography, and ecology upon a loom whose frame consists of correlations based on evolutionary conclusions gained from sound taxonomic and morphologic studies. Analysis of the constituent threads making up this fabric is awkward because of their interdependence.

SOME DEFINITIONS

1. The term "platform," as I intend to use it, refers to a relatively level, flat structure made up of basement complex type rocks. Since we are concerned with the early Paleozoic, the basement happens to be of pre-Cambrian age. On top of this relatively level ancient basement complex are deposited a variety of sedimentary sequences:

A. Carbonate sequences consisting chiefly of dolomite and dolomitic shale surrounded by a relatively narrow rim of interbedded limestone and calcareous shale. The stratigraphic thicknesses present in these carbonate sequences are relatively small, normally a few hundred feet for a period like the Silurian and seldom, even locally, exceeding a few thousand feet for such a time interval.

B. Platform mudstone sequences consisting chiefly of interbedded siltstone and mudstone with minor additions of calcareous, impure units. The stratigraphic thicknesses in these platform mudstone sequences are relatively small for a period like the Silurian, normally from a few hundred to a thousand feet or at most three to five thousand feet, but never reaching the tens of thousands of feet commonly encountered within the geosynclinal suite of stratified rocks.

C. Non-marine sequences consisting chiefly of mudstone and siltstone of variable thicknesses which seldom exceed a few thousand feet for a period like the Silurian. These non-marine rocks, deposited above sea level, are derived from sources upon the platform for the most part.

2. The term "Geosyncline," as used here, refers to a linear feature of the earth's surface, situated lateral to a platform or between two platforms, or portions of two platforms. It consists of a trough receiving relatively large amounts of sediment as contrasted to the adjacent platform or platforms. This trough may be the site of local volcanic sources arranged in either a linear or an irregular manner. It may also contain local islands or land masses made up of pre-Silurian rocks, with relatively small greatest dimension, which provide a certain amount of clastic debris to the adjacent deeper parts of the geosyncline. The maximum width of the Silurian geosynclines appears to be about 300 km (Fig. 1), and the average width would appear to be about one to two hundred kilometers. The lengths of these features are measured in thousands of kilometers, and their trace, while linear, may be very sinuous. The stratified filling of the geosynclines of Silurian age is characterized by an overwhelming preponderance of non-calcareous clastic rocks in which detrital quartz and clay-to-silt-size clastic debris make up the bulk of the rocks. The most abundant and widespread rock types are siltstone, mudstone, and argillaceous sandstone with dark-colored bedded chert, with volcanic rocks being of local importance. Limestone and calcareous shale are relatively minor constituents, and dolomite and evaporites virtually unknown. Black shale is a relatively minor constituent, as are sedimentary iron ores and conglomerates. The average stratigraphic thicknesses for geosynclines during the Silurian are of the order of ten to twenty thousand feet, but local deviations both above and below the average are known.

3. The "shore line," as the term is used here, is essentially the locus of a series of

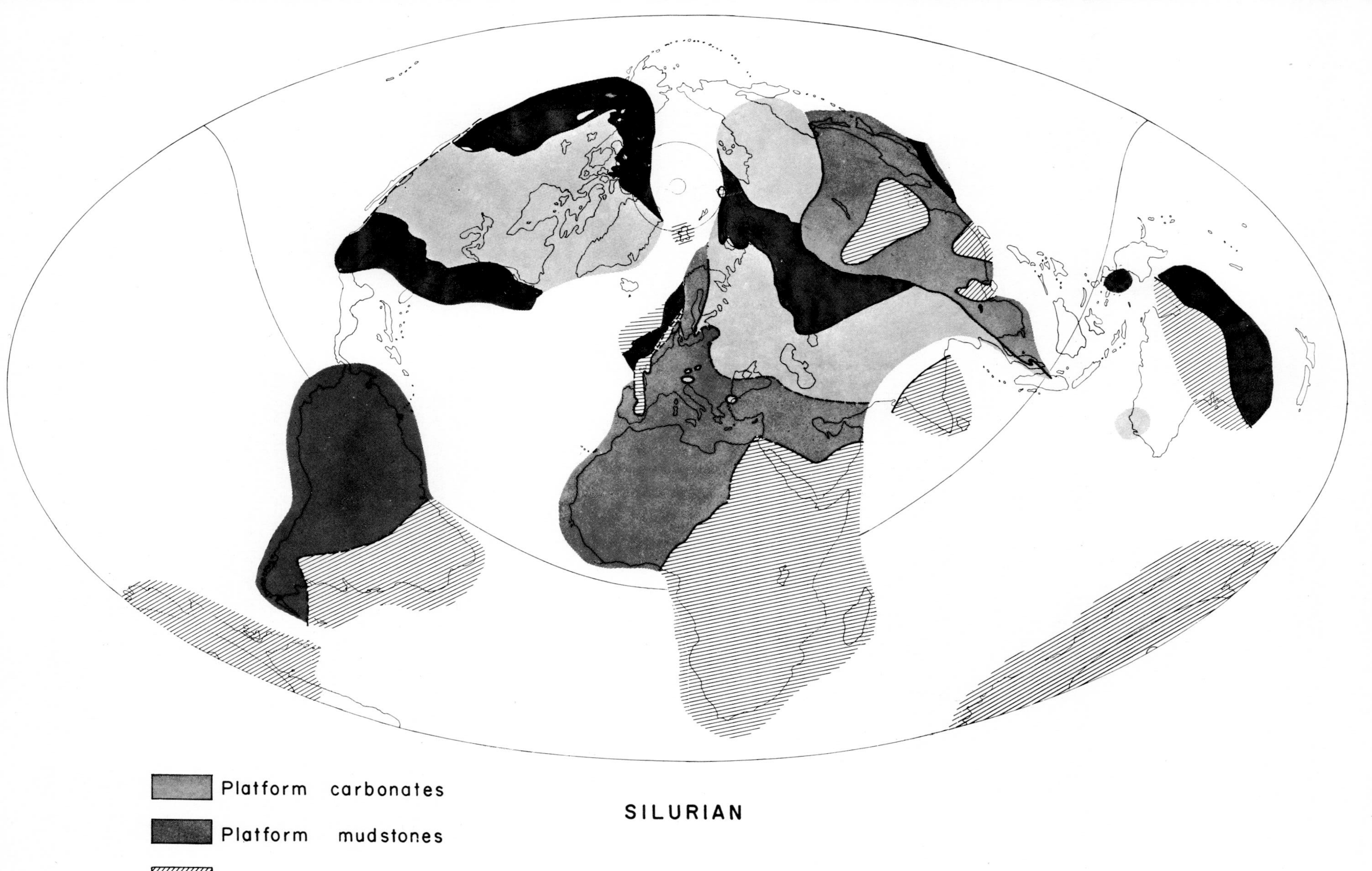

Figure 1. Silurian lithofacies, compiled by W. B. N. Berry and A. J. Boucot.

points obtained from both physical and biological considerations that indicate a change from marine to non-marine deposits. In a limited number of cases the precise point where this change takes place is known, but in most cases it is approximated on the basis of disjunct exposures of marine and non-marine type correlated with each other by both physical and biological means. The accuracy of this method is completely dependent upon the quality of the available data. In many cases it is necessary to extrapolate the information over great distances for particular time intervals by appealing to the argument that it is allowable to take into account the intermediate position of a particular shore line if data are available for the preceding and following time intervals.

4. The term "non-marine rocks" is used here essentially to refer to strata having an overall similarity, for the Silurian, to those of the British Old Red Sandstone. This similarity is considered to incorporate both physical and faunal criteria. The physical criteria consist chiefly of the presence of a sequence of mudstone, siltstone, sandstone, and conglomerate, with rare limestone and chert, in variable proportions. A respectable proportion of the strata may be characterized by a red color, but this feature is not an invariable attribute of the type. (The thickness of these beds may also be variable.) These units commonly lack marine fossils. Locally, they may be characterized by the presence, in varying degrees and abundance, of plant remains, fish, eurypterids, and a limited assemblage of pelecypods, gastropods, and ostracods.

5. The term "marine rocks" refers to all those beds, of both platform and geosynclinal type, which possess physical and biological features consistent with their having been deposited below sea level. For strata which contain fossils, the nature of the organisms is normally sufficient to make the marine category convincing.

DESCRIPTION OF THE SILURIAN PLATFORMS AND GEOSYNCLINES

During the preparation of Silurian correlation charts for the continents we have made accompanying index maps on a scale of 1:5,000,000. Upon these index maps is plotted the known extent of both Silurian surface exposures and subsurface occurrences. Despite the fact that post-Silurian erosion has removed a large percentage of the Silurian strata from the continents, it is possible from lithofacies, animal community, and a variety of physical arguments to make reasonable inferences about the original extent of the Silurian beds. The presently preserved scatter of Silurian strata is relatively random in spite of the involved post-Silurian history, a fact which greatly facilitates inferences about the original extent of the Silurian strata. Figure 1 (taken from Berry and Boucot, in press) is a summary of the data provided by the various index maps combined with the inferences derived from them and associated data concerning the original extent of the Silurian seas, the relations of platform and geosyncline during the Silurian, and the probable areas which were above sea level during a large part of the Silurian.

It is worth emphasizing, in keeping with the title of this report, that the information synthesized by Berry and Boucot for the Silurian as regards the relationships of land, sea, platform, and geosyncline also appears to be relatively valid for the preceding

Ordovician and Cambrian systems. During the Early Devonian, however, this situation breaks down very rapidly, Devonian paleogeography and lithofacies not being, in many important instances, a carbon copy of Silurian relationships. It may turn out that in some ways the rapidly changing Devonian geography and platform-geosyncline relationships are the turning point from the relative stability of the Cambro-Silurian to the more mercurial situation present in the Permo-Carboniferous. However, until the data for the late Paleozoic and Devonian have been synthesized in a manner similar to that done for the Silurian it is premature to draw any far-reaching conclusions except for those which are self evident from the Silurian data.

North America during the Silurian was characterized by a broad platform (Figure 1), bordered on the east by the Appalachian geosyncline, on the north by the Franklinian geosyncline, on the west by the Cordilleran geosyncline, and on the south by the Ouachita geosyncline. The only conspicuous break in the continent's rim of geosynclinal deposits is to be found on the southwest, where platform deposits are present up to the vicinity of the San Andreas–San Jacinto–Gulf of California lineament, but not to the southwest where, indeed, Paleozoic fossiliferous rocks of any type are not known. Inboard from the geosynclinal rim is found a sub-circular belt, characterized by an abundance of limestone and calcareous shale, which borders the dolomitic strata covering the bulk of the North American platform. The Cambro-Ordovician story for the North American platform is a modification of that for the Silurian. At the beginning of the Cambrian a large portion of the platform's pre-Cambrian basement was above sea level, and it became progressively submerged during the remainder of the period. Total submergence was probably not achieved, however, until some time during the Middle Ordovician. In any event the maximum relief and elevation of the North American platform during the Cambro-Ordovician was probably very low judging by the relatively small volume and the physical characteristics of the clastic sediment derived from exposed pre-Cambrian rocks. The only conspicuous Silurian landmass which acted as an important source area for clastic debris during the Silurian was situated on the boundary between the platform and the Appalachian geosyncline in the central and northern Appalachians. This linear, sinuous landmass had a maximum area approximately equivalent to two Cubas laid end to end. Consideration of reasonable rates of uplift and erosion for this landmass, as well as of its steadily diminishing area during the course of Silurian onlap, shows conclusively that it was not the major source of the clastic debris present in the Appalachian geosyncline proper, although it was capable of providing the clastic debris present to the west in the Valley and Ridge Province of the central Appalachians. The importance of volcanic rocks of Silurian age as sources of clastic debris for the geosynclines of North America is hard to evaluate. Information available for the northern Appalachians suggests that volcanics were relatively unimportant. For the other geosynclines the available sampling pertinent to this question is inadequate at present to provide a meaningful answer. In any event, the North American story is one of a platform covered by carbonates during the bulk of Silurian time, surrounded by a geosynclinal rim, the contents of which raise thorny problems as to source area. Sources outside the present North American shore line can be invoked to satisfy the deficit of clastic debris for the North American geosynclines, but their presence and availability are strictly hypothetical.

During the Silurian, Eurasia and Africa act essentially as one vast platform bordered on the northwest by the Caledonian geosyncline and split in the north-central region by the Ural-Kazakstan geosyncline, which terminates against the Angara Shield on the southeast (Fig. 1). There is a possibility of another geosynclinal area in Japan, although the sampling is not yet sufficient to make this possibility clear. The distribution of sedimentary rocks upon this vast Eurasian-African platform is not simple and will be considered as going from west to east.

It is convenient first to consider Europe (between the Caledonian geosyncline on the northwest and the Uralian on the east) and Africa as a unit. Europe itself may be divided into a western half joined with North Africa, characterized by platform mudstones during the Silurian, and an eastern half, the "Russian platform," characterized by a cover of Silurian carbonates. The carbonates of the Russian platform include a limestone and calcareous shale rim, similar to that present on the North American platform, and a dolomitic inner core which is areally important. The limestone and calcareous shale on the western rim extends from Gotland, and possibly Jemtland on the northwest, east through the Baltic States, and then south to Podolia and Moldavia before turning easterly to be last recorded in the subsurface just west of Odessa. To the east this rim extends from the southern part of the North Island of Novaya Zemlya southeasterly through Vaigatch and Pai-Hoi to the western slope of the Urals as far south as the central Urals, from where it is presumably displaced westerly into the subsurface of the Russian platform. The dolomitic rocks of the core are known in the easterly portions of the Baltic states, in Podolia on the west, in the Kanin Peninsula area on the north and to the west of the limestone and calcareous shale rim on the flanks of the Urals as far south as the central Urals. It should be emphasized that proven Silurian strata have not thus far been recognized over the bulk of the Russian platform. This absence can be ascribed largely to the effects of pre-Middle Devonian erosion, which in some places cuts down through the Cambro-Ordovician cover to the pre-Cambrian. However, there are areas of dolomitic rocks, known in the subsurface of this region, which when more carefully studied for their content of microfossils may prove to be of Silurian age. The homologies between the lithofacies pattern present on the North American and Russian platforms are striking in their geometry. To the southeast, the carbonates of the Russian platform strike into the poorly known and widely scattered exposures of similar rock types in southwest Asia in the region which includes Turkestan, Iran, Afghanistan, Pakistan, and India, regions for which our sampling is too poor to permit an evaluation of whether or not the division into a dolomitic core region bordered by a limestone and calcareous shale rim still holds.

The western half of Europe and adjacent North Africa, southeast of the Caledonian Geosyncline, was mantled during the Silurian by a blanket of platform mudstone which extended from the central Sahara, Libya, and portions of the Near East, north through most of France on the west, into Scandinavia and the Baltic region. This blanket of mudstone has a thickness of five hundred to a thousand feet, fifteen hundred feet at most, and does not contain volcanic rocks (with one minor exception in the Prague region), graywacke, or other geosynclinal rock types. In places, a fair amount of very impure limestone may be present. All of the available physical and biological evidence suggests the presence of a region of relative stability and relatively slow deposition during the Silurian. To the southeast in the central Sahara as well as to the northeast in Libya, parts of the Near East, and also regions to the southwest, there is a major facies change from marine to non-marine type rocks characterized by the presence of fine-grained sandstone. The southerly limits of these non-marine clastic rocks are not well known. However, for the rest of Africa to the south, marine beds of Silurian age are not known; in fact marine Paleozoic rocks are completely unknown except for widespread beds of late Early Devonian age and a few occurrences of Permian age. This information permits the conclusion that Africa south of the central Sahara was above sea level during the early Paleozoic.

Between the Caledonian geosyncline to the northwest and the Eurasian platform there is abundant evidence in parts of Scandinavia, Britain, northwestern France, and parts of Spain for an irregularly shaped landmass (Fig. 1) or series of islands. This landmass is reminiscent of a somewhat similar, mirror image mass present in the Appalachians between the geosynclinal and platform regions, and it also has too limited an area, even when maximized, to account for the bulk of the clastic debris needed to account for the adjacent Silurian sediment of the platform mudstones and Caledonian geosyncline.

The Cambro-Ordovician story for Europe and Africa is relatively similar to that for the Silurian, except that a more complete cover of carbonates is known on the Russian platform, and the platform mudstone region of the Silurian is occupied by somewhat coarser-grained clastic debris in the medium-to-coarse-grained sand-size range rather than the clay, silt, and fine-sand-size range present during the Silurian.

The Caledonian geosyncline on the northwest side of the Eurasian and African platform is a relatively simple structure from a Silurian point of view. It extends from the northern part of the British Isles to the north through portions of Norway and western Sweden. To the northwest in Britain it is bounded by a landmass of unknown seaward extent and to the southeast by the landmass or landmasses which separate it from the platform mudstones. The thickness of the filling is very large in most places, on the order of tens of thousands of feet. Volcanic rocks are present in a few areas, but Silurian volcanics are not the predominant rock types, whereas mudstone and graywacke are very abundant. The source or sources for the bulk of this clastic debris are not well known at present.

The Ural-Kazakstan geosyncline to the east of the Russian platform is a relatively complex structure. Its northern extent in the northern half of North Island, Novaya Zemlya, northeastern Vaigatch, and eastern Pai-Hoi, as well as the northern half of the Taimyr Peninsula, consists of mudstone and siltstone for the most part. Further to the south, on the east slope of the Urals from the polar to the southern parts, volcanic rocks form the major element of the sequence together with clastic rocks, in large part of volcanic origin. The west slope of the southern Urals also belongs in this same category. The logical continuation of the Uralian volcanics is found to the southeast in the volcanics of Silurian age in Kazakstan and adjacent areas. To the north these volcanics are bounded by an area rich in mudstone and siltstone, as well as some carbonate rocks in which the thicknesses (of the order of 10,000 feet) are intermediate between the geosynclinal and the platform, being assigned here to the geosyncline. On the west and the north sides of the Siberian platform there is a facies change into graptolitic mudstone, which has been interpreted (Fig. 1) as the eastern limit of geosynclinal sedimentation. Due to our ignorance of Silurian rocks under the West Siberian Lowland this last conclusion may be in error, and these relatively thin graptolitic mudstones may actually belong to a belt of platform mudstones intermediate between the carbonates of the Siberian platform and the volcanics of the eastern Urals. It is notable that the Ural-Kazakstan geosyncline ends to the southeast against the pre-Silurian massif, which includes some Cambro-Ordovician as well as pre-Cambrian rocks of the Angara Shield.

Next to be considered is the less-well-known eastern half of the Eurasian-African platform. The southern portion consists of a relatively linear belt, characterized by the presence of carbonate rocks, which extends from, and is presumably continuous with, the carbonates of the Russian platform. This belt carries through the Himalayan region, where our information is very scanty, to the Shan states in Burma and portions of adjacent Yunnan before turning south into western Thailand, the Isthmus of Kra, and those portions of Malaya bordering the Straits of Malacca until about the latitude of Kuala Lumpur, south of which further evidence is lacking at present. A division into dolomitic rocks versus limestone and calcareous shale cannot yet be made for this region, probably owing in large part to the poor density of the available information. To the north in China and to the east in southern China and Indo-China, as well as in Thailand and Malaya, there is a major facies change into a region of platform mudstones. The stratigraphic thicknesses of these rocks, where known, are much greater than those of the large area of platform mudstones known in Europe, the Near East, and Africa, averaging about 3,000 to 5,000 feet. To the north of this vast area of platform mudstones occurs the Angara Shield, which appears for the most part to have been above sea level during most if not all of the Silurian, and to have separated the area of marine sedimentation occurring on the Siberian platform and points to the east from those to the south in China and adjacent regions. Facies relations on either side of the Angara Shield suggest that it was an area of relatively low relief which shed a large volume of relatively fine-grained debris during the Silurian after having been strongly affected by orogeny of Late Ordovician age. On the south side of the Angara Shield there is evidence for progressive offlap during the Silurian, with shallow-water sediments containing a shelly fauna being, in time, deposited further and further to the south over somewhat deeper-water platform mudstones. To the north on the Siberian platform and on the adjacent mountain ranges to the east there was initial deposition of a widespread sheet of very Early Silurian platform-type mudstone followed by carbonate-type deposition for the remainder of the Silurian. This picture is complicated, however, by progressive onlap of the Silurian sea to the south over the Angara Shield, so that younger and younger Silurian strata rest upon the pre-Silurian basement of the Shield as one progresses in a southerly direction. This picture is further complicated by a vertical change to non-marine conditions as one approaches the boundary between the Silurian and the Devonian. From a Silurian point of view, the Siberian platform, not the present Siberian Shield, extended way to the east of the present area of relatively flat-lying rocks of early Paleozoic age. It is worth pointing out here that the Silurian relationships of Alaska and adjacent parts of the Soviet Union are not yet well enough known to be of any great significance. The presence of geosynclinal rocks in Japan is based on a limited amount of data: essentially some volcanic rocks associated with a few scraps of fossiliferous limestone in a structurally very complex region. In southeastern China there are several areas which are best interpreted as having been regions of non-marine-type deposition during the greater part of the Silurian (Fig. 1).

The Australian Silurian is relatively easy to deal with. On the east there is good evidence for a Tasman geosyncline from Tasmania through New South Wales; further north in Queensland there is scattered evidence consistent with the presence of a Tasman geosyncline. To the west of the Tasman geosyncline there is inadequate evidence for the presence of a possible shore line abutting a major land area occupying most of the present continent. Cambro-Ordovician strata are scattered about over the interior of Australia, but Silurian beds are unknown, the available facies information suggesting that marine Silurian beds may never have been present. A single bore hole occurrence of Silurian beds is known to the west on Dirk Hartog Island. A single occurrence of geosynclinal Silurian is known to the north in New Guinea in the Vogelskap region.

South America, from the Silurian point of view, is unfortunately the land of mystery. We have reasonable data in parts of Bolivia and northern Argentina, a little bit in the Precordillera of San Juan, a bit in the Amazon Basin in outcrop and subsurface, a few exposures in the Andes of Venezuela, plus a little in Paraguay. These limited data enable us to extrapolate a limit between platform mudstones to the west and non-marine rocks of possible Silurian age to the east. We have no proof that the rocks to the east are actually of Silurian age. The elegant fit of this boundary with the corresponding one of somewhat similar type in Africa may be more apparent than real because of the poor quality of much of the data for South America. Probably the most critical thing to emphasize for South America is the total absence of deposits which could be categorized as geosynclinal, particularly in the Andean region.

Silurian rocks are unknown in Antarctica, where at present the best information about the early Paleozoic leads to the conclusion that there is a major break, both disconformable and angular, depending upon the locality, between an older basement including rocks as young as Late Cambrian, and an overlying unit including rocks as old as late Early Devonian.

The foregoing is a summary of our present knowledge of Silurian lithofacies for the world. From it a number of conclusions can be made, some of which have great significance in any arguments concerning the crust of the earth:

1. The platforms of the Silurian, and essentially the Cambro-Silurian, of the world were relatively large as contrasted with those of the late Paleozoic, Mesozoic, and Tertiary, with minimum dimensions measured in thousands of kilometers.
2. The geosynclines of the Silurian, and essentially the Cambro-Silurian, are relatively few in number (Fig. 1), and limited in extent.
3. Most post-Silurian geosynclines are situated upon, rather than peripheral to, the platforms of the Cambro-Silurian. These include the bulk of the Hercynian, Mesozoic, and Tertiary geosynclines of the world.
4. Most post-Silurian mountain ranges are situated upon, rather than peripheral to, the platforms of the Cambro-Silurian.
5. Large portions of the Cambro-Silurian platforms, with the exception of the bulk of the North American platform, border the present oceans without the known intervention of belts contemporaneous with Cambro-Silurian geosynclinal rocks.
6. In view of the above conclusions it is unreasonable to postulate continental growth by accretion since the beginning of the Phanerozoic; in fact for large portions of all the continents except North America one could postulate a theory of continental diminution.
7. The Mediterranean Sea from a Cambro-Silurian point of view is merely a somewhat larger Red Sea type which has had a somewhat longer and more complex history.
8. Eurasia and Africa have formed a single block since at least the beginning of the Cambrian.
9. The presence of Cambro-Silurian platform-type rocks adjacent to the present oceans in many localities makes a theory of continental oceanization or of continental drift reasonable from a purely lithofacies point of view, but does not provide any data favoring one conclusion over the other.
10. Differential vertical movements since the Silurian would appear to be the most reasonable explanation for the relatively disordered location of post-Silurian geosynclines and mountain belts upon, for the most part, the platforms of the Cambro-Silurian.

Early Silurian and Devonian Animal Geography and Its Consequences

Animal geography is a discipline whose rewards for the historical geologist are virtually uninvestigated. If we assume that the shelly marine faunas of the past have been largely characterized by planktonic larvae, like those of the present, we can learn the following kinds of things by delineating the zoogeographic provinces of the past:

1. An indication for the past of those regions—the zoogeographic provinces—in which environmental conditions were favorable for an effectively interbreeding population.
2. The provision of clues as to potential paths which ocean currents of the past may

have taken as evidenced by the distribution of shelly marine organisms, the majority of which may be safely concluded to have had planktonic larvae.

3. We can identify sharp environmental barriers by locating the boundaries between zoogeographic provinces of the past.

4. The recognition of those cosmopolitan taxa whose environmental tolerances or modes of distribution were such as to remove them from the provincial category. Needless to say, information regarding the evolution of these taxa is invaluable, as it forms the firmest basis for correlation between zoogeographic provinces.

5. The possible conclusions regarding the distributions of climates during the past as evidenced by the position on a worldwide basis of provincial boundaries. Any significant conclusions regarding climatic distributions during the past may afford insights into such problems as polar wandering, continental drift, and continental stability.

6. Last, but by no means least for the practicing paleontologist, is the provision of a firmer zoogeographic background against which to assess the possibilities of consanguinity of closely related taxa occurring in the same or differing zoogeographic provinces. A better understanding of zoogeographic situations can afford a far better opportunity of assessing the possibilities inherent in parallel evolution and convergent evolution of related groups—the question of whether or not closely related stocks can be safely concluded to have developed independently of each other from common stocks.

The practicing paleontologist venturing for the first time into the essentially unexplored thicket making up the problem of Silurian and Early Devonian animal geography must try to consider and evaluate the following questions on a worldwide basis:

1. The definition of zoogeographic provinces and subprovinces on a basis of adequate sampling which avoids the pitfall of mistaking markedly different and distinctive environments with unique faunas within a single zoögeographic entity for separate provincial entities.

2. The identification and delineating of boundaries between zoogeographic entities, together with the critical recognition of those restricted boundary regions in which the faunas of adjoining zoogeographic entities are intermixed.

3. The recognition of disjunct taxa whose distribution patterns on either a worldwide or multi-provincial basis provide important clues for correlation between provinces as well as to possible worldwide dispersal routes independent of the environmental restrictions characterizing each zoogeographic entity.

In the present state of the art this work can best be carried out by first obtaining a familiarity with the faunas on a worldwide basis and then pondering deeply upon the various geographic and stratigraphic possibilities inherent in the faunal data. Basically one is faced with facts, in this instance an almost endless number of fossil collections, published illustrations, and faunal lists, which must be analyzed and filtered to provide the maximum number of meaningful generalizations regarding distribution patterns at every level, both geographic and stratigraphic. It is soon apparent that no two faunal lists are ever identical in all particulars. This introduction should make clear that we face a problem in data analysis fully as complex as that faced in most physical sciences, with the additional complication that sample control is so variable as initially to produce despair among the mathematically sophisticated. Despite the absence of mathematical rigor from our analyses, we have no more cause to give up the search than has the farmer ignorant of the biophysical and biochemical implications of planting and reaping to abandon his livelihood. We fall back upon a large body of information, available to all interested parties, which enables us to make significant progress towards our zoogeographic goal at this stage.

SILURIAN BRACHIOPOD ZOOGEOGRAPHY

Almost twenty years of ever-increasing familiarity with the Silurian brachiopods of the world has given the distinct impression that provinciality is not one of their outstanding characteristics. This observation stands in marked contrast to the situation in the Early Devonian. Basically, Silurian shells from Siberia or Gotland, New York or New South Wales, Malaya or Michigan, Wales or Wisconsin have so much in common that distinctions at the generic level are commonly as marked between adjacent localities within the same region as between regions. In more professional terms, correlation between Silurian rocks far removed from each other geographically is relatively easy because of the large number of taxa common to most areas. This cosmopolitan relationship is in marked contrast to that obtaining during the Late Ordovician or the Early Devonian. The conclusion is unavoidable that planktonic larvae during the Silurian were faced with far fewer barriers to effective dispersal than during the previous and subsequent time intervals.

As with all generalizations certain exceptions must be noted. The following exceptions for the Silurian brachiopod faunas of the world are relatively insignificant, but appear to be consistent.

1. The *Clarkeia* fauna of the northern Argentinian and Bolivian Silurian appears to be unique as regards most of the genera recognized at present within it.
2. The smooth atrypacean genus *Dayia* is recognized at present with certainty only in Europe, where its abundance at certain localities is noteworthy and puzzling in connection with its apparent absence, at least in any abundance, elsewhere in the world.
3. The Wenlockian and Ludlovian brachiopods of Japan and China appear to include at least a few taxa unique to that region, of which *Nikiforovaena tingi* is one of the most easily identified.

Information concerning other groups of Silurian marine invertebrates has not been adequately analyzed from the point of view of cosmopolitanism versus provinciality. One exception to this statement concerns Berry's work on the Silurian graptolites, which display a mild degree of provinciality. The provinciality of the Silurian graptolites contrasts markedly, in Berry's opinion, with their highly provincial character during the Ordovician. Early Devonian graptolites are still too poorly known on a worldwide basis to afford a firm basis for conclusions regarding animal geography.

DEVONIAN BRACHIOPOD ZOOGEOGRAPHY

The Early Devonian was a period of rapid transition for the brachiopods of the world from the taxonomically cosmopolitan condition characterizing the Silurian to the conditions of high provinciality exhibited during the later part of the Early Devonian. The Middle Devonian is a time interval, for the world as a whole, of marked drift back to a cosmopolitan condition which, by the end of the Devonian, became fully established for the brachiopods. The information we have regarding other phyla is not inconsistent with these conclusions. Our data for the Early Devonian are most widespread for the Early Emsian (Fig. 2), which, with the preceding Siegenian, happens to be the time interval of greatest provinciality.

We have defined (Boucot, Johnson, Talent, in press) the following provinces and subprovinces for the Early Emsian of the world:

1. Appalachian Province (the region included between Gaspé, Quebec, and central Texas for North America and the boundary region of Colombia and Venezuela).
2. Malvinokaffric Province (includes South Africa, South America from Bolivia south to the Falkland Islands, and Antarctica).
3. Old World Province:

A. Rhenish-Bohemian Subprovince (includes the Annapolis Valley of Nova Scotia, Europe, except for the Urals and the Carnic Alps, North Africa, and at least portions of the Near East).

B. Uralian Subprovince (includes the Carnic Alps, the Urals, and Asiatic Russia).
C. Tasman Subprovince (includes eastern Australia as a minimum).
D. New Zealand Subprovince (includes at least part of South Island).
E. Cordilleran Subprovince (includes the Great Basin of the western United States).

The taxa characterizing each of the provinces and subprovinces have antecedents in the Silurian, with the exception of a few whose ancestry is largely unknown. From this Silurian base the various provincial faunas have diverged. All of the provincial faunas share a certain background of both Silurian holdovers and cosmopolitan taxa in varying percentage. Both the Appalachian and Malvinokaffric provinces have a relatively low percentage of both Silurian holdovers and cosmopolitan taxa, whereas the Old World Province is characterized by a large percentage of Silurian holdovers. In many respects the Old World Province brachiopods are a Silurian fauna which has persisted through the Early Devonian and acquired Devonian characteristics both by the incoming of entirely new taxa and by the gradual development from Silurian stocks of additional new taxa. The subprovinces of the Old World Province are defined on the basis of the distribution of locally endemic taxa outstanding after the filtering out of both Silurian holdovers and cosmopolitan elements.

All of the provinces except for the Malvinokaffric incorporate a great variety of environmentally controlled faunas of a very distinctive nature. For example, the Rhenish

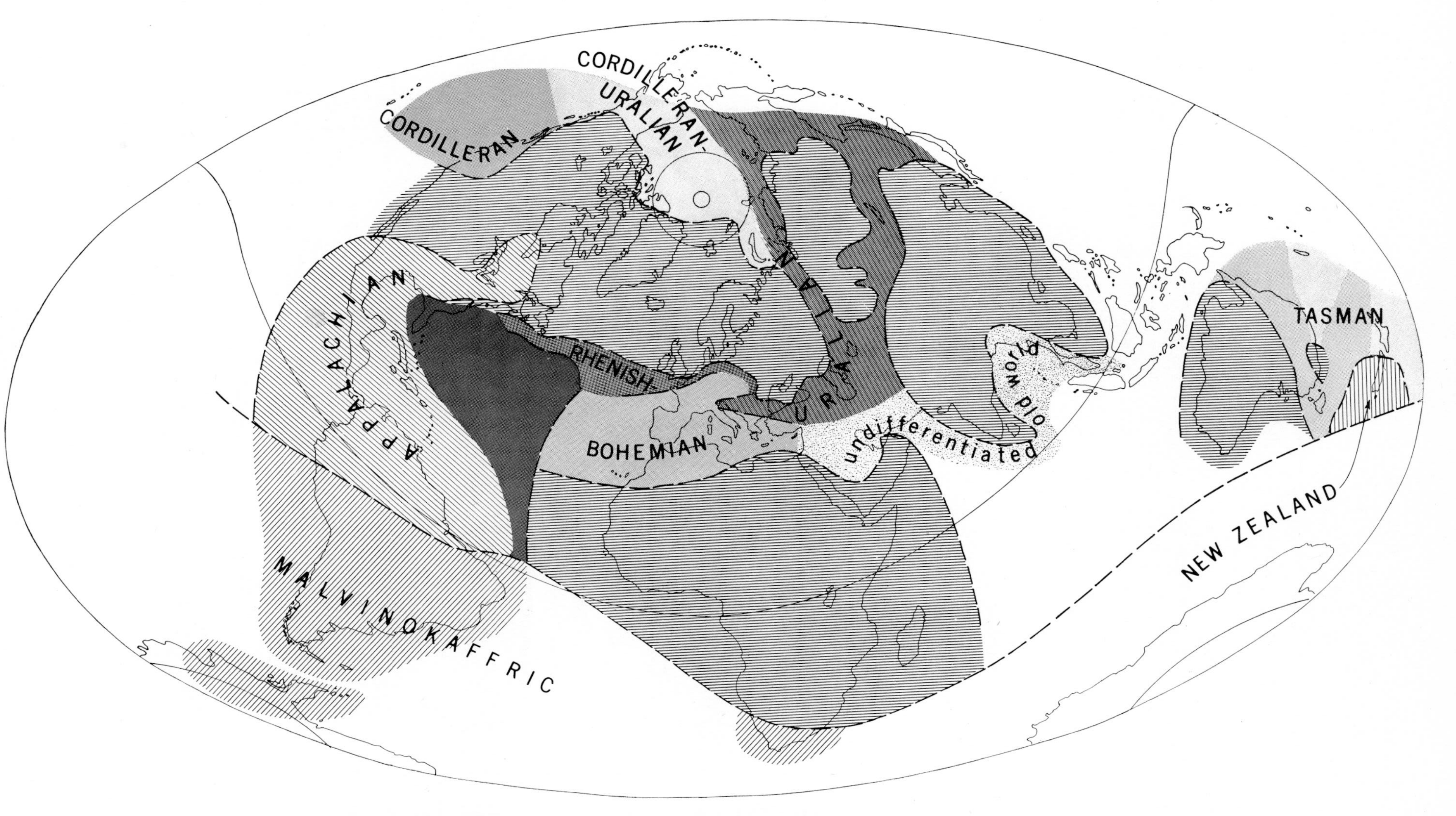

Figure 2. Early Emsian animal geographic provinces, from Boucot, Johnson, and Talent (in press).

and Bohemian facies of the Rhenish-Bohemian Subprovince are so distinctive insofar as brachiopods are concerned as to have almost no taxa in common, the situation regarding other phyla being in some cases even more extreme. However, the occurrence in certain scattered areas of Europe, North Africa, and the Near East of Rhenish-Bohemian mixtures is excellent evidence for environmental factors rather than basic isolating mechanisms being at the root of the distinction between the two facies.

The monotonous uniformity of the Malvinokaffric fauna, together with the relatively small number of taxa recognized within it, bespeaks a very uniform environment. Available ecologic information regarding the significance of homalonotids, the bellerophontid genus *Plectonotus*, and abundant mutationellinid brachiopods suggests the presence of a very widespread, shallow, relatively uniform environment.

Several examples of intermixed faunas occurring in provincial boundary areas during the Early Emsian interval are known. The most well-defined of these are Bolivia, where an intermixing of Malvinokaffric and Appalachian elements has been detected; and Gaspé, where an intermixing of Rhenish-Bohemian and Appalachian elements is known. The interbedding in the Great Basin of both Old World and Appalachian faunas marks that region as lying on the boundary between the two provinces for much of Early Devonian time, although by Emsian time such a high degree of endemism had developed that the setting aside of the region as a separate province has essentially become necessary. New Zealand during the Early Emsian is of particular interest, as the New Zealand Subprovince is essentially an area in which Tasman Subprovince, Malvinokaffric Province, and New Zealand endemic taxa mingle, in other words, a subprovince near the boundary between the Old World and Malvinokaffric provinces isolated enough from both regions to have developed certain distinctive local faunal elements.

The occurrence of a limited number of Appalachian Province taxa in eastern Australia, New Zealand, Manchuria, and some of the eastern regions of Asiatic Russia makes clear a possibility for a circum-North Pacific migration route available to a limited number of taxa whose environmental tolerances were greater than the average.

The foregoing brief discussion of Early Devonian zoogeography permits only a limited number of significant conclusions to be drawn:

1. The Old World Province is largely restricted to the northern hemisphere with the exception of Australia and New Zealand.

2. The Malvinokaffric Province of the Early Emsian is purely a southern hemisphere affair except for portions of Bolivia.

The boundaries between these two provinces, as well as the included Appalachian Province, are too scantily based at present to provide an adequate basis for determining whether the latitudinal component is significant or not.

3. The high degree of provinciality and the high percentage of land area on the northern hemisphere continents contrast strongly with the Silurian situation.

4. The environmental factors responsible for the relatively restricted distribution patterns of the Early Devonian may be ascribed to such a variety of factors that it would be rash to invoke the distribution patterns in support of any of the present controversies regarding continental drift, polar wandering, and the like.

It is possibly disappointing to the geologist and geophysicist to realize how little which is of immediate service to them can be concluded at the present time from the data of biogeography. There are so many ways to distribute things around today—and presumably in the past the available distributive mechanisms were not wildly different or fewer in numbers. But it does rain fish and frogs, and other unlikely things in a variety of accidents; given five to ten millions of years many improbable accidents and conjunctions become near certainties. One can only conclude therefore, that biogeographic data provide a very shaky foundation at best for any conclusions relative to crustal considerations.

New Possibilities: The Feasibility of Obtaining Geodetic Data from Ancient Stratified Rocks

The importance of obtaining data relative to changes in elevation over intervals of geologic time is well illustrated by McConnell's recent paper (1965, pp. 5171-5188). The possibility for securing this type of information readily from ancient stratified rocks does not appear to have been investigated. Initially the possibility of establishing level lines over extensive portions of the continents during, for example, the Silurian would seem to be an absurdity. However, owing to the recent work of Berry and Boucot on North America (in press), and on South America plus the Old World we now have available, on a 1:5,000,000 scale, data about the position of shore lines during the Silurian for extensive portions of the continents for about six discrete time intervals. In addition, Ziegler (1966) has established the utility of mapping shallow-water animal community boundaries for the Silurian. Ziegler has shown that these animal communities are, in large part, depth-controlled. Ziegler and Boucot (in Berry and Boucot, in press) have mapped such animal community boundaries, which are essentially isobaths, for five intervals during the Silurian for North America. The position of these animal community boundaries and shore line positions is a measure of the relative vertical stability of the regions in question. An important additional consideration now under investigation is the thickness of stratified rock present at any station under consideration plus a reasonable coefficient for compaction of same.

Qualitatively there are the following broad configurations for the Silurian:

1. For certain regions, including the western border of the North American platform, the northern border of the North American platform, and the western and eastern borders of the Russian platform, relative stability in elevation has been maintained during the Silurian. This is to say that in these regions there has been no significant displacement of community boundaries during the Silurian and that the thickness of marine sediment is relatively small.

2. In North Africa and the north side of the Angara Shield there has been a steady movement of the shore line over distances of about 1,000 km for the Silurian, indicating either a gradual tilting of these large blocks or gradual rising or lowering of the entire block. In the case of the Angara Shield tilting is indicated, as both shore line and community displacements on the north and south sides are inverse.

3. There is no systematic relationship between amount of sediment deposited during the Silurian, or for that matter during the succeeding Devonian interval, and the relationships outlined above or below. Shore lines moved relatively rapidly during the Silurian across both North Africa and the Angara Shield, whereas in the Caledonian regions of Britain and also in the Appalachians they moved relatively slowly. Yet in the Appalachian and Caledonian regions, relatively large amounts of sediment were being deposited, whereas in North Africa and the Angara regions rather small amounts of sediment were being deposited. Therefore, it is safe to conclude that sedimentary load superimposed on these regions during the Silurian, and also the succeeding Devonian, does not correlate or seem to be an important factor in trying to explain the observed rapid or slow movements of shore line.

4. For the Appalachians from the Gulf of Saint Lawrence to Alabama, shore line and community boundary displacements amount to several thousand kilometers northeast-southwest and only fifty to a hundred kilometers east-west. Similar scale displacements are manifest for the Caledonian region in Britain and southern Scandinavia.

5. Finally, relatively rapid differential displacement of animal communities has taken place within the Williston Basin, the Prague region, the Istanbul region, and the Holy Cross Mountains; all of them smaller regions situated on the stable platforms.

These qualitative comments make clear that when isopach and volumetric data are integrated for the time intervals into which the Silurian can be reliably divided we will have available an additional category of geophysical data capable of throwing additional light on the nature and behavior of the crust and shallow mantle during this interval of time. Needless to say, similar syntheses for non-Silurian portions of the geological record should prove invaluable to the geophysicist interested in crustal and near crustal problems.

Division of Geological Sciences
California Institute of Technology (A. J. B. and J. G. J.)
and
University of California, Berkeley (W. B. N. B.)

Discussion

DR. WASSERBURG: I fail to see precisely what connection can be drawn between these observations and the possibilities of continental drift.

DR. BOUCOT: The chief conclusion that I have is that from the point of view of lithofacies and source of terrigenous debris, the Old World—Africa, Asia, and Europe—have made up one block since the Cambrian. This is not in agreement with everybody's view of the relationship of the continents since the Cambrian.

As far as the New World relationships are concerned, the data we have developed can be interpreted more or less as you will. The relationships of Australia and Antarctica to the major land masses are not things we can provide much help with. Peninsular India we would much rather have in its present position relative to Asia than drifting around somewhere.

DR. BULLARD: Could you enlarge on that?

DR. BOUCOT: We need a source of sediment for the terrigenous debris we get in a good bit of the Near East, Africa north of the central Sahara, and Europe. Peninsular India fits in much more nicely with Arabia and Africa south of the central Sahara as a potential source region.

DR. BULLARD: It is an awfully small place to provide sediment for the whole area.

DR. BOUCOT: One thing we would like to do eventually is to consider the volume of terrigenous sediment involved in these regions and then try to understand something about reasonable rates of uplift and erosion, so that we can come up with some idea of the area necessary to provide this debris. Possibly something can be learned about its location, or at least the direction in which it is located.

DR. IMBRIE: Can you make any conclusions by focusing on reef lithologies and their worldwide distribution in the Silurian? Can you pick an equatorial belt?

DR. BOUCOT: I don't think we can during the Silurian. We have so-called Silurian reefs—reef-like masses rich in stromatoporoids and some corals—the sort of things you get in parts of Nevada and the Michigan basin, Gotland, and a few places in the Urals. The distribution of these reef-like structures is so limited in at least the present state of our knowledge, and in many cases so poorly studied, that I think it would be very presumptuous to use these data to arrive at any conclusion. The same, I think, is true of evaporites and red beds during the Silurian. It would be a very rash man who would take our present knowledge of the distribution of these phenomena in the Silurian and arrive at any conclusion relative to the equatorial belt.

DR. ANDERSON: You don't seem particularly impressed by the fact that when you shove South America into Africa the lithofacies patterns match.

DR. BOUCOT: Naturally I am impressed. I loaded the dice that way myself. On the other hand, I am realistic enough to understand that there may be an alternate explanation of this seeming good fit. I feel it is much wiser to be conservative in these matters than to stick your neck way out and prejudge things. I would rather say the data we have are consistent with continental drift, but certainly don't prove it. It is also consistent with a lack of continental drift.

DR. BULLARD: Why is it more conservative not to believe in continental drift?

(*Laughter*)

DR. BOUCOT: Possibly it is because I am afraid I might be wrong. I am not sure which side to come out for.

DR. DEWEY: There are many places in your diagram where major structures are striking right into the edges of continents. What does it suggest to you?

DR. BOUCOT: It suggests to me that either pieces of the continents have been oceanized or that continental drift has occurred.

DR. DEWEY: Can you think of a process in connection with that?

DR. BOUCOT: That is another man's job.

DR. IMBRIE: This is one type of conclusion which may have been buried in your talk. You can point out from lithologic evidence that something has happened to blocks—either drift or some other process.

DR. BOUCOT: Yes. This is why I feel very strongly that the paleontologist should be encouraged to do this sort of thing. If we had a time sequence series of maps of this type from the Cambrian to the Tertiary, I think there are some questions we would be asking now which we are not asking, and some which have been discussed here today which we will not have to discuss any more. The effort involved is not a very intellectual one. It is mostly gritting your teeth and plowing through an awful lot of dull literature.

DR. WASSERBURG: Are there no forbidden rules of juxtaposition of colors as you have used them? Are there any rules which one can make out which would prohibit the juxtaposition of certain facies or certain sedimentary sequences with another that might be valid most of the time?

DR. BOUCOT: Not really, because you have so many types of situation. You have land areas. Adjacent to them you will get shallow-water sediments, but the width of the band occupied by the shallow-water sediments will depend on the slope. You may not see this band at all.

On the platforms you can get marine conditions with carbonate sedimentation, marine conditions with mud and silt type sedimentation, or non-marine sedimentation. In the geosynclines you can get islands with non-marine sedimentation occurring on them as commonly today as in the past. You can get carbonates present next to volcanics and graywackes, in thin-platform and thick-platform sequences. From my point of view the geosyncline incorporates a welter of environments. It is not a simple thing.

DR. FAIRBRIDGE: Dr. Boucot deserves our commendation for pointing out the dangers of the misuse of paleontological data, which has been indulged in to a very large degree over the last 50 years. There are one or two rules which you can apply—one you mentioned yourself—that you can't take a major land mammal across a large body of water. In other words, the fossil elephants of the Celebes have to have got there on foot because even a living elephant doesn't swim very far. Only the major organisms are involved in such rules. Most of the smaller organisms can go as passengers in some way—spores are blown in the wind, seeds are carried in mud and stick to branches and float. Seeds were found adhering in the mud of a duck's foot.

DR. BULLARD: They can take a ride on the backs of elephants, too.

(*Laughter*)

VOICE: I did read that an elephant was once seen swimming in the middle of the Bay of Bengal, 50 miles from land, which rather shook me. Maybe that will change some of these ideas.

(*Laughter*)

DR. IMBRIE: Was it pregnant?

DR. FAIRBRIDGE: That is a point. I understand in the rutting season the major mammals do travel in very peculiar ways.

REFERENCES

Berry, W. B. N., and A. J. Boucot (in press), *Silurian of North America,* Geological Society of America.

Boucot, A. J., and J. G. Johnson (in prep.), Brachiopods of the Bois Blanc Formation in New York, *U.S. Geol. Surv. Prof. Pap.*

Boucot, A. J., J. G. Johnson, and J. A. Talent (in press), *Lower and Middle Devonian Faunal Provinces Based on Brachiopoda,* Geological Society of America.

McConnell, R. K., Jr. (1965), Isostatic adjustment in a layered earth, *J. Geophys. Res.* 70:5171-88.

Walliser, O. H. (1964), Conodonten des Silurs, *Abh. Hess. L.-Amt. Bodenforsch.* 41:1-106., 32 pls.

White, E. I. (1950), The vertebrate faunas of the Lower Old Red Sandstone of the Welsh borders; *Pteraspis leathensis* White, a Dittonian zone-fossil, *Bull. Brit. Mus. Nat. Hist., Geol.,* 1, no. 3, 51-67.

4. CLOSING REVIEW

CONFERENCE ON THE HISTORY OF THE EARTH'S CRUST

E. C. BULLARD

This conference has been about the history of the crust of the earth but has only covered part of the subject. The crust has many features which have a history and which we might have discussed but have not. We have concentrated on the great division between continents and oceans and especially on the history of the oceans and of the mid-ocean ridges. It is in these matters that the great advances have been made in the last year or two, advances that seem to me to herald a new precision in our knowledge about the earth and a real revolution in geology. The main subject we have left out is the structure and history of the fold mountains of the continents. The change of emphasis is remarkable; until a few years ago the fold mountains were at the center of thought on structural geology, and it would have been inconceivable that a conference on the crust of the earth could have left them out. The great efforts made at sea during the last 20 years have now borne fruit, and we understand what has happened beneath the oceans better than we understand the continents. Perhaps at another conference in ten years' time we shall be able to explain why the Urals are in the middle of a continent, why the Andes are at the continental edge and their extension in the Rockies is not. Already we have ideas, but the subject has been considered, perhaps rightly, either too obscure or too speculative for the conference. There are other things too that we have left out; we have only referred briefly to island arcs and their ocean deeps and not at all to those remarkable non-magnetic features of the ocean floor, the 90° ridge and the Lomonosov ridge.

Our central theme has been the demonstration that the continents have moved large distances in the last 200 million years. Like many people I had been uncertain about the reality of such movements ever since I was an undergraduate. The thing that convinced me was the work of E. Irving on the magnetization of Australian rocks. The discrepancy with the European and American results was so great that there were only two possibilities: either Australia had moved by 40° or so relative to the northern continents or the earth's magnetic field during the Mesozoic was grossly different from that of a dipole. There is some direct evidence that the field was not far from a dipole field, and it should be possible to put the matter beyond doubt; in the meantime the assumption of a non-dipole field seems a quite arbitrary and unattractive hypothesis; J. C. Briden's contribution is relevant to this question.

Now we have something new: N. Opdyke's measurements of the magnetization of sediments have confirmed every detail of Cox, Doell, and Dalrymple's account of the history of the earth's magnetic field during the last 2½ million years. This agreement of the results from lavas in California with those from red clays in widely separated places settles many doubts. The reality of reversals of the earth's field is put beyond doubt. The reversals are world wide and sudden, the field does not swing round through 180°;

it oscillates with an amplitude of 10° or so for something of the order of a million years about the value appropriate to the latitude, then in the relatively short period of a few thousand years it flips to the opposite direction, passing through a small value while changing direction. It seems certain that the main dipole has reversed. Here we have not only a challenge to the magnetodynamic theorists but also a stratigraphic tool which may prove of great importance.

But this is not all: F. Vine has shown that we have the history of the magnetic field spread out horizontally on the sea floor in the pattern of magnetic anomalies. The first of the great linear patterns of magnetic anomalies was discovered by Mason, Raff, and Vacquier off the west coast of North America, but their widespread occurrence and true nature has only very recently become apparent, largely through the work of J. Heirtzler. The pattern runs parallel to the mid-ocean ridges and is very accurately symmetrical about them. It is not a reflection of either the topography of the sea floor or of a buried basement beneath the sediments. The hypothesis of Matthews and Vine that it is associated with the spreading of the ocean floor, postulated by Hess, Dietz, and others, seems very likely to be correct. They have suggested that new ocean floor is formed along the axial crack of the ridge and is magnetized in the direction of the contemporary field. The magnetic pattern is then formed by the carrying away of these magnetized rocks to each side of the axis of the ridge, the rocks being magnetized now in one direction, now in the other, as the field reverses.

It appears that the magnetic patterns from widely separated places in all the oceans can be explained by the sequence of field reversals established quite independently. This agreement of three lines of evidence, the field reversals established from lavas on land, the same reversals found in the magnetization of ocean sediments, and the magnetic patterns on the ocean floor, is perhaps the most remarkable and unexpected advance in earth science of this generation. We not only get an explanation of the magnetic pattern but a timetable for the development of the oceans. By fitting the oceanic magnetic pattern to the times of reversals found from the lavas we get the velocity of spreading of the ocean floor. It turns out to be remarkably linear and to give velocities of from 1 to 6 cm/yr. These ideas are strikingly confirmed by L. Sykes's study of the first motion in earthquakes along the mid-ocean ridges. These occur either beneath the central valley or on the transverse fracture zones. The ones along the valley show that it is in fact cracking open, and those on the fracture zones show that they are, in Tuzo Wilson's nomenclature, "transform faults."

It is possible to believe in the spreading of the ocean floor without believing in continental drift; the two ideas do, however, go very naturally together. If one accepts the spreading of the ocean floor it is natural to seek drift explanations of a wide range of phenomena. The geometric and geological fits of the opposing coast lines on the two sides of the Atlantic are striking. The agreement of Marshall Kay's two fits, one made on geometrical grounds and the other from geological evidence, is remarkable, as is the fit of the recent age determinations by P. Hurley on rocks from Nigeria and Brazil. Another important general rule very suggestive of a separation of the continents in the past is the truncation of all pre-Tertiary structural lines at the continental edge. An example that has been studied in detail is the WSW Hercynian structures of the English Channel. These appear in southern Ireland, in the granites of Cornwall, in Brittany, and in the pattern of magnetic anomalies on the south side of the Channel. The magnetic anomalies extend to the continental edge and there disappear, to be replaced over the deep ocean by a quite unrelated system running almost at right angles and associated with a pattern in the center of the Bay of Biscay. Many similar examples of the truncation of pre-Tertiary structures could be given. Tertiary structures, on the other hand, frequently extend across the continental edge; obvious examples are the continuation of the Andes in the island arc joining South America to Antarctica, the continuation of the mountains of Burma in the Andaman islands, and the continuation of the line of volcanos in the Cameroons as a line of sea mounts running across the floor of the Atlantic to St. Helena. All this, and much else, is compatible with the Atlantic Ocean having been formed by a separation of the continents.

These ideas are important partly because they are very likely correct but also because they suggest things that we might do. The world is so wide that we badly need a means of choosing what it will be most instructive to do. Perhaps the most important thing is to drill some holes through the sediments beneath the ocean floor. Maurice Ewing has pointed out how important it is to know how long a time interval is represented by the great piles of sediment in the equatorial Pacific and also to find the pattern of sediment ages in the Atlantic. Lower Tertiary sediments near the axis of the ridge would be of great significance and perhaps fatal to the hypotheses of the spreading of the ocean floor and of continental drift.

All the things I have mentioned so far are facts or near facts, at any rate they are things about which one can plan rational programs. It is otherwise when we consider mechanisms; here we are talking about materials and processes of which we have no direct knowledge. We are making guesses guided, to some extent, by seismology and using extrapolations of laboratory experience to deduce their consequences. There are, so far as I know, only two theories as to why the Atlantic Ocean opened. One view supposes that the sea floor and the continents are carried away, as on a conveyor belt, by convection currents rising under the mid-ocean ridge and spreading out on either side. The other view supposes that the interior of the earth has swollen and that the surface has therefore been obliged to crack. The latter hypothesis seems to me unattractive in that it involves an increase in radius by over 10% in the last 2% of geological time, which would imply a rate of lengthening of the day greater than is observed at present and which is also probably incompatible with estimates of the radius from palaeomagnetism.

It seems simplest to assume provisionally that the movements are associated with convection currents in the mantle. If this is so, the materials of at any rate the upper part of the mantle must have a very low strength and not too high a viscosity. It has always seemed to me likely that this is so. Great efforts have been made by engineers to obtain materials that do not creep at temperatures between 500° and 1000°C, but even the best of them do creep at rates which would be significant to us. D. Anderson, R. McConnell and D. P. McKenzie discussed this question; in particular McKenzie gave what seemed to me a very convincing exposition of the theory and concluded that the upper mantle would behave like a liquid with a low enough viscosity to allow convection.

What form would one expect the convection cells to have? The answer usually given is that there will be a cellular pattern of the kind known as "Rayleigh-Benard convection." This seems to me not to be what one should expect. Convection takes the form of Rayleigh-Benard cells if it has no reason to do anything else; to observe these cells the circumstances must be carefully arranged and, in particular, the boundaries must be at very uniform temperatures. If horizontal temperature gradients are present they will control the pattern. Now all discussions of temperatures within the earth show that at depths of a few hundred kilometers the temperatures under the oceans are 100°C or so higher than they are under the continents. Whatever doubts one may have about the details of these calculations, the rocks beneath the oceans and continents are certainly different and have differing rates of generation of heat, and it would be expected that if convection occurs in the upper mantle it will be controlled by the resulting differences in temperature. Thus the form of the convection currents will be controlled by the distribution of oceans and continents, and the distribution itself will be altered by the convection currents; we therefore have a coupled system of some complexity without the rigid geometrical pattern of Rayleigh-Benard cells.

The principal difficulty arises from the equality of the mean continental and oceanic heat flows. The rocks at the surface of the continents are about an order of magnitude more radioactive than those beneath the oceans. If the continents do not move the explanation is simple: the radioactivity from the mantle beneath the continents has worked its way upward into the continental rocks, but under the oceans it is still in the mantle; the total amount beneath oceans and continents is the same, but that beneath the oceans is spread over a greater range of depth. If the continents move this account cannot be correct; the heat flow through a continent advancing over an ocean should be that derived from the over-ridden ocean plus that from the continental crust and should be greater than normal. The situation is complicated by the thermal effects of the convection currents and by the very long thermal time constants of crustal blocks, but there seems no reason to expect equal heat flows on land and at sea. Detailed calculations on various hypotheses are required; the work of X. Le Pichon provides an example.

In many ways the picture drawn by the proponents of continental drift is lamentably vague. Do we believe that the break-up of the continents in the late Mesozoic was a once-for-all event or is it merely a stage in a process that has always been going on, in which continents sometimes approach each other and collide and sometimes separate? In particular, what do we believe about the Appalachian geosyncline; was it a narrow arm of the sea throughout the Palaeozoic, or did two continental blocks approach each other much as India is supposed to have traveled north and collided with the main bulk of Asia? If North America was ever in contact with Europe, its west side cannot have been in contact with Asia, we need a gap somewhere. The reconstructions usually show the gap at the Bering Straits, which is, I understand, not very plausible on geological grounds. Perhaps Asia was split somewhere; perhaps at the Urals; there is some palaeomagnetic evidence for Asia being a mosaic of assembled pieces. The answers to these and many similar questions are quite unknown but can, in principle, be found. Detailed studies in which particular areas are viewed in the light of the new ideas are of great importance; the papers of J. Dewey and H. W. Menard are examples.

It is to be hoped that we can have another meeting in a few years' time at which many of our present problems will be found nearer solution and at which we can discuss the great problems of the continental crust which we have largely neglected at this meeting.

Department of Geodesy and Geophysics
University of Cambridge

AUTHOR INDEX

Page numbers in roman type refer to text references and those in italics refer to items in the bibliography following an article.

Adams, W. M., 123, *149*
Ade-Hall, J.M., 73, *87*
Airy, G., 4
Aldrich, L. T., 153, *159*
Allen, C. R., 133, 141, *148*, *149*
Allen, R. O., 19, *27*
Allsopp, H. L., 101, 102, *108*
Almeida, F.F.M. de, 153, 157, *159*
Anders, E., 18, *26*
Anderson, D. L., 6, *11*, 21, *26*
Anderson, O. L., 34, *42*
Armstrong, W. M., 35, *42*
Auskern, A. B., 35, *42*
Austerman, S. B., 35, 39, *42*
Axelrod, D. I., 190, *193*

Baird, D. M., 162, 163, *166*
Balakina, L. M., 147, *148*
Barmore, W. L., 35, *42*
Baron, J. G., 82, *88*, 90, 91, *100*, 134, *149*
Bassot, J. B., 153, *159*
Bath, J., 85, 86, *88*
Bell, W. A., 163, *166*
Belle, J., 35, *42*
Beloussov, V. V., 6, *11*
Belt, E. S., 163, *166*
Bence, E. A., 15, *26*
Bentle, G. G., 34, *42*
Bernazeaud, J. J., 153, *159*
Berry, W.B.N., 209, 212, *228*
Bessoles, B., 154, *159*
Biehler, S., 141, 142, *148*
Birch, F., 17, 19, 24, *26*, 34, *42*
Black, R. F., 190, *193*
Blackett, P.M.S., 6, *11*, 169, *176*, 179, *193*
Bloomfield, K., 140, *148*
Bodvarsson, G., 134, *148*
Boelrijk, N.A.I.M., 155, *160*
Boer, J. de, 168, *176*, *177*
Bogard, D. D., 20, *27*
Boissonnas, J., 154, *159*
Bollinger, G. A., 132, *150*
Bonhomme, M., 153, 154, *159*
Boucot, A. J., 6, 209, 212, 221, *228*
Brace, W. F., 38, *43*
Briden, J. C., 6, 170, 174, *177*, 182, 184-88, *193*, 196, 197
Brodie, J. W., 144, *148*
Brown, D. A., 186, 187, *193*
Brueckner, W. D., 163, *166*
Brunhes, B., 101, *107*
Budinger, T. F., 85, *87*
Bullard, E. C., 5, 6, *11*, 37, *42*, 153, *159*, 161, *166*
Bullen, F. P., 38, *42*
Bunce, E. T., 130, 131, 133, *149*
Burckle, L. H., *12*, 85, *89*

Cahen, L., 153, 154, *159*
Cameron, A.G.W., 20, *26*
Cameron, B.E.B., *205*, 206, 207
Campbell, N. J., 136, *148*
Cann, J. R., 74, 83, *88*, 120, *148*
Cannon, R. T., 155, *160*
Carey, S. W., 6, *11*
Chamalaun, F. H., 76, *88*, 101, 102, *107*, *108*, 191, *193*
Chandler, B. A., 35, *42*
Chase, R. L., 155, *159*
Choubert, B., 155, *159*
Christoffel, D. A., 84, *88*
Clark, S. P., 15, 23, *26*
Cleary, J., 10, *11*
Clegg, J. A., 169, *176*
Clifford, P. M., 162, *166*
Cloud, P. E., *205*, 206, 207
Coble, R. L., 31, 35, *42*, *43*
Collinson, P. W., 168, *177*
Compston, W., 15, 17, *26*, *27*, 155, *160*
Coogan, A. H., *205*, 206
Cook, A. H., 7, *11*
Cooke, R.J.S., 144, *148*
Cooper, G. A., 198, 199, *205*, 206, 207
Cooper, J. A., 22, *26*
Cooper, J. R., 163, 164, *166*
Cordani, U. G., 153, 157, *159*
Cosson, J., 154, *159*
Cottrell, A. H., 29, *42*
Couch, R. W., 80, *88*
Cox, A., 5, 8, *11*, 61ff, *72*, 73, 76-78, *88*, 101, 102, 104, 105, 107, *108*
Creer, K. M., 168-73, *177*, 182, 191, *193*
Crittenden, M. D., 36, *42*
Crockford, M.B.B., *205*, 206
Cummerow, R. L., 35, *42*
Cumming, L. M., 163, *166*
Curtis, G. H., 101, *108*

Dagley, P., 85, *88*
Dalrymple, G. B., 8, *11*, 61ff, *72*, 73, *88*, 101, 102, 104, 105, *107*, *108*
Damon, P. E., 20, *26*
Darwin, G. H., 4
Davies, K. A., 145, *148*
Davis, G. L., 153, *159*
Dawson, E. W., 144, *148*
De Bruin, H. J., 35, *42*
Dehlinger, P., 80, *88*
Delafosse, R., 154, *160*
Dickson, G. O., 66, *72*, 76, *88*, 103-105, *108*
Dietz, R. S., 5, *12*, 120, *148*
Dixon, G. O., 90

Doe, B. R., 24, *26*
Doell, R. R., 5, 8, *11*, 61ff, *72*, 73, *88*, 101, 102, 104, *107*, *108*
Donn, W. L., 169, *177*
Dorn, J. E., 34, *43*
Drake, C. L., 90, 99, *100*, 136, 137, *148*
Dunbar, C. O., 163, *167*
Duplan, L., 154, *159*
Dutro, J. J., *205*, 206

Eberhardt, P., 20, *26*, 153, 154, *159*
Ehlers, E. G., 133, *149*
Eldredge, N., 164, *167*
Elsasser, W. M., 6, *12*
Enbysk, B. J., 85, *87*
Engel, A.E.J., 16, 18, *26*
Engel, C. G., 18, *26*
Epstein, S., 174, *177*
Ericson, D. B., 49, *57*
Everett, J. E., 6, *11*, 153, *159*, 161, *166*
Everitt, C.W.F., 176, *177*
Evernden, J. F., 101, *108*
Ewing, J., 7, *12*, 85, *88*
Ewing, M., 5, 7, *12*, 49, *57*, 85, *88*, *89*, 90, *100*, 130, 134, 136, 143, 145, 146, *148-50*, 169, *177*

Fairbairn, H. W., 153-55, 157, *159*
Farrand, W. R., 48, *57*
Favre-Mercuret, R., 154, *159*
Ferrara, G., 153, 154, *159*
Fisher, R. L., 140-42, *148*, *150*
Folweiler, R. C., 30, 35, *42*
Foster, J., 9, *12*, 61, 64-66, *72*, 76, 77, *88*, 103-105, *108*, 180, *194*
Fowler, W. A., 15, 19, 21, *27*
Friedel, J., 29, *42*
Fritz, J. N., 15, *26*
Fryxell, R. E., 35, *42*
Funnell, B. M., 61, *72*, 103, *108*

Gale, N. H., 85, *88*
Gaskell, T. F., 182, *193*
Gast, P. W., 10, 15, 17, 19, 22, *26*
Geiss, J., 20, *26*
Gemperle, M., 80, *88*
Gerard, R. D., 130, *149*
Girdler, R. W., 137, 139, 142, 145, *148*
Girty, G. H., *205*, 206
Glangeaud, L., 153, 154, *159*
Glass, B., 9, *12*, 61, *72*, 103, 104, *108*, 180, *194*
Goldich, S. R., 16, *26*
Goodell, H. G., 103, 104, *108*
Gordon, R. B., 8, *12*, 37, 39, *43*
Gough, D. I., 168, 169, 173, *177*
Grant, R. E., 198, 199, 206
Grassaud, G., 154, *159*
Gravelle, M., 153, 154, *159*
Green, R., 168, 170, *177*, 191, *193*
Griggs, D. T., 33, 39, *42*, *43*
Grimbert, A., 153, *159*
Gromme, C. S., 101, 102, 104, *107*, *108*
Guerard, Y. H., 35, *42*
Gutenberg, B., 135, 140, 142, *148*

Hall, W. E., *205*, 206
Hamilton, E. I., 22, *26*
Hamilton, W., 80, *88*, 141, *148*
Harding, S. T., 133, *149*
Harker, P., *205*, 206, 207
Harland, W. B., 169, 174, *177*
Harper, J., 34, *43*
Harrison, C.G.A., 61, 65, *72*, 103, *108*
Hart, S. R., 153, *159*
Haskell, N. A., 36, *43*, 49, *57*
Hatherton, T., 144, *149*
Hausner, H., 34, *43*
Havens, R. G., 18, *26*
Hawkes, D. D., 155, *160*
Hay, R. L., 101, 102, *108*
Hayes, A. D., 163, *166*
Hays, J. D., 9, *12*, 62, 67, *72*, 103, 104, *108*, 180, *194*
Heard, H. C., 33, 39, *43*
Hebeda, E. H., 155, *160*
Hedge, C. E., 15, 16, 22, *26*
Heezen, B. C., 61, 64-66, *72*, 77, 82, *88*, 90, *100*, 103-105, *108*, 115, *118*, 120, 130-34, 136, 138, 139, 143, 145, 146, *148*, *149*
Heier, K. S., 17, 18, 25, *26*
Heirtzler, J. R., 9, 10, *12*, 74, 78-80, 82, 83, *88*, *89*, 90-92, 96, 97, 99, *100*, 104-106, *108*, 116, *118*, 120, 134, 141, *149*, *150*
Hekinian, R., 73, 74, *88*
Helsley, C. E., 182, *194*, 196, *205*, 206, 207
Heroy, P. B., 191, *194*
Herring, C., 30, 31, 39, *43*
Herron, E., 90
Hersey, J. B., 130, 131, 133, *149*
Hess, H. H., 5, *12*, 39, *43*, 73, 74, 80, 85, *88*, 109, *118*, 120, 130, *149*
Hilten, D. van, 168, *177*
Hirshman, J., 90, 99, *100*
Hodgson, J. H., 123, 133, *149*
Holmes, A., 9, *12*, 80, *88*, 145, *149*, 153, 154, *159*
Honda, H., 147, *149*
Hopkins, D. M., 102, *107*
Hospers, J., 101, *108*
Howie, R. D., 163, *166*
Hoyle, F., 15, 19, 21, *27*
Hull, D., 29, *43*
Hurley, P. M., 9, *12*, 16, 22, *26*, 153-55, 157, *159*, *160*
Hutchinson, R. D., 166, *166*

Ichikawa, M., 147, *149*
Irvine, W. R., 35, *42*
Irving, E., 139, *149*, 168, 170, 171, 173, 175, *177*, 180-88, *193*, 196, 197, 202, *205*
Isacks, B., 10, *12*, 147, *149*

Jacobson, R.R.E., 154, *159*
James, G. T., 101, *108*
Jeffreys, H., 4, 5, *12*, 28, *43*
Johnson, H., 163, *166*
Johnson, J. G., 209, 221, *228*
Jones, R. B., 34, *43*

Kamb, W. B., 39, *43*
Kaula, W. M., 37, *43*

Kawai, N., 202, *205*
Kawashita, K., 153, 157, *159*
Kay, M., 161, 163, 164, *166*, *167*
Khramov, A. N., 101, *108*, 168, *177*
Kigoshi, K., 19, *27*
Kindle, C. H., 163, *167*
Kindle, E. M., *205*, 206
King, L. C., 170, *177*
King, R. E., *205*, 206
Kingery, W. D., 35, *43*
Kolbe, P., 153, *159*
Komissarova, R. A., 168, *177*
Kovach, R. L., 21, *26*, 141, *149*
Krogh, T. E., 153, *159*
Kronberg, M. L., 35, *43*
Kulp, J. L., 18, 20, *26*, *27*, 184, *194*

Lachance, G. R., 164, *167*
Laing, E. M., 154, *159*
Lamb, H., 4
Lambert, I. B., 17, 18, *25*, *26*
Landisman, M., 130, 136-39, *150*
Langseth, M. G., 137, 139, *150*
Larimer, J., 21, *26*
Larochelle, A., 73, *88*
Lasserre, M., 154, 155, *159*
Laughton, A. S., 136-39, *149*
Lay, C., 153, 154, *160*
Ledent, D., 153, 154, *160*
Lensen, G. J., 144, 147, *149*
Le Pichon, X., 10, *12*, 74, 80, 82, 83, *88*, *89*, 90, 91, 97, *100*, 120, 141, *150*
Leutwein, F., 154, *160*
Li, C., 191, *194*
Lilly, H. D., 163, *167*
Link'ova, T. L., 61, *72*
Lochman, C., 163, *167*
Lomnitz, C., 33, *43*
Loncarevic, B. D., 85, *88*
Loupekine, I. S., 130, *149*
Love, A.E.H., 4
Lowden, J. A., 164, 165, *167*
Lowenstam, H. A., 174, *177*
Lucas, J., 154, *159*
Ludwig, W. J., 7, *12*

Maack, R., 170, *177*
MacDonald, G.J.F., 5, 7, *12*, 15, 21, 23, *26*, *27*, 47, *57*
MacDougall, I., 15, *26*
Maisonneuve, J., 154, *159*
Malahoff, A., 109, *118*
Marsh, S. P., 15, *26*
Mason, C. S., 85, *88*
Mason, R. G., 9, *12*, 74-76, 79, 82, *88*, *89*, 90, *100*, 113, *118*
Matthews, D. H., 9, *12*, 83, 85, 86, *88*, *89*, 91, *100*, 120, 136, 138, 139, *149*, *150*
Matuyama, M., 101, *108*
Maxwell, L. H., 35, *43*
McAlester, A. L., 182, *194*
McConnell, R. B., 155, *160*
McConnell, R. K., Jr., 45, 46, 49, 50, *57*, 224, *228*
McDougall, I., 61, *72*, 76, *88*, 101, 102, 104, *107*, *108*, 155, *160*
McElhinny, M. W., 168, 169, 173, *177*
McEvilly, T. V., 133, *149*
McKenzie, D. P., 8, 10, *12*, 37, 39, 43, 47-49, *57*
McLellan, A. G., 39, *43*
McManus, D. A., 80, *88*
McQueen, R. G., 15, *26*
Melcher, G. C., 153-55, 157, *159*
Melson, W. G., 74, *88*
Menard, H. W., 109-11, 113-15, *118*, 120, 133, 140-43, *149*
Mercanton, P. L., 101, *108*
Merriam, C. E., *205*, 206
Miller, E., 153, *159*
Miller, J. A., 85, *88*, 153, *160*
Millot, G., 154, *159*
Misra, A. K., 33, *43*
Moffit, F. H., *205*, 206, 207
Mohorovicic, A., 4
Mohr, P. A., 140, *149*
Moorbath, S., 15, *27*, 85, *88*
Morgan, W. J., 10, *12*, 82, 85, *88*, *89*
Morley, L. W., 73, *88*
Mott, N. F., 29, *43*
Muehlberger, W. R., 16, *27*
Muir, I. D., 74, 85, *88*
Mumme, W. G., 174, *177*
Munizaga, F., 153, *159*
Munk, W. H., 47, *57*
Murrell, S.A.K., 33, *43*
Murthy, V. R., 17, *27*
Myers, W. B., 80, *88*

Nabarro, F.R.N., 29, 36, *43*
Nafe, J. E., 34, *42*, 136, *148*
Nagata, T., 61, *72*
Nairn, A.E.M., 182, *193*
Nash, W. A., 164, *167*
Neale, E.R.W., 163, 164, *167*
Neiman, A. S., 35, *43*
Ninkovich, D., 64-66, *72*, 77, *88*, 103-105, *108*
Nishamura, M., 19, *27*
Norton, F. H., 31, 35, *43*

Oishi, Y., 35, *43*
Olehy, D. A., 19, *27*
Oliver, J., 10, *12*
Opdyke, N. D., 9, *12*, 61, 62, 64-67, *72*, 73, 74, 77, *88*, 103-105, *108*, 168-71, 173, *177*, 180, 182, *194*
Orowan, E., 6, *12*, 28, *43*, 86, *88*
Ostenso, N. D., 73, 74, 85, 86, *89*

Passmore, E. M., 31, 35, *43*
Parry, L. G., 170, 171, *177*
Patterson, C., 22, *27*
Pavoni, N., 80, *89*
Pepin, R. O., 20, *27*
Peter, G., 82, *89*
Phillips, J. D., 191, *194*
Phillips, R. P., 142, *149*
Picard, M. D., 176, *177*
Picciotto, E., 154, *160*

Pinson, W. H., 17, 19, *27*, 153, 154, *159*, *160*
Pitman, W. C., III, 74, 78, 79, *89*, 90, 92, 96, *100*, 104-106, *108*, 120, *149*
Plumstead, E. P., 179, *194*
Poldervaart, A., 17, *27*
Popova, A. V., 191, *194*
Posadas, V. G., 155, *160*
Press, F., 37, *43*, 141, *149*
Priem, H.N.A., 155, *160*

Quon, S. H., 133, *149*

Raff, A. D., 9, *12*, 74-76, 79, 82, 84, *89*, 90, *100*, 109, 113, 115, *118*
Rand, J. R., 153-55, 157, *159*
Rayleigh, J.W.S., 4
Read, W. T., 29, *43*
Reed, G. W., 19, *27*
Richards, J. R., 22, *26*
Richter, C. F., 135, 136, 140, 142, *148*
Riley, G. C., 162, *167*
Rimsaite, R.Y.J., 164, *167*
Rinehart, W., 133, *149*
Ringwood, A. E., 15, 18, 23, *26*, *27*
Ritsema, A. R., 147, *150*
Rodgers, J., 161, 163, *167*
Rodionov, V. P., 168, *177*
Romney, C., 133, *150*
Ronov, A. B., 190, *194*
Roques, M., 153, 154, *159*
Ross, D. I., 84, *88*
Rothé, J. P., 136, *150*
Rothwell, W. S., 35, *43*
Roubault, M., 154, *160*
Rowe, M. W., 20, *27*
Rudwick, M.J.S., 174, *177*
Runcorn, S. K., 5, 6, 8, *11*, *12*, 168, *177*, 182, *194*
Rusnak, G. A., 141-43, *148*, *150*
Rutten, M. G., 101, *108*

Saito, T., 7, *12*, 85, *89*
Sandell, E. B., 19, *27*
Sander, G., 136, *148*
Saul, J. M., 153, *159*
Savage, D. E., 101, *108*
Schmitt, R. A., 19, *27*
Schnetzler, C. C., 154, *160*
Schuchert, C., 3, *12*, 163, *167*
Schumacher, E. E., 31, *43*
Sclater, J. G., 137, 139, *150*
Sewell, R.B.S., 136, *150*
Shand, S. J., 133, *150*
Shepard, F. P., 141-43, *148*, *150*
Sherby, O. D., 31, 33, *43*
Shirokova, H. I., 147, *148*
Short, K. C., 155, *160*
Simons, J., 85, *88*
Smales, A. A., 17, 19, *27*
Smith, A. G., 6, *11*, 153, *159*, 161, *166*
Smith, R. H., 19, *27*
Snelling, N. J., 154, 155, *159*, *160*
Somayajulu, B.L.K., 103, *108*
Sonet, J., 154, *160*
Squires, R. L., 34, *43*
Stauder, W., S. J., 132, 133, *150*
Steenken, W. E., 155, *160*
Stehli, F. G., 6, 182, 187, *194*, 198, 200, 203, *205*, 206, 207
Stevens, R. D., 164, *167*
Stover, C. W., 139, *150*
Strange, W. E., 109, *118*
Stubbs, P.H.S., 169, *176*, 191, *194*
Sutton, J., 163, *167*
Swainbank, I. G., 22, *26*
Sykes, L. R., 9, *12*, 39, *43*, 80, *89*, 120, 121, 130, 131, 134-42, 144, 147, *149*, *150*

Talent, J. A., 209, 221, *228*
Talwani, M., 74, 80, *89*, 91, *100*, 120, 134, 141, *150*
Tarling, D. H., 61, *72*, 101, *108*, 139, *149*
Tatsumoto, M., 18, 22, *27*
Taylor, F. B., 161, *167*
Taylor, S. R., 17, 19, *27*
Teichert, C., 181, *194*
Teng, H., 191, *194*
Tharp, M., 82, *88*, 120, 130-34, 138, 139, 143, *149*
Thorsteinsson, R., *205*, 206, 207
Tilley, C. E., 74, 85, *88*, 133, *150*
Tilton, G. R., 22, 24, *26*, *27*
Tobin, D. G., 80, *89*, 130, 142, *150*
Tolstoy, I., 133, 134, *150*
Tongiorgi, E., 153, 154, *159*
Truswell, J. F., 154, *159*
Turekian, K. K., 18, 20, *27*
Turkevich, A., 19, *27*
Turner, F. J., 33, 39, *42*, *43*

Uffen, R. J., 37, *43*, 63, *72*
Urey, H. C., 21, *27*

Vachette, M., 153-55, *159*, *160*
Vacquier, V., 79, 82, *89*, 90, *100*, 109, *118*, 120, *150*
Valiev, A. A., 107, *108*
Van Andel, T. H., 74, *88*
Vandervoort, R. R., 35, *42*
Vandoros, P., 153, 157, *159*
Vasilos, T., 35, *43*
Vening Meinesz, F. A., 49, *57*
Vershure, R. H., 155, *160*
Vialette, Y., 154, *159*
Vine, F. J., 73, 74, 76, 78, 80, 82, 83, 85, *89*, 91, *100*, 104, *108*, 109, 111, 113, *118*, 120, 136, 137, 141, 142, *148*, *150*
Vogt, P. R., 73, 74, 85, 86, *89*
Von Herzen, R. P., 80, *89*, 137, 139, 142, *150*
Vvedenskaya, A. V., 147, *148*

Wachtman, J. B., 34, 35, *43*, *44*
Wagner, J. W., 39, *42*
Walker, G.P.L., 15, *27*, 85, *88*, 134, *148*
Walliser, O. H., 209, *228*
Walthall, F. G., 15, 16, *26*
Wang, T., 191, *194*
Wanless, R. K., 164, *167*
Warren, P. S., *205*, 206
Warren, R. E., 79, *89*, 90, *100*, 109, *118*

Warshaw, S. I., 31, 35, *43*
Washington, H. S., 133, *150*
Wasserburg, G. J., 15, 19, 21, 22, *27*
Watkins, N. D., 103, 104, *108*
Watson, G. M., 35, *42*
Watson, J., 163, *167*
Weber, F., 154, *159*
Wedepohl, K. H., 18, *27*
Weertman, J., 29, 31, *44*
Wegener, A., 3, *12*, 168, *177*, 181, *194*
Wensink, H., 61, *72*, 101, 104, *108*
White, E. I., 209, *228*
Whittington, H. B., 163, 166, *167*
Williams, E., 155, *160*
Williams, H., 161, 164, *167*
Wilson, J. T., 9, *12*, 74, 85, *89*, 92, *100*, 109, 110, 113, 115, *118*, 120, 121, 124, 135, 138, 139, 141, 142, *150*, 153, *160*
Wilson, R. L., 176, *177*
Wilton, J. T., 80, *89*
Windisch, C., 85, *88*
Wiseman, J.D.H., 133, 136, *150*
Wohlenberg, J., 130, *150*
Wollin, G., 49, *57*
Woollard, G. P., 5, 109, *118*
Worzel, J. L., 85, *88*

Yajima, S., 35, *44*
Yeh, S., 191, *194*
Yochelson, E. L., *205*, 206

Zeigler, A. M., 224, *228*
Zharkov, V. N., 37, *44*

SUBJECT INDEX

Acadian orogeny, in Newfoundland, 164
accretion of continents, Phanerozoic, 218
accumulation time of the earth, 20
activation energy, for self-diffusion and creep, 30ff
Africa, age data, 153-60; Silurian, 214
age data, in Africa and South America, 153-60
animal geography, Silurian and Devonian, 218ff
Angara Shield, 214ff
annealing, 29
Appalachian Province, Early Emsian, 221ff
Appalachians, evidence for continental drift in, 161-67
Ar^{40}, atmospheric abundance of, 20
Arabian peninsula, movement of, 139
Arabian Sea, 136
Arctic Basin, earthquake mechanism in, 135
asthenosphere, 6
Atlantic Ocean, 233; fit of continents around, 6; formation of, 169ff
Atlantic Ocean, equatorial, fracture zones in, 130; offset ridges in, 147
Atlantic Ocean, North, earthquake mechanisms in, 133; evidence for continental drift in, 161-67
Atlantic Ocean, South, magnetic anomalies in, 95
Avalon Belt, in Newfoundland, 162ff
axial dipole, 178-94, 195-207

Baffin Bay, seismicity of, 135
Baja California, reconstruction, 142
bending apparatus for creep experiments, 34
biogeographic relationships, in Lower Paleozoic, 208ff
brachiopods, 208ff; Silurian and Devonian, 220ff; terebratuloid, taxonomic diversity, 207; "Tethyan," taxonomic diversity, 206
Brazil, basement age provinces in, 153ff

Caledonian geosyncline, 214-15
Caledonides, evidence for continental drift in, 161-67
Cambrian paleogeography and lithofacies, 213ff
carbonate sequences, 210ff
carbonates, paleolatitude time variations of, 185; world distribution of, 190
Carboniferous, continental movements in, 168ff; Lower, magnetic field reversals in, 176; in Newfoundland and Maritimes, 163
Caririan Orogenic Episode, 157
Carlsberg Ridge, 139
Central America, earthquake epicenters, 140
Central Mobile Belt, Newfoundland, 161ff
ceramics, creep in, 34
chemistry of upper mantle, 15-27
circum-Pacific belt, tectonics, 10
climate, 178ff
coalfields, paleolatitudes of, 186
coals, Carboniferous and Permian, 186
composition of earth, mean, 25
compression, in first motions, 125ff
continental crust, and ice ages, 168-77
continental drift, 5ff, 73, 120-50, 146, 153-60, 168ff, 178ff, 196, 225, 232; at the end of the Paleozoic, 174; evidence for in the North Atlantic, 161-67
continental evolution, by vertical segregation, 23-24
continents, 151-228
convection, 5ff, 28, 110; asymmetric, 113ff; in the lower mantle, 8
convection cell, 86
convection currents, 119, 233
corals, hermatypic, taxonomic diversity, 201
cores, deep sea sediment, 61-72, 76. *See also* inclination
covalent bonding, 33
creep, in crystalline solids, 6; equation of state, 33; high-temperature, 10, 28-44; steady-state, 28; transient, 28
creep experiments, high-temperature, 33ff

creep mechanisms, 29ff
creep rate, stress dependence, 34
Cretaceous latitudes, 174-75
crust, distinction between continental and oceanic, 4; evolution of, 15-27; mass of, 17; mean Sr^{87}/Sr^{86} ratio in, 16
Curie point, in cores, 67

dating errors, reversal time scale, 102
deformation, long and short wavelength, 45ff
degassing of the earth, 20
density variations in the mantle, 7
Devonian, Early, 208ff
Devonian, Middle, lithofacies, 190ff
diffusion, 30ff
diffusion creep, 30ff
dilatation, in first motions, 125ff
dipole geomagnetic field, 178-94, 195-207
dislocation climb, 31
dislocation creep, 29ff; instability due to, 38
dislocations, 29ff

earthquake mechanisms, 121ff
earthquakes, 120ff; deep, 38; first motion of, 121ff, 232; stress release of, 37
East Africa, earthquake mechanism in, 136
East Pacific Rise, 110ff, 119, 140; magnetic profile on, 79
Eburnean orogeny, 153ff
ecology, Lower Paleozoic, 208ff
edge dislocations, 29ff
elephants, 228
elevation, over geologic time, 224
Emsian, Early, zoogeography, 221ff
equation of state for creep, 33
Eurasia, Silurian, 214
evaporites, paleolatitude time variations of, 184
Ewing-Donn theory, 169ff

fault plane, 124
faunas, as paleoclimatic indicators, 178ff; Silurian and Devonian, geography of, 218ff
Fennoscandia, isostatic rebound of, 45ff
finite strength, 36
first motion, of P and PKP, 122ff; quadrant distribution of, 124ff
floras, as paleoclimatic indicators, 178ff
fluid intergranular material, 41
fold mountains, 231
fracture mechanisms, 29
fracture zones, 120ff; termination of, 109ff
free energy of dislocations, 29

gastropods, taxonomic diversity, 201
geoid, flattening, 45
geomagnetic field, 178ff
geomagnetic reversal, 8, 61ff, 76ff. *See also* polarity reversal
Geomagnetic reversal time scale, 101-108; statistical properties of, 104
geosynclines, 210ff
glaciation, 168-77; Ewing-Donn theory of, 169ff; in southern hemisphere, 169ff
Glossopteris, 189
Gondwanaland, disruption, 173
Gorda Ridge, 110, 111; magnetic anomalies on, 79
grain boundary sliding, 31
graptolites, 208ff; Ordovician and Silurian, 220
gravity field, external harmonics, 37, 47; nonhydrostatic, 5; decay, 37, 39
Grenville ages in Newfoundland, 163
Guayana Shield, 155
Guianas, basement age provinces, 153ff
Guinean Shield of West Africa, 153
Gulf of Aden, earthquake mechanism in, 136; heat flow in, 139
Gulf of California, 140-41; crustal structure of, 142; heat flow in, 142
Gutenberg low velocity zone, 6

heat flow, 6, 15ff; average oceanic and continental, 7, 234; its dependence on source distribution, 24-25; distribution of, 5, 10; in the Gulf of Aden, 139; in the Gulf of California, 142; oceanic, 119
heat sources, radioactive, 6
Hess-Dietz hypothesis, 9
Highland Boundary Fault Belt, in Britain, 165

ice ages, 168-77. *See also* glaciation
ice rafting, 70
imperfections in solids, 29ff
impurities, 29ff
inclination of magnetic field in sediments, 61ff, 70, 176, 180; depth of reversals of, 66
Indian Ocean, magnetic anomalies in, 97
instability in dislocation creep, 38
interstitial atoms, 29ff
intensity, magnetic, 64, 70
invertebrates, marine, taxonomic diversity in Permian, 199
island arcs, 231; earthquake mechanisms in, 146. *See also* oceanic trench
isostasy, 4ff
isostatic rebound of Fennoscandia, 45

Juan de Fuca area, ocean floor spreading and magnetic anomalies, 79ff
Juan de Fuca Ridge, 110ff
Jurassic, latitudes of southern continents in, 172-73

K, terrestrial abundance of, 20
K-Ar ages, 153ff
K-Ar dating, as basis of the reversal time scale, 101ff
K/U ratio, in terrestrial materials and chondritic meteorites, 19ff
Kiaman Magnetic Interval, 176

labyrinthodonts, paleolatitude distribution of, 186-87
latitude changes of continents, 169ff
Liberian orogenic episode, 154
line defects, 29ff
lithofacies, Lower Paleozoic, 208ff
lithophile elements, abundance of, 17ff
lithosphere, 10ff; thickness of, 10; viscosity of, 36

M discontinuity, 4, 136
Macquarie Ridge, earthquake mechanism on, 143
magnetic anomalies, oceanic, 9, 73-89, 119, 120
magnetic field, history, 232
magnetic reversal time scale, *see* geomagnetic reversal time scale
magnetization in sediments, 231. *See also* inclination, intensity
Malvinokaffric Province, Early Emsian, 221ff
mantle, 13-57; bulk composition of, 15-22; convection currents in, 233; mode of deformation of, 28; melting temperature of, 37; temperature distribution in, 37; viscosity of, 45-57. *See also* upper mantle
mantle convection, 120
mass of earth's crust, 17
melting, in upper mantle, 42
melting temperature, in the mantle, 37
metallic bonding, 33
Mexico, earthquake epicenters in, 140
Mid-Atlantic Ridge, heat flow in, 119
mid-ocean ridge system, 5
mid-ocean ridges, 119; earthquake mechanisms on, 121ff; magnetic anomalies on, 73-89
Mohorovicic, *see* M discontinuity
mountains, density, 4
mudstone sequences, 210ff

Nabarro-Herring creep, *see* diffusion creep
Newfoundland, pre-Carboniferous, 161ff
nodal planes, of first motions, 126ff
nonhydrostatic gravity field, 47. *See also* gravity field
non-marine sequences, 210ff
normal faulting, 125; on ridge system, 145
North American platform, Silurian, 213
North Atlantic, *see* Atlantic Ocean

ocean basins, 61-150
ocean floor spreading, *see* sea floor spreading
oceanic cores, paleomagnetism, 61-72, 103
oceanic crust, age, 79; composition and bulk magnetic properties of, 74; upper layers of, 7
oceanic trench, 7, 10. *See also* island arcs
oceanic volcanic rocks, K-Ar ages of, 101ff; Sr isotope composition of, 15; Pb isotope composition of, 22
Old Red Sandstone, 212
Old World Province, Early Emsian, 221ff
Ordovician paleogeography and lithofacies, 213ff
Owen fracture zone, 138

P, first motions, 122ff
Pacific Antarctic Ridge, 92ff
paleoclimates, 178-94, 195-207
paleogeography, Lower Paleozoic, 208ff
paleomagnetism, 5, 11, 178ff; and age dating, 101ff; and continental drift, 168ff; of oceanic cores, 61-72; Permian data, 195ff
Paleozoic, late, drift events, 169ff; Lower, 208-28; paleomagnetic poles, 170
Paleozoic-Mesozoic boundary, continental drift at, 168
Pan-African Orogenic Cycle, 154, 158
Pb isotope, composition in upper mantle, 15, 22ff; fractionation, 23; from oceanic islands, 24
pelecypods, taxonomic diversity, 201
Permian, paleoclimates, 182, 195ff; paleolatitudes, 171; paleomagnetism, 168, 195ff; taxonomic diversity in, 203
Phanerozoic paleolatitudes, 180ff
plasticity, of rocks, 38
platform, 210ff
Plio-Pleistocene, in oceanic sediments, 69
plutonic intrusives in Newfoundland, 164
polar wandering, 196
polar wandering paths, Carboniferous, 168
polarity epochs, 61, 76, 101ff; normal, 78
polarity reversal, ages of, 104-105; origin of, 101. *See also* geomagnetic reversal
polarity in sediments, *see* inclination
point defects, 29ff
pre-Cambrian age provinces, 153-60
pre-Cambrian metamorphic rocks in Newfoundland, 161
pre-Carboniferous rocks in Newfoundland, 161ff
pre-Miocene reconstruction of Gulf of Aden, 138
Proterozoic, in Newfoundland, 162ff

Quaternary time scale, 8. *See also* geomagnetic reversal time scale

radioactivity, in oceans and continents, 6, 234
radiolarian zones, in Antarctic cores, 62-63
radiometric reversal time scale, 101ff. *See also* geomagnetic reversal time scale
Rayleigh-Benard convection, 8, 234
Rayleigh stability criterion, 8
Rb-Sr ages, 153ff
Rb/Sr ratio of the earth, 17
Recent benthonic marine invertebrates, taxonomic diversity, 201, 203
Red Sea, 136
reef lithologies, Silurian, 226
reflection profiling, 7
reflector A, 7
refractory elements, enrichment, 21
relaxation time, as a function of wavenumber, 45ff
remanent magnetization of oceanic crust, 73ff
reversal of the geomagnetic field, 120, 231; frequency of, 65, 176. *See also* polarity reversal; geomagnetic reversal time scale
Reykjanes Ridge, 80-81, 90ff
rotation of the earth, and spreading, 87
Russian platform, Silurian, 214

S wave polarization, 131
San Andreas fault, 80, 141
screw dislocation, 29ff
sea-floor spreading, 5, 9ff, 73ff, 90-100, 109-18, 119-50, 232; in Juan de Fuca area, 80; rate of, 99, 106
sediment thickness, distribution, 7, 233

sedimentary rocks, as paleoclimatic indicators, 178ff
sedimentation rates, in cores, 70
sediments, inclination of magnetic field, *see* inclination
seismicity, along fracture zones, 120ff
self-diffusion, 30ff
serpentinites, in Newfoundland, 164
Silurian, 208ff; platforms and geosynclines, 212ff
solar nebula, condensation from, 21
South America, age data, 153-60
South Atlantic, *see* Atlantic Ocean
spreading rates, 99, 106
Sr^{87} in earth and meteorites, 15ff
statistics, 87
steady-state creep, 30
stratigraphy, Lower Paleozoic, 208ff; magnetic, 62; using reversal time scale, 107
strength, finite, 36; of upper mantle, 233
stress, critical, 32; in the lithosphere, 11
stress differences, 4, 36
strike-slip fault systems, 10
strike-slip motion, on ridge system, 145

Tasman geosyncline, 217
taxonomic diversity, 187, 198ff
temperature, mean, variation with latitude, 179
temperature distribution in the mantle, 37
Tertiary, paleomagnetic poles, 202
Tertiary, Upper, latitudes, 174-75
thrust faulting, on Macquarie Ridge, 143
time scale, for geomagnetic reversals, 78, 101-108
trace element content of common rocks, 18
transcurrent faulting, 10, 120ff
transform fault, 9, 80, 109ff, 120-50, 232
trench, oceanic, 7, 10
twinning, 33

U, possible distribution in crust and mantle, 25
U/Pb ratio, 23
upper mantle, chemistry, 15-27; partial melting in, 42; strength and viscosity of, 233; viscosity gradient in, 50. *See also* mantle
Ural-Kazakstan geosyncline, 214-15

vacancies, 29ff
vacancy diffusion, 31ff
Venezuela, basement age provinces, 153ff
Vine-Matthews hypothesis, 9, 74ff, 82, 92, 99
viscosity, 30ff; effective, 40; of lithosphere, 36; of mantle, 45-57, 233; Newtonian, 46
viscosity gradient in upper mantle, 50
volatile elements, depletion, 21
volcanic rocks, K-Ar ages, 101ff
volcanoes, on Juan de Fuca Ridge, 113, 114
von Mises' criterion, 32

West African Craton, 154
Western Belt of Newfoundland, 161ff

yield stress, 28

zoogeography, Silurian and Devonian, 218ff